Telemedicine Technologies

Big Data, Deep Learning, Robotics, Mobile and Remote Applications for Global Healthcare

Telemedicine Technologies

Big Data, Deep Learning, Robotics, Mobile and Remote Applications for Global Healthcare

EDITED BY

HEMANTH D. JUDE
Karunya University, Coimbatore, India

VALENTINA EMILIA BALAS
Aurel Vlaicu University of Arad, Arad, Romania

ACADEMIC PRESS
An imprint of Elsevier

Publisher: Mara Conner
Acquisition Editor: Chris Katsaropoulos
Editorial Project Manager: Mariana Kuhl
Production Project Manager: Punithavathy Govindaradjane
Cover Designer: Christian Bilbow

Academic Press is an imprint of Elsevier
125 London Wall, London EC2Y 5AS, United Kingdom
525 B Street, Suite 1650, San Diego, CA 92101, United States
50 Hampshire Street, 5th Floor, Cambridge, MA 02139, United States
The Boulevard, Langford Lane, Kidlington, Oxford OX5 1GB, United Kingdom

Working together
to grow libraries in
developing countries

www.elsevier.com • www.bookaid.org

List of Contributors

Surajudeen Abiola Abdulrahman
Emergency Medicine Department
James Paget University Hospital
Great Yarmouth, Norfolk, United Kingdom

R. Amutha
Department of Electronics and Communication
   Engineering
SSN College of Engineering
Chennai, India

T. Anitha
Department of Electronics and Instrumentation
   Engineering
Sri Ramakrishna Engineering College
Coimbatore, India

J. Anitha
Department of Electronics and Communication
   Engineering
Karunya Institute of Technology and Sciences
Coimbatore, India

Mohammed Imran Basheer Ahmed
Department of Computer Science
College of Computer Science and Information
   Technology
Imam Abdulrahman Bin Faisal University
Dammam, Saudi Arabia

Sujitha Juliet Devaraj
Department of Computer Science and Engineering
Karunya Institute of Technology and Sciences
Coimbatore, India

Khalid ElBahnasy
Faculty of Computer and Information Sciences
Ain Shams University
Cairo, Egypt

Ahmed Elbokl
Neurology Department
Ain Shams University
Cairo, Egypt

Tamer Emara
Neurology Department
Ain Shams University
Cairo, Egypt

Andrei Gabor
"Vasile Goldis" Western University of Arad
Arad, Romania

Kurubaran Ganasegeran
Clinical Research Center
Seberang Jaya Hospital
Ministry of Health Malaysia
Seberang Jaya, Penang, Malaysia

G. Gopu
Department of Bio Medical Engineering
Sri Ramakrishna Engineering College
Coimbatore, India

G. Indumathi
Mepco Schlenk Engineering College
Sivakasi, India

R. Jansi
Department of Electronics and Communication
   Engineering
SSN College of Engineering
Chennai, India

Antoanela Naaji
"Vasile Goldis" Western University of Arad
Arad, Romania

M. Nagarajapandian
Department of Electronics and Instrumentation
   Engineering
Sri Ramakrishna Engineering College
Coimbatore, India

Soulakshmee Devi Nagowah
Department of Software and Information
   Systems
Faculty of Information, Communication and Digital
   Technologies
University of Mauritius, Réduit, Mauritius

Anand Nayyar
Duy Tan University
Da Nang, Vietnam

Gaurav Pareek
National Institute of Technology
Goa, India

Florin Popa-Iovanut
"Vasile Goldis" Western University of Arad
Arad, Romania

Marius Popescu
"Vasile Goldis" Western University of Arad
Arad, Romania

M. Prabhakaran
Department of Electronics and Instrumentation
   Engineering
Sri Ramakrishna Engineering College
Coimbatore, India

Pijush Kanti Dutta Pramanik
National Institute of Technology
Durgapur, India

Sahar Qazi
Department of Computer Science
Jamia Millia Islamia
New Delhi, India

S. Radha
Department of Electronics and Communication
   Engineering
SSN College of Engineering
Chennai, India

Atta-ur-Rahman
Department of Computer Science
College of Computer Science and Information
   Technology
Imam Abdulrahman Bin Faisal University
Dammam, Saudi Arabia

Khalid Raza
Department of Computer Science
Jamia Millia Islamia
New Delhi, India

J. Revathi
Department of Biomedical Instrumentation
   Engineering
Avinashilingam Institute for Home Science and Higher
   Education for Women
Coimbatore, India

V. Sathananthavathi
Mepco Schlenk Engineering College
Sivakasi, India

Raginee Sooklall
Department of Software and Information Systems
Faculty of Information, Communication and Digital
   Technologies
University of Mauritius
Réduit, Mauritius

Khushnuma Tanveer
Department of Computer Science
Jamia Millia Islamia
New Delhi, India

Muhammad Ali Javed Tengnah
Department of Software and Information Systems
Faculty of Information, Communication and Digital
   Technologies
University of Mauritius
Réduit, Mauritius

Ramjeevan Singh Thakur
MA-NIT Bhopal
Bhopal, India

Vivek Tiwari
IIIT Naya Raipur
Naya Raipur, India

K. Uma
Department of Biomedical Instrumentation
    Engineering
Avinashilingam Institute for Home Science and Higher
    Education for Women
Coimbatore, India

Hoda MF Wahba
Geriatrics Department
Ain Shams University
Cairo, Egypt

Contents

CHAPTER 1

Mobile Application for Medical Diagnosis

ANDREI GABOR • MARIUS POPESCU • FLORIN POPA-IOVANUT •
ANTOANELA NAAJI
"Vasile Goldis" Western University of Arad, Arad, Romania

1 INTRODUCTION

Mobile applications continue to be a challenge for software developers, although the first attempts yielded results decades ago. Technological evolution has moved this industry from an exclusive area to a purely commercial one, which appeals to large hardware and software companies in the field [1–3].

Digital pager technology dominated the market until the early 1990s, when the emergence of mobile phones dramatically reduced pager sales [4].

Today, the use of smart mobile devices such as tablets and cell phones is steadily increasing. This is why mobile device manufacturers are forced to improve their hardware and come up with improved technologies [5,6]. It is also well-known that the volume of hardware sales from giant companies such as Apple, Samsung, BlackBerry or Nokia are mainly due to the number and quality of software that each platform offers to its users.

Mobile applications are divided into two main categories: native applications and hybrid applications. Native applications are those that are written in the programming language specific to the platform for which they are developed. This is because native applications must retain the behavior and the design of the running operating system. They must integrate into the platform chosen by the end user. For a "native" product to run on multiple operating systems, it usually requires a programmer for each platform based on workload, deadlines, and other factors [7–9].

Hybrid applications are those that are typically developed using WEB technologies (JavaScript, HTML, CSS). Nowadays, hybrid applications are starting to grow stronger because many development environments are becoming more stable and provide access to more and more hardware features of devices. Hybrids typically consist of a native web browser component that can access various features of the running platform (e.g., video camera, contact list, etc.). As a rule, hybrid applications have the same graphic design on all platforms that have been published, regardless to the user interface implementation guide of each platform running the application [10–12].

In this chapter we present a mobile application which was designed to diagnose users' illnesses following a set of questions. Any element in the application can be modified by using Firebase. The software is developed in the Xcode programming environment using the Objective-C programming language.

The second section of the chapter contains the description of the application (the structure, the architecture, the description of the modules, the structure of the database, the features, etc.). The third section describes how the application works. At the end of the chapter some conclusions are drawn and future work is identified.

2 DESCRIPTION OF THE APPLICATION

The application for medical diagnosis was designed to help reluctant people who, because of being afraid of going to a doctor when experiencing health problems, prefer to look up on the internet for possible causes and find a diagnosis.

To further simplify matters, MedicalDiag offers its users, free of charge, a presumptive medical diagnosis service guided by their answers to various questions. The responses are "Yes" and "No". Finally, to make sure that the diagnosis is as accurate as possible, they

Telemedicine Technologies. https://doi.org/10.1016/B978-0-12-816948-3.00001-5

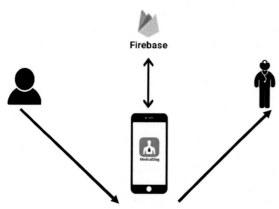

FIG. 1.1 The MedicalDiag block diagram.

have the opportunity of contacting a specialist doctor from a chosen city by phone.

The block diagram of the application is shown in Fig. 1.1.

2.1 Application Features
The application has two modules: the administrative one and the client one.

The administrator module has the following features:
- Change any question for any illness;
- Change user data (name, email, PNC — personal numeric code);
- Export/import the database;
- Statistics for any access;
- Setting permissions for users.
The client module has the following features:
- Account creation and authentication;
- Ability to change your password;
- Displaying the human body (male/female, front and back) and choosing the main areas of the body;
- Identifying a possible disease and displaying a diagnosis based on the user's answers;
- Providing the contact details of a specialist in the field of the user's possible illness, within his or her geographical area;
- Statistics in a pie chart with the most affected areas (data taken from all users in real time);
- The possibility to change personal data (PNC, gender, name).

2.2 Gantt Chart
For a better organization and planning of the development stages of the application, the Gantt chart shown in Figs. 1.2 and 1.3, created with the SmartSheet application, was used. The Gantt chart is a tool used to

plan projects, events and work in general. It follows the stages of running a project according to their duration. It is a common method in project management. It was invented by Henry L. Gantt (American engineer and sociologist) in 1917 and was considered a revolutionary technique at the time, its author being rewarded for his contributions to management [13]. Specifically, the Gantt chart graphically illustrates an activity schedule that helps plan, coordinate, and monitor project-specific tasks. It has the form of a matrix with two axes: a horizontal one indicating the amount of time required for each activity (with the start and end dates of the task); and a vertical one, indicating the tasks to be performed.

2.3 Architecture of the Program
The architecture of the application is presented in Fig. 1.4. The software is developed in the Xcode programming environment using the Objective-C programming language. Google's web and mobile application development platform called Firebase was used for storing and synchronizing the data.

The navigation through the application interfaces is done through a navigation viewer controller that contains, after login, a tab bar viewer controller for managing the other tabs ("statistics" and "my account").

2.4 Design of the Modules
The application was built on the model presented by Apple, in which applications must keep a uniform design, so that the user could see where the application's functions are. The application also follows an auto-layout, an Apple standard for scaling the app on any screen size. The application structure is presented in Fig. 1.5.

The application icon was designed using Adobe Photoshop CC2017, considering current design requirements, the need for fluid integration among other iOS icons, and the need to be user-friendly (Fig. 1.6). This is performed vectorally to scale on any device running an iOS operating system.

The description of the modules is presented in the next paragraphs.

2.4.1 Logging in module
In order to be able to use the application, the user will need to log in, which is also the first screen encountered (start view controller). After logging in from Firebase, the function that authenticates the user is called, with the contents of the email and password fields in the application, and when a response is received from the server, the user and password are saved in a separate native file (NSUserDefaults), otherwise they are the error encountered is displayed.

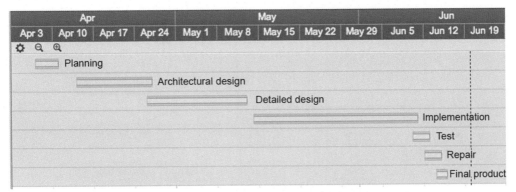

FIG. 1.2 Gantt chart (part 1).

Task Name	Start Date	End Date	Days	% Complete
Detailed design	04/07/17	04/10/17	3	100%
Architectural design	04/14/17	04/26/17	12	100%
Detailed design	04/26/17	05/12/17	16	100%
Implementation	05/14/17	06/10/17	27	100%
Test	06/10/17	06/12/17	2	100%
Repair	06/12/17	06/14/17	2	100%
Final product	06/14/17	06/15/17	1	100%

FIG. 1.3 Gantt chart (part 2).

The next function will allow the user to use the "Next" button on the keyboard in order to jump to the next field, and if it is the last one, the application will "press" the "LOGIN" button on its own. If the user will press anywhere else on the screen, the keyboard will disappear.

If the users have an account, they enter their email and password here. These are saved so that next time when the users wish to use the application, their login is made automatically.

2.4.2 Registration module

If the user does not have an account already created, s/he can register by pressing the top right button, which will send him/her to the next screen, i.e. the Register screen. When the user has entered all the required data and pressed the registration button, the *validateInput* function will be called to validate all fields. For a cleaner and easier way to debug layout, a separate Boolean return function was used for each validation. The email will have to contain the "@" and the "." signs, and the name and password must contain at least 5 characters.

Otherwise, the users will see a warning message on the screen informing them of what they have entered incorrectly.

In order for the PNC to be considered valid, it must start with 0 if the user has selected "Male", or 1 if "Female" is selected. The selection of the gender is made by using the two buttons located at the top of the registration form.

If all these conditions are met, a dictionary containing the name, PNC, and gender will be created; these will be created separately from email and password (encrypted and decrypted by Firebase) in the database user table. If the registration has been successful, the user will receive a message on the screen and will be redirected to the login screen, where the email and password fields will be automatically filled in.

2.4.3 The human body module

The human body consists of two images: the front- and rear-view body. Fig. 1.7 represents the male body. When the user is a woman, the images change.

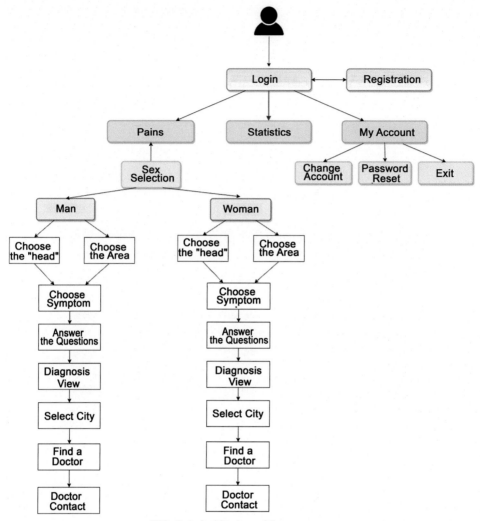

FIG. 1.4 Architecture of the program.

For the head we have an interim class that works according to the same principle, and a separate image will be uploaded containing only this part, to better select other related areas (forehead, eye, nose, mouth, chin, ears). Over the images, buttons have been set with unique identifiers that are related to the Firebase database. When selecting a specific area (pressing a button on the body), it will be shaded to highlight it. This is what the "setupBodyActions" feature does, which cycles through the entire collection of buttons, sets them to be hidden, adds an action that will go to the "goToNextStep"

function and will shade the selected area for better highlighting.

The goToNextStep function verifies if the user has pressed the head-marker area (has element_cap as the identifier) to know which screen to go forward into (the head seen from the front or back, both having separate classes defined), or else it will lead directly to the symptoms section (QuestionManager ViewController) for the area that was pressed. The first query is made in the QuestionManagerViewController class, where the existing malfunctions are listed. If the user chose

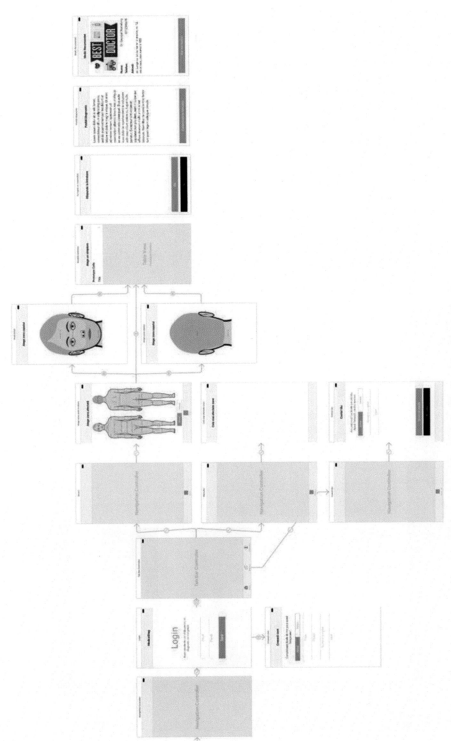

FIG. 1.5 Application structure in Xcode.

FIG. 1.6 Creating the MedicalDiag icon.

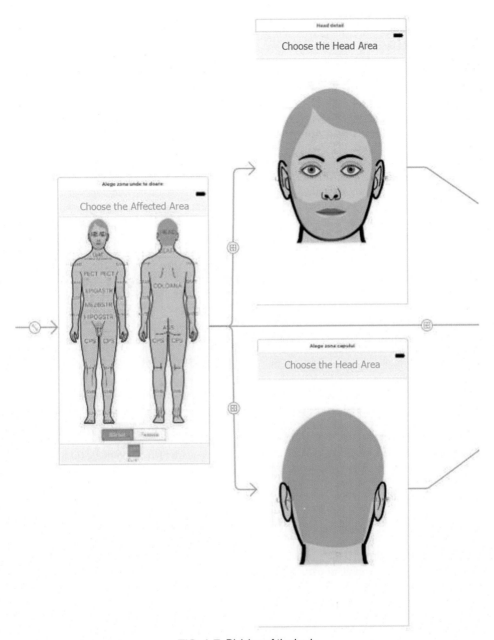

FIG. 1.7 Division of the body.

the head, where there are several sub-zones, then this fact will be taken into account when making the query. Also, this initial query increments the number of hits for that area.

Once we have the illness chosen by the user, we move to the QuestionViewController class, which makes a query, and we take the first set of questions for the chosen illness. Depending on the contents of the "yes" and "no" keys, we open the same class but with a new query, and the query path retains its position. This is done until the "yes" or "no" keys no longer contain dictionaries, but values, meaning that that is the diagnosis.

Next, a new class − DiagnosticViewController - is presented, which will display the diagnosis in a scrollable text view next to a doctor contact button. We also display a button so that the user can return to the human body page, without having to press "back" repeatedly. The "contact a specialist" button opens a native action sheet that gives the user a list of counties in the country to choose a physician who is closer to their location. The list of counties is read from a text file located in the external file folder.

A new class opens, *RecDoctorViewController*, where we display the recommended details of the doctor in the selected county (data read from text files). Here on the top right corner there is a button where the doctor can be called directly from the application or a button that opens a web page containing a list of all doctors in the country.

2.4.4 My account module
My account gives the user:
- the possibility to change the gender, name and PNC. It contains the same checks as the registration module. In order to change the name, PNC, and gender, a query in Firebase will be taken to retrieve this data and display it in the corresponding fields in the application interface.
- the possibility of logging out (removes the email and password, saves the changes and sends us to the login screen)
- the possibility to reset the password. The sendPasswordResetWithEmail function is called in Firebase, giving it the current user's email as parameter, and when a response is received, the users will receive a message informing them that the password has been requested and that they will need to access the reset link received via email.

2.4.5 The statistics sub-module
The statistics tab makes a query through all parts of the body in the database, then extracts the "hits" key to generate a body part dictionary and the number of hits.

To create the pie chart, we use an external library called *AAChartView* that helps us build different types of charts very easily, just by moving the data; it deals with computing, building and interacting with charts.

In order to populate the chart with the required data from Firebase (the number of hits for each affected body part), we created the configChart function that takes this data from the database as the library expects (a key and a dictionary value). Each area on the chart will show the name of the affected area and

the number of hits. This area can be hidden by pressing the name of the condition in the legend.

2.5 Structure of the Database
Firebase has been chosen because it allows easy access to in-app data, is easy to scale to a large number of users, increased security is provided by Google, and APIs are tailored for mobile systems. The database was designed with Google's recommendations for real-time firebase databases. Its structure is in the form of questions:
- head body part: hits - value, illness (question - value, answer 1 - value or another dictionary of the question&answer1&answer2, answer 2 - value or another dictionary of the type question&answer1&answer2
- part of the body: hits - value, illness (question - value, answer 1 - value or another dictionary of the question &answer1&answer2, answer 2 - answer 2 - value or another dictionary of the type question&answer1 &answer2
- users: user idcnp - valuesex - valuename - value

The structure of the Firebase database is illustrated in Fig. 1.8.

3 USAGE OF THE APPLICATION
3.1 Registration Process and Finding a Presumptive Diagnosis
The first step which is necessary for using the app is registration. In order to benefit from all of the application's utilities, the user has to create an account (Fig. 1.9), selecting gender (to differentiate body areas and possible diagnoses) and then filling out a form containing email address, password, full name and Personal Numeric Code (PNC).

All these fields will have to be to be completed by following a set of rules:

The email will have to be in the form of [user] @ [domain]. [TLD], where [domain] is the address of the website that provides the email service and [TLD] is the geographic location extension of the domain (*Top Level Domain*);

The password must be longer than 6 characters;

The name must be longer than 5 characters;

The PNC will have exactly 13 characters and will have to begin with the number 1 if the user selected "Male", or 0 if they selected "Female" (this rule is available in Romania, but it can be changed). Otherwise, a pop-up warning will be displayed specifying what needs to be corrected. After completing this form and thus creating the account, the user is redirected to the Login page, where s/he will enter the email

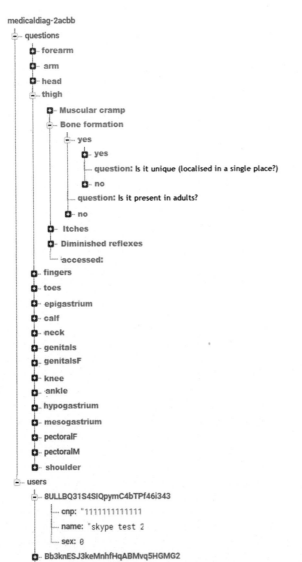

FIG. 1.8 Structure of the database.

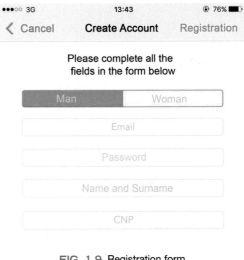

FIG. 1.9 Registration form.

address and password followed by the *"Enter"* button. Once logged in, the human body will be displayed according to the gender chosen by the two buttons located at the bottom of the screen: male and female (Fig. 1.10).

The next step will be to select the affected area from the human body (differing by gender) by tapping on it. To make sure that tapping is done exactly in the desired area, the application will highlight the area with a shadow, as it can be seen in Fig. 1.11. The available body parts are:

- Head — the most complex part of the body; when selected, the application will zoom in on it to facilitate easy selection;
- Feet, thigh, knees, fingers, hands;
- Shoulder, arm, forearm, fingers;
- Body: pectorals/breast, hypogastrium, mesogastrium, epigastrium and genital organs.

After selecting the affected area, the user will select the symptom (Fig. 1.12) and respond to a set of successive YES or NO questions, after which s/he will receive a diagnosis, indicating that this is a presumptive one, and a medical check is recommended for the diagnosis to be as accurate as possible.

With the receipt of the presumed diagnosis (Fig. 1.13), the user will have a button located at the bottom of the screen, to search for a specialist doctor. When pressed, the user will be presented with a list of the country's 41 counties to narrow searches to the selected region (Fig. 1.14).

Finally, after displaying the recommended physician in the selected region, along with his/her address and phone number, the user can contact him/her via the button illustrated by a handset at the top right of the screen.

3.2 Statistics Area

Apart from the above functionalities, the user can access a statistics area (by clicking on the "Statistics"

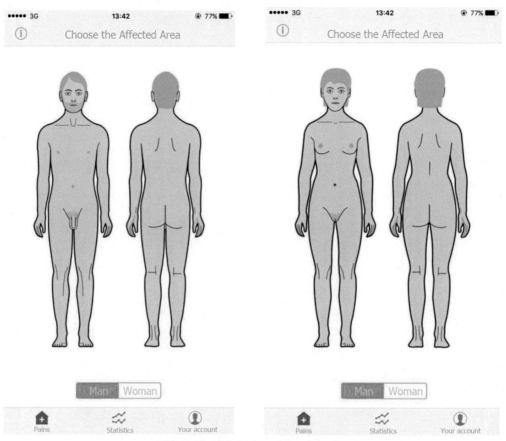

FIG. 1.10 Male body and female body.

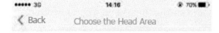

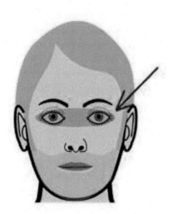

FIG. 1.11 Selection of affected area.

tab located in the menu at the bottom of the screen) where s/he can see with a pie chart, the most affected areas in all users. Body parts are represented on the diagram with different colors and when they are pressed, the number of affected users in the area is shown (Fig. 1.15).

3.3 Update/Modify the Personal Data

One of the key features of the app is the possibility to change personal data (gender, name or PNC), this being done in the "My Account" menu, accessible from the lower-menu menu bar.

If the user wants to change the password for personal reasons, the "Change Password" button is pressed (Fig. 1.16), at which time a pop-up will inform them that a mail has been sent to their email account with which they created the account (Fig. 1.17).

In order to complete the password change process, it will be necessary to access the email link, where the user will need to enter the new password.

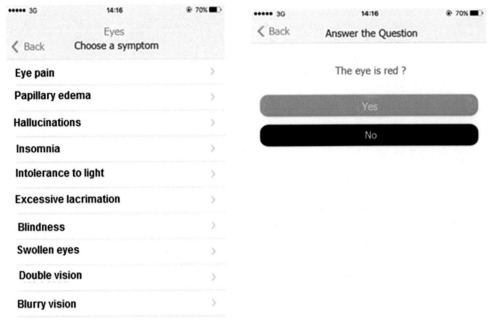

FIG. 1.12 Symptom list.

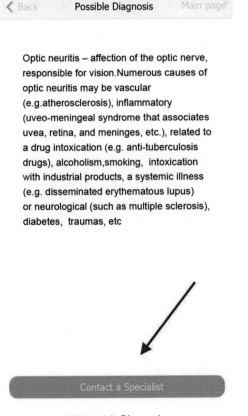

FIG. 1.13 Diagnosis.

FIG. 1.14 List of counties.

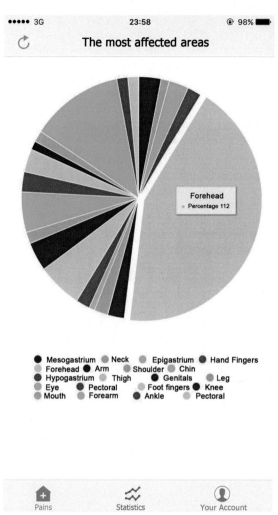

FIG. 1.15 Affected area percentage.

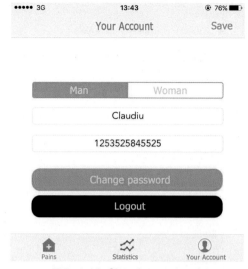

FIG. 1.16 Changing password.

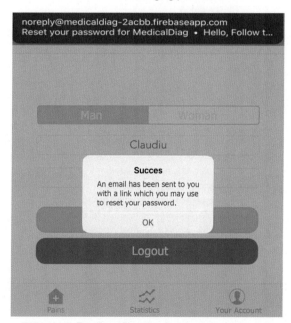

FIG. 1.17 Email confirmation for changing password.

4 CONCLUSIONS

The MedicalDiag application was built according to the specifications provided, eventually reaching version v.1.3.2, bringing new functionalities and improvements to it during the development. MedicalDiag is available on any device running the iOS operating system, designed to diagnose users' illnesses following a set of questions.

The biggest challenge of this application was to create a simpler and more logical structure for arranging questions, answers and diagnostics. The current solution is not just "memory friendly", but also scalable. If one should want to add one or 100,000 questions later, this will not affect the application at all. Any element in the application is scalable and any question, answer, or diagnosis can be modified without interfering with the application, all being changed in Firebase.

As an application for a vast domain, it has many development directions and will be periodically upgraded, bringing new functionalities and even solving existing problems. The app will be placed as soon as possible in Apple's AppStore, and new features will be brought to it, such as the ability to contact a physician in real time with a live chat, or search for it using the embedded GPS module.

These two new functionalities are already under development and another is in plan, namely the possibility of receiving a diagnosis after uploading a photograph (if the disease has visual signs, for example varicella).

REFERENCES

[1] R. Ford, J. Wierdermann, The App&mobile Case Study Book, Indexed edition, TASCHEN, 2011.
[2] W.J. Yi, W. Jia, J. Saniie, Mobile sensor data collector using android smartphone, in: Circuits and Systems, IEEE 55th International Midwest Symposium, 2012, pp. 956–959.
[3] A. Székelya, R. Talanow, P. Bágyi, Smartphones, tablets and mobile applications for radiology, Eur. J. Radiol. 82 (May) (2013) 829–836.
[4] M. Panhale, Beginning Hybrid Mobile Application Development, first ed., Apress, 2015.
[5] W.J. Yi, W. Jia, J. Saniie, Mobile sensor data collector using android smartphone, in: Circuits and Systems, IEEE 55th International Midwest Symposium, 2012, pp. 956–959. September.
[6] R. Pryss, M. Reichert, J. Herrmann, B. Langguth, W. Schlee, Mobile crowd sensing in clinical and psychological trials — a case study, computer-based medical systems, in: IEEE 28th International Symposium, June, 2015.
[7] R. Pryss, M. Reichert, B. Langguth, W. Schlee, Mobile crowd sensing services for tinnitus assessment, therapy, and research, mobile services, in: IEEE International Conference, August, 2015.
[8] K.F. Braekkan Payne, H. Wharrad, K. Watts, Smartphone and medical related App use among medical students and junior doctors in the United Kingdom (UK): a regional survey, BMC Med Inf. Decis. Mak. (2012) (October).
[9] M.N. Faruk, J. Mallipeddi, An optimal healthcare self-diagnosis system using cloud framework, in: International Conference On Big Data Analytics and Computational Intelligence, 2017.
[10] L. Briz Ponce, J. Antonio, J. Méndez, F.J. García Peñalvo, First approach of mobile applications study for medical education purposes, in: TEEM '14, Salamanca, October, 2014.
[11] M.N. Kamel Boulos, A.C. Brewer, C. Karimkhani, D.B. Buller, R.P. Dellavalle, Mobile medical and health apps: state of the art, concerns, regulatory control and certification, Online J. Public Health Inform. 5 (February) (2014).
[12] L. Csiki, A. Gabor, M. Popescu, A. Naaji, Software application for disabled students, Brain 8 (4) (2017) 32–40 (December).
[13] http://www.scritub.com/stiinta/informatica/Ce-este-diagrama-Gantt1032201711.php, accessed in January 2017.

CHAPTER 2

Emerging Paradigms in Transform-Based Medical Image Compression for Telemedicine Environment

SUJITHA JULIET DEVARAJ
Department of Computer Science and Engineering, Karunya Institute of Technology and Sciences, Coimbatore, India

1 INTRODUCTION

This chapter focuses on the challenges in affording remote health care through telemedicine and presents the importance of image compression methods for effective storage/transmission. With the widely dispersed populations, efficient medical services and health care have scarcely touched the lives of the people in rural areas. India, being the second-most populated nation in the world with a total area of 3.28 million Sq. Km., sustains 16.7% of the world population with over 1.35 billion residents [1]. Around 70% of India's population is residing in rural areas in which nearly 800 million people living in the residential area on the outskirts of a city have no direct access even to primary health care. The ready availability of health care services still has not kept pace for these low-income and under-served individuals in the suburbs. Telemedicine provides quality medical services by bringing the knowledge and experience of health professional closer to patients with less healthcare cost.

The four categories of telemedicine systems are:

a. Asynchronous Telemedicine
b. Synchronous Video Conferencing or Interactive telemedicine
c. Remote Patient Monitoring
d. Mobile Health

1.1 Asynchronous Telemedicine

Asynchronous telemedicine system as shown in Fig. 2.1 involves the transmission of acquired medical images and patient's record to the healthcare provider at a suitable time for detailed inspection. This system is often used in non-emergent situation and doesn't require the simultaneous attention of medical specialist and primary care practitioner [2]. Specialized medical fields such as radiology, tele-pathology and tele-dermatology rely on this type of telemedicine.

1.2 Synchronous Video Conferencing or Interactive Telemedicine

The interactive telemedicine system also known as synchronous video conferencing provides two-way real time interactions between patients and specialists through telecommunications technology with excellent network infrastructure. The live conferencing system facilitates the quick sharing of patient's critical information and further improves the overall level of treatment. Advanced medical system like tele-radiosurgery that requires visualization of patient's treatment depend on interactive video conferencing system (Fig. 2.2).

1.3 Remote Patient Monitoring

Remote patient monitoring system as shown in Fig. 2.3 assists the medical specialist to monitor patients' bio-signals from a far, through specialized equipments. This system is particularly useful for aged patients with chronic diseases to monitor their vital health signals and alert the medical experts during emergency [3]. It offers immense comfort to elderly and disabled since it minimizes the time spend in the clinic and provides a much better level of care.

1.4 Mobile Health or m-Health

m-Health provides a new phase of healthcare reform. This system plays a vital role in modern healthcare monitoring/alerting systems and clinical data storage and maintenance systems. This system uses smart mobile devices that utilize 3G and 4G mobile networks

Telemedicine Technologies. https://doi.org/10.1016/B978-0-12-816948-3.00002-7

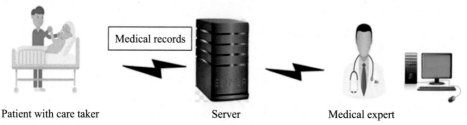

Patient with care taker   Server   Medical expert

FIG. 2.1 Asynchronous telemedicine.

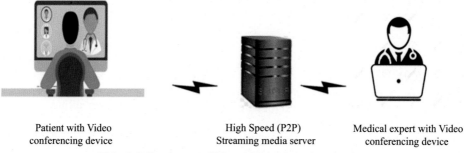

Patient with Video   High Speed (P2P)   Medical expert with Video
conferencing device   Streaming media server   conferencing device

FIG. 2.2 Synchronous Video conferencing telemedicine.

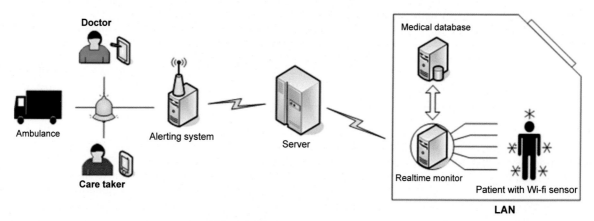

FIG. 2.3 Remote patient monitoring.

for data transmission. The specialist and the patient can access the medical registers conveniently regardless of time and place and patient may contact the specialist for any emergency [2].

2 TECHNICAL CHALLENGES IN TELEMEDICINE

Despite the telecommunication technology afford greater services in improving the quality of medical services through telemedicine, there are many technical challenges faced by telemedicine technology. This section addresses those challenges:

- **Security:** In telemedicine services, highly sensitive medical information about the patient is exchanged through wireless channels which need to be preserved from public notice. Providing security measures like verification code, critical security question and behavior-based security algorithms can result in beneficial outcome on the quality of health care [4]. Therefore, security is one of the critical issues to be addressed in telemedicine.

- **Patient monitoring:** Patient monitoring is one of the most important requirements in establishing improved health care especially for disabled people who require personal attention. Real time monitoring with adaptable equipments to measure bio-signals and to diagnose the severity is crucial in telemedicine [5]. The system should assist the medical specialist to observe and oversee the health conditions of patients with chronic sickness.

- **Scheduling:** Scheduling process identifies and allocates resources between varieties of possible tasks. In asynchronous telemedicine, the clinical information about the patient in remote site is evaluated offline by the medical specialist. Therefore scheduling and resource allocation are easily achievable. However, in synchronous or interactive consultation, an optimal scheduling based on priority is essential to avoid the signal transmission delay and to improve clinic accessibility [6]. The tracking of differently abled people and their clinical signals using real-time tracking system is also the need of the hour.

- **Digital Health Record:** In providing enhanced medical support through telemedicine, personal e-health record is one of the key elements, which result in effective diagnosis, shared decision making, reduced duplication of medical tests and abating inappropriate drug combinations [7]. Even though EHRs enable quality orientated health care, the implementation challenges need to be addressed.

- **Network Access:** Telemedicine system needs to transfer large amount of medical information through a secured and high speed network with shortest possible time. The telemedicine network often encounters challenges toward bandwidth management and network congestion. A strategy needs to be planned for optimizing the telemedicine network utilization upon considering the image file size and traffic forecast [8].

- **Bigdata analysis:** An emergent Bigdata analytics in healthcare delivery systems has become progressively more important in the recent years. The Big data in health care domain is gathered from diverse sources such as clinical data, sensor information and bio-signals. The analysis of huge volumes of medical data facilitates personalized clinical services which further leads to improved care delivery [9].

- **Knowledge Based Systems:** Development of telemedicine technology with intelligent decision support system that substitutes a medical specialist with virtual specialist is one of the great challenges in telemedicine [10]. The expert system should ensure

that the patient is treated properly and consultation by medical specialist is provided if needed.

- **Medical Image Compression:** In recent years, digital medical imaging schemes are widely used in clinical investigations to monitor the bio-signals and to calculate the physiological cross-sectional area of patient's body. Since CT and MRI modalities offer detailed images with superior quality, they are extensively employed in telemedicine for efficient diagnosis. The colossal volume of digital images needs to be preserved and transmitted over the network for diagnosis. However, preserving medical information for many years may become unaffordable for majority hospitals owing to the storage requirements. In order to improve the overall clinical care, it is crucial to reduce the demand for storage and transmission by utilizing efficient medical image compression methods. Thus, developing a compression method that results in reduction of redundancies in images as well as preserving the vital information is one of the great solutions to address this issue in telemedicine [11].

All the issues pointed out here are extensively significant to be solved in telemedicine. However, compression of medical images while maintaining acceptable visual quality is the critical challenge in telemedicine for efficient transmission of medical data [12,13]. Generally, image compression is achieved by transforming the input data, projecting it on a basis of functions, and then encoding the transform coefficients. This chapter discusses and analyses the emerging paradigms in transform-based medical image compression, particularly suitable for telemedicine.

2.1 Motivation of the Chapter

In recent years, medical imaging modalities such as CT and MRI are extensively being utilized in clinical investigations through telemedicine. Handling the colossal volume of image datasets increases the need for efficient image compression methods. During the last two decades, several image compression methods such as run length coding, arithmetic encoding, JPEG 2000, LZ compression and Huffman coding have been proposed in order to meet the increasing demands for medical images. Out of the existing compression methods, the transform based methods are widely used due to their decorrelation, linearity and orthogonality properties. In transform based compression, the transformation process converts a large set of highly correlated pixels into a smaller number of decorrelated transform coefficients, so that energy is concentrated into as few coefficients as possible. Among the existing compression methods, much interest has been focused on attaining the desirable characteristics such as high compression ratio, and little work has been done on hierarchical representation of images that may lead to improvements in compression performance. Also, an efficient compression method that utilizes less computational time to compress the medical images is required in telemedicine. Moreover, a compression method that helps the healthcare provider to communicate to the medical specialist by locating the region of interest at different scales and angles is much required in telemedicine.

Grounded on these motivations, this chapter describes a paradigm of transform based medical image compression for telemedicine applications. The main objective of the chapter is to investigate the performances of a variety of image transforms on a set of medical images for different compression evaluation parameters.

3 EMERGING PARADIGMS IN TRANSFORM BASED CODING

This section discusses on seven different transforms (Ripplet, Bandelet, Radon, Contourlet, Curvelet, DCT, Haar Wavelet and RST transforms) for medical image compression and investigates their performances for a set of medical images. All the above mentioned transforms are encoded using SPIHT (Set Partitioning in Hierarchical Trees) encoder to provide better and quality compression.

3.1 Ripplet Transform

The ripplet transform [14] is effective in capturing directional features (edges and curves) in medical images. It generalizes curvelet transform by introducing arbitrary support c and degree d. The arbitrary components c and d provide anisotropy capability of representing singularities along randomly shaped curves. Ripplet transform is capable of representing 2D images at different scales and directions. Since ripplet transform is optimal for representing images with C^2 singularities, the image edges represent sparsity in ripplet space. The image region satisfies the property for its length and width as $width \approx c \times length^d$ where c represents the support and d defines the degree of ripplets.

The ripplet transform of an $M \times N$ image $f(a,b)$ is given as

$$R_{j,\overrightarrow{k},l} = \sum_{a=0}^{M-1} \sum_{b=0}^{N-1} f(a,b) \cdot \overline{\rho_{j,\overrightarrow{k},l}(a,b)} \qquad (2.1)$$

where $R_{j,\overrightarrow{k},l}$ denotes the ripplet coefficients and $\rho_{j,\overrightarrow{k},l}(a,b)$ are the ripplets of scale j at position index k with angle index l in discrete domain, $\overline{(.)}$ represents the conjugate operator.

The input medical image $f(a,b)$ is initially decomposed into a set of multi-resolution subbands $P_0, (\Delta_s, s > 0)$ through biorthogonal 9/7 wavelet transform. The decomposed $f(a,b)$ is given as

$$f(a,b) \mapsto (P_0 f(a,b), \Delta_1 f(a,b), \Delta_2 f(a,b), \ldots) \qquad (2.2)$$

where $P_0 f(a,b)$ represents lowest frequency component and $\{\Delta_1 f(a,b), \Delta_2 f(a,b), \ldots\} \in \Delta_s f(a,b)$ denote high frequency components. $\Delta_s f(a,b)$ contain details about 2^{-2s} wide. The frequency domain is partitioned into 3 subbands as $s = 1, 2, 3$ and the decomposed wavelet bands j are partially reconstructed into ripplet subbands as $j \in \{2s, 2s + 1\}$.

The high frequency subbands $\Delta_s f(a,b)$ are smoothly dissected into dyadic squares and then renormalized to the unit square $[0,1] \times [0,1]$. The renormalized squares are denoted as ripplets in spatial domain. Finally, the ripplet coefficients and the lowest frequency components are further coded using SPIHT encoder to get the compressed image.

3.2 Bandelet Transform

The bandelet transform [15] builds sparse representation that employs small number of coefficients to represent a signal. This representation utilizes the benefit of geometric regularity of edges in images. The image gray levels have regular variations in the direction of geometric flow. In bandelet based compression, the input medical image $b(x,y)$ is initially divided into spatial frequency sub-bands using biorthogonal wavelet transform. The approximation low-pass sub-band, a_j, is a coarser version of the input image, while the other sub-bands, d_j^H, d_j^V, d_j^D represent the high frequency details in the horizontal, vertical and diagonal directions, respectively. The subbands are given as

$$a_j(x,y) = \langle b(x,y), \phi_j(x)\phi_j(y) \rangle$$
$$d_j^H(x,y) = \langle b(x,y), \psi_j(x)\phi_j(y) \rangle$$
$$d_j^V(x,y) = \langle b(x,y), \phi_j(x)\psi_j(y) \rangle$$
$$d_j^D(x,y) = \langle b(x,y), \psi_j(x)\psi_j(y) \rangle$$

where $\phi_j(x,y)$ represents the scaling function and $\psi_j(x,y)$ denotes the wavelet function. j represents the scale factor. The decomposed image is recursively segmented into different dyadic squares S of equal size 2^k. In each sub-square, the discrete wavelet coefficients "warp-around" the edges rather pass through them. The warped wavelet coefficients are then computed with subband filtering that uses reconstruction filter and sub-sampling method.

The bandelets are inserted in warped wavelet basis to obtain bandelet coefficients and are given as

$$\left\{\begin{array}{l} \psi_l^{m_1}(x)\psi_j^{m_2}(y-c(x)), \\ \psi_j^{m_1}(x)\phi_j^{m_2}(y-c(x)), \\ \psi_j^{m_1}(x)\psi_j^{m_2}(y-c(x)) \end{array}\right\}_{j,l>j,m_1,m_2} \quad (2.3)$$

The resultant bandelets are encoded using SPIHT encoder to acquire better compression. Researchers [16] have utilized the characteristics of Bandelet transform for the compression of medical images.

3.3 Radon Transform

The Radon transform estimates the projection of the image intensity along specific angles. The resulting projection is the total intensities of the pixels in each direction. It maps the spatial domain coordinates (x,y) to projection domain coordinates (t,θ). Radon projections are computed at various orientations and the pixel intensities are added [17]. For $f(x,y)$, the projection is formed by performing a line integration of the image intensity along a line L at a distance t from the origin and at angle θ. The projection at distance t and the vector v in the (x,y) plane are represented as $t = x\cos\theta + y\sin\theta$ and $v = -x\sin\theta + y\cos\theta$.

The Radon transform over integrable region is defined as

$$r(t,\theta) = \int_{-\infty}^{\infty} f(x,y)dv, \quad (x,y)\in L_t \quad (2.4)$$

and the radon transform of $f(x,y)$ is given by $r(t,\theta)$ as

$$r(t,\theta)\overset{\Delta}{=}\int_{-\infty}^{\infty}\int_{-\infty}^{\infty} f(x,y)\delta(x\cos\theta + y\sin\theta - t)dxdy \quad (2.5)$$

where δ denotes the Dirac function and $t \in (-\infty,+\infty)$ and $\theta \in [0,\pi]$.

More the number of projections and the points measured on each projection improve image quality with reduced artifacts. Researchers have utilized Radon transform with projections for medical image compression [18]. The radon transformed coefficients are coded using SPIHT encoder to attain improved compression performance.

3.4 Contourlet Transform

Contourlet transform introduced by Do et al. [19] is able to capture inherent geometrical image structure. It achieves optimal approximation performance for 2-D piecewise smooth functions except for discontinuities along C^2 curves. Contourlets uses discrete filter bank structure for images with smooth contours to obtain sparsity. In this structure, the Laplacian pyramid captures the point discontinuities and the directional filtering relates point discontinuities into linear structures. Contourlets have supports at various aspect ratios, scales and directions. Since Contourlet transform is directly formulated in discrete domain, it does not require conversion from continuous time-space domain. Researchers have investigated the application of contourlet transform for image compression [20].

The input image $f(x,y)$ is decomposed into multiscale sub-bands through Laplacian pyramid $(C_0f(x,y),\Delta_1f(x,y),\Delta_2f(x,y), ...)$. Where $C_0f(x,y)$ represents coarser component of the image and $\{\Delta_1f(x,y), \Delta_2f(x,y),...\} \in \Delta_df(x,y)$ denote the detailed components. Each bandpass channel $\Delta_df(x,y)$ is provided into Directional Filter Bank (DFB) to capture the directional information and to compute the coarse component. The combination Laplacian pyramid and DFB offers multiscale and directional subbands which recommends a flexible contourlet transform.

3.5 Curvelet Transform

Starck et al. [21] proposed curvelet transform, based on the properties of multi-scale ridgelet transform. The curvelet transform is a high-dimensional generalization of wavelet transform and is efficient in representing the signal functions with high directionality. Curvelet is a suitable basis for representing images with smooth curve discontinuities. It provides sparsity by establishing orthogonality through subband filtering. Also the subband decomposition enforces anisotropic relationship between the width and length of the frames and observes $width = length^2$.

The curvelet transform utilizes the benefit of dyadic sequence of scales and subband filtering. The input image $f(x,y)$ is decomposed into a set of multi-resolution sub-bands $(P_0f(x,y),\Delta_1f(x,y),\Delta_2f(x,y), ...)$. The passband filter $\Delta_sf(x,y)$ is concentrated near the frequencies $[2^{2s},2^{2s+2}]$. Each subband $\Delta_sf(x,y)$ is smoothly divided into squares of appropriate scale and then renormalized to unit scale which results in the ratio of $width = length^2$. Each square is analyzed through curvelet transform and the curvelet coefficients with coarsest subband coefficients are further coded using SPIHT encoder to get the compressed image.

3.6 Discrete Cosine Transform (DCT)

DCT is one of the known and widely used transform for image compression. It transforms the image from the spatial domain $f(x,y)$ to the frequency domain $F(u,v)$.

At low bitrates, DCT offers lower image deformation and better visual quality. Most of the critical detail about the image is concentrated in few coefficients of the DCT. This property makes DCT to be at the heart of the famous lossy image compression JPEG.

The two dimensional DCT of $M \times N$ matrix S is defined as

$$D_{ab} = \alpha_a \alpha_b \sum_{m=0}^{M-1} \sum_{n=0}^{N-1} S_{mn} \cos \frac{\pi(2m+1)a}{2M} \cos \frac{\pi(2n+1)b}{2N}, \quad (2.6)$$

where

$$\alpha_p = \begin{cases} \dfrac{1}{\sqrt{M}} & p = 0 \\ \sqrt{2/M} & 1 \leq p \leq M-1 \end{cases} \quad \text{and}$$

$$\alpha_q = \begin{cases} \dfrac{1}{\sqrt{N}} & q = 0 \\ \sqrt{2/N} & 1 \leq q \leq N-1 \end{cases}$$

The coefficients D_{ab} are called as DCT coefficients of S. The decomposed DCT coefficients are encoded using SPIHT to improve the compression performance.

3.7 Haar Wavelet Transform

Haar transform, introduced by Alfred Haar in 1910 is one of the simplest and oldest transform. Due to its low computing requirement and wavelet-like structure, Haar transform is mainly used in signal and image compression and to analyze the localized aspects of the input signal [22]. The Haar functions represent the signal in different scales and frequencies.

The Haar's mother wavelet function $\psi(h)$ is defined as

$$\psi(h) = \begin{cases} 1 & 0 \leq h \leq \dfrac{1}{2} \\ -1 & \dfrac{1}{2} \leq h \leq 1 \\ 0 & \text{otherwise} \end{cases} \quad (2.7)$$

and the scaling function $\phi(h)$ is given as

$$\phi(h) = \begin{cases} 1 & 0 \leq h \leq 1 \\ 0 & \text{otherwise} \end{cases}.$$

Haar wavelet transform obtains a set of subclasses of images by decomposing the input image into appropriate subsamples using pyramidal architecture. The decomposition is performed along the vertical and horizontal directions to preserve constant number of pixels to represent the image [23]. The basic concept of Haar is to transfer the input image into a square matrix in which each element represents a pixel in the image. This matrix includes information about global, horizontal, vertical and diagonal details embedded in the image. The decomposed wavelet coefficients are vector quantized and encoded using SPIHT to yield compressed image.

3.8 Geometric (RST) Transform

In telemedicine, there may be a necessity for healthcare provider at rural medical center to communicate to a medical specialist at a distance, by positioning the infected region or region of interest at different scales and angles. The compression of region of interest without compromising the quality of the medical image is one of the major requirements in providing quality medical service through telemedicine.

Geometric transformation, a unique class of affine transformation maps the pixel intensity values in an input image into new variables in an output image by performing linear operation such as rotation, scaling and translation. Geometric scaling changes the visual appearance of an image by altering the quantity of pixel information in a particular region. When scaling a medical image with adjustable scaling factor, a new image with a higher amount of pixels with no loss of image quality is produced. Image scaling also regards the association between neighboring pixels through bilinear interpolation to achieve a best approximation of pixel intensity values based on surrounding pixels values [24].

Consider an image $g(x,y)$ that undertakes geometric deformation to result in a transformed image $g'(x',y')$. Here, $x' = p(x,y)$ and $y' = q(x,y)$, where $p(x,y)$ and $q(x,y)$ are the spatial relationships that yield geometrically deformed image $g'(x',y')$. The geometrical deformation is obtained through the operations such as rotation, scaling and translation [25] applied on an image.

Scaling performs image resizing with a trade-off between sharpness and efficiency. Image scaling is also termed as image re-sampling, image enhancement or image interpolation. Let the (x,y) coordinates are multiplied by scaling factors r and s to achieve (x',y') transformed coordinates. If r and s are given similar value, the transformation results in uniform scaling on both x and y directions. If $0 < r < 1$, it results in image compression or sub-sampling in x direction. The image reduction is obtained by the pixel replacement or by interpolating between pixel values in local neighborhoods. Image zooming or expansion is achieved in x direction through pixel replication when $r > 1$. A similar transformation is performed in y direction by scaling (x,y) coordinates with s factor.

The scaling transformation is denoted as

$$\begin{aligned} x' &= x \cdot r \\ y' &= y \cdot s \end{aligned} \quad (2.8)$$

The image repositioning is achieved by rotating the image $g(x,y)$ by angle θ in a counterclockwise direction around its center point. The new coordinates of a point along the z direction is derived as

$$x' = x \cos \theta + y \sin \theta$$
$$y' = -x \sin \theta + y \cos \theta \qquad (2.9)$$

The image translation shifts or repositions the image $g(x,y)$ along a straight-line path from one location to another through translation vector (t_x, t_y). The translated coordinates (x', y') is obtained by adding t_x and t_y to image coordinate (x,y). The transformation of coordinates is expressed as

$$x' = x + t_x$$
$$y' = y + t_y \qquad (2.10)$$

Hence the geometric transformations are demonstrated in matrix form as

$$\begin{bmatrix} x' \\ y' \end{bmatrix} = \begin{bmatrix} r & 0 \\ 0 & s \end{bmatrix} \cdot \begin{bmatrix} \cos \theta & -\sin \theta \\ \sin \theta & \cos \theta \end{bmatrix} \cdot \begin{bmatrix} x \\ y \end{bmatrix} + \begin{bmatrix} t_x \\ t_y \end{bmatrix} \qquad (2.11)$$

where $\begin{bmatrix} r & 0 \\ 0 & s \end{bmatrix}$ defines the scaling matrix, $\begin{bmatrix} \cos \theta & -\sin \theta \\ \sin \theta & \cos \theta \end{bmatrix}$ refers the rotation matrix and (t_x, t_y) represents the translation or shift vector.

3.8.1 Image interpolation

Image interpolation tries to attain a best estimation of a pixel's intensity based on neighboring pixel values on proximity basis. Interpolation is particularly essential when resizing or resampling of image is performed to meet the specifications of the transmission channel or to present the final image with no visual loss. Image interpolation is generally achieved through one of three methods: nearest neighbor, bilinear interpolation, or bicubic interpolation. Since each method has its own merits and challenges, the choice of appropriate method is based on state of affairs.

The basic method for interpolation is nearest neighbor where the output pixel value is estimated by considering the nearest pixel's value to the specified input coordinates. This method is simple and easy to implement and no artificial data is introduced in the final output. However the output image contains blocking artifacts which may have effect on error calculations in performance evaluation. Bilinear interpolation estimates the appropriate intensity pixel values by finding the distance weighted average of the four nearest pixels, located in diagonal directions. Bicubic interpolation determines the pixel value from the weighted average

of the 16 closest neighboring pixels. Bicubic methods produce smoother output with a better representation of the input image. However, the efficiency is achieved at the expense of complexity to perform mathematical evaluation. Based on the requirements for efficient interpolation and compression with less complexity, Bilinear interpolation is selected in this research for image resampling.

The region of interest specified by the medical specialist is extracted through image scaling by assigning the scaling factor more than 1. Image scaling enlarges the input image about the origin in x and y direction by the scaling factors r and s, denoted by the scaling matrix $\begin{bmatrix} r & 0 \\ 0 & s \end{bmatrix}$. A uniform scaling is obtained by assigning g and h the same value of 1.2. The scaled image is then up sampled using bilinear interpolation to determine the pixel value of the output coordinates. While performing rotations, neither rows nor columns of the image are reproduced unless the rotation angle is a multiple of 90. Consequently, the resampled image is then translated by including translation distances (t_x, t_y) with (x, y) coordinates that result in (x', y') transformed coordinates.

4 SPIHT ENCODER

Encoders improve the quality of the images at high compression rates. Many entropy encoders are proposed to encode the medical images with excellent result. Few of the encoders presented by researchers are Huffman coding [26], Embedded Zerotree Wavelet (EZW) [27], Set Partitioning in Hierarchical Trees (SPIHT) by Said et al. [28]. Context-based, Adaptive, Lossless Image Codec (CALIC) proposed by Wu et al. [29]; Embedded Block Coding with Optimized Truncation (EBCOT) algorithm proposed by Taubman [30] and Set Partitioned Embedded block coder (SPECK) proposed by Pearlman et al. [31]. While many encoders are proposed for image compression, SPIHT is considered as one of the efficient encoders owing to its significant characterizes such as intensive progressive capability, scalability, low computational complexity and variable bit rate. This coder exploits the spatial relationships between the decomposition subbands. It works as spatial orientation tree, where each subband coefficient is encoded based on the significance of coefficients.

SPIHT employs three sets of lists based on the significance of coefficients. (i) The list of insignificant pixels (LIP), (ii) the list of insignificant sets (LIS) and (iii) the list of significant pixels (LSP). LIP and LSP holds the single pixel nodes and LIS consists of descendants nodes. The maximum number of bits needed to represent the highest coefficients in the spatial orientation tree is obtained and expressed as n_{max} where $n_{max} = [\log_2(\max_{x,y}\{|T_{x,y}|\})]$.

The encoding method uses two stages of operation: Sorting and Refinement stages. The encoder segments the transformed coefficients into significant and insignificant sections based on the significance test $s_n(T)$ as in Eq. (2.12).

$$s_n(T) = \begin{cases} 1, \max_{(x,y) \in T}\{|T_{x,y}|\} \geq 2^n \\ 0, otherwise \end{cases} \tag{2.12}$$

where $T_{x,y}$ represents the resultant coefficients at coordinates (x,y). The sorting and refinement procedures are repeated until all the transformed coefficients are evaluated completely.

All the above mentioned transformed coefficients (Ripplet, bandelet, Radon, Contourlet, Curvelet, Haar Wavelet, DCT and RST transforms) are encoded using SPIHT encoder to improve the compression performance.

5 EXPERIMENTS AND DISCUSSIONS

The experimental investigations on the transform based image compression methods are carried out on gray intensity CT and MRI images represented in Axial, Coronal and Sagittal planes/views. The performance measures employed to analyze the behavior of the transforms for image compression are Peak signal to Noise ratio (PSNR), Structural Similarity Index Measure (SSIM), Computational Time (CT), Compression Ratio and Subjective Assessment.

PSNR is a pixel-wise error metric employed to arbitrate the visual image quality based on mean square error. Higher the PSNR, better the compressed image quality. The visual image quality of the transform based compression method is analyzed for a range of bitrates varying from 0.07 to 1.5 and the average PSNR is considered for performance analysis.

$$MSE = \frac{1}{P * Q} \times \left[\sum_{x=0}^{P-1} \sum_{y=0}^{Q-1} (f(x,y) - F'(x,y))^2 \right] \tag{2.13}$$

$$PSNR = 10 * \log_{10}\left(255^2 / MSE \right) \tag{2.14}$$

SSIM is an objective image quality metric that measures the structural similarity between the original and compressed images based on the human visual system. It is defined as

$$SSIM(x,y) = \frac{(2\mu_x\mu_y + A_1)(2\sigma_x\sigma_y + A_2)}{\left(\mu_x^2 + \mu_y^2 + A_1\right)\left(\sigma_x^2 + \sigma_y^2 + A_2\right)} \tag{2.15}$$

where x and y denotes the spatial windows, μ_x and μ_y represent average intensities and σ_x^2 and σ_y^2 represent variations of x and y respectively and A_1 and A_2 are constants.

The computational time measures the time utilized by the processor to compress the medical image. It depends on the complexity of compression algorithm, implementation efficiency and processor speed. Compression ratio specifies the ratio of uncompressed (original) file size to the compressed file size. It is computed as the number of bits per pixel before compression, divided by the number of bits engaged per pixel after compression. Compression ratio for gray scale image is $(1/bpp)*8$.

The compression methods are validated through MATLAB software executed in Intel(R) i5 processor with 4 GB RAM. The test images of size $(256 \times 256,$ 8 bpp) are obtained from KG Hospital & Postgraduate Medical Institute, Coimbatore, India. Fig. 2.4 shows the sample set of medical images employed for investigation.

5.1 Investigations on Ripplet Transform

The performance measures of ripplet transform based image compression with SPIHT encoder is shown in Table 2.1.

From Table 2.1 it is inferred Ripplet transform based compression method result in relatively higher PSNR for test medical images with an average value of 46.47. Ripplet transform holds anisotropy behavior that leads to efficient representation of edges in images. Since ripplet transform effectively confines the 2D singularities more precisely along the curvatures [32], the SSIM value reaches nearly to 1. The average time utilized to compress the medical images is 0.60 s. This is due to the redundant behavior of ripplet transform that evades boundary effect. It is also inferred that ripplet yields greater compression ratio of 11.83 and performs well on the compression of both MRI and CT images.

5.2 Investigations on Bandelet Transform

The performance measures of Bandelet transform based image compression with SPIHT encoder is shown in Table 2.2.

Table 2.2 shows that the Bandelet transform based compression results in fairly higher PSNR for test medical images with an average value of 46.32. For Axial T2 Weighted MRI brain image, Bandelet transform

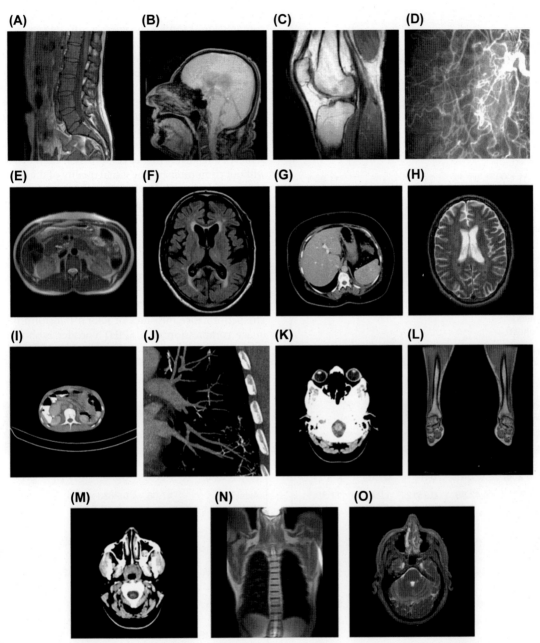

FIG. 2.4 Sample test images employed for analysis. (A) Sagittal T1 Weighted MRI Spine, (B) Sagittal T1 Weighted MRI brain, (C) Sagittal MRI knee, (D) Sagittal MRI cerebral, (E) Axial Haste T2 Weighted MRI abdomen, (F) Axial T1 Weighted MRI brain, (G) Axial CT abdomen, (H) Axial T2 Weighted MRI brain, (I) Axial CT abdomen with oral contrast, (J) Axial CT Chest pulmonary arteries, (K) Axial CT skull, (L) Coronal T1 Weighted MRI leg, (M) Axial CT CVJ brain, (N) Coronal T1 Weighted MRI chest, (O) Axial T1 Weighted MRI brain.

TABLE 2.1
Performance Measures of Ripplet Based Medical Image Compression.

Images (256 × 256)	PSNR	SSIM	CT (s)	CR
Axial T2 Weighted MRI brain	48.75	0.9796	0.62	12.72:1
Axial CT abdomen	46.93	0.9788	0.55	12.17:1
Axial CT CVJ brain	47.22	0.9687	0.65	11.89:1
Axial CT skull	47.25	0.9777	0.69	11.49:1
Sagittal T1 Weighted MRI brain	45.59	0.9732	0.59	11.48:1
Sagittal MRI Cerebral	42.61	0.9389	0.56	10.42:1
Coronal T1 Weighted MRI chest	46.93	0.9610	0.57	12.65:1
Average	**46.47**	**0.9783**	**0.60**	**11.83:1**

achieves a higher PSNR of 47.89. Bandelet based compression captures geometrics in images which further leads to sparse representation with high PSNR. It is also shown that the SSIM value is close to "1" with increased bitrates. The SSIM of 0.9889 for Sagittal MRI Cerebral confirms that the structural shape of compressed image is very close to the original image. In Bandelet transform, the decomposition of two dimensional wavelet coefficients is done through adaptive segmentation and hence reduces blocking artifacts. Therefore it achieves a relatively higher compression ratio of 10.58 without compromising the visual quality of the images. However the excellent visual quality images are obtained at the cost of a significant raise in the computational time. Since an in-depth search method is needed for searching the best directions of pixel scale, the computational time utilized by bandlet transform is relatively high with an average value of 1.43 s when compared with other compression methods.

5.3 Investigations on Radon Transform

The performance measures of Radon transform based image compression with SPIHT encoder is shown in Table 2.3.

TABLE 2.2
Performance Measures of Bandelet Based Medical Image Compression.

Images (256 × 256)	PSNR	SSIM	CT (s)	CR
Axial T2 Weighted MRI brain	47.89	0.9512	1.27	11.38:1
Axial CT abdomen	46.52	0.9188	1.55	10.14:1
Axial CT CVJ brain	46.17	0.9523	1.59	11.08:1
Axial CT skull	46.51	0.9678	1.38	9.79:1
Sagittal T1 Weighted MRI brain	47.61	0.9632	1.41	10.37:1
Sagittal MRI Cerebral	42.49	0.9889	1.39	10.45:1
Coronal T1 Weighted MRI chest	47.06	0.9772	1.47	10.86:1
Average	**46.32**	**0.9599**	**1.43**	**10.58:1**

TABLE 2.3
Performance Measures of Radon Based Medical Image Compression.

Images (256 × 256)	PSNR	SSIM	CT (s)	CR
Axial T2 Weighted MRI brain	47.06	0.9633	0.41	12.68:1
Axial CT abdomen	47.82	0.9788	0.36	11.83:1
Axial CT CVJ brain	46.53	0.9756	0.42	12.49:1
Axial CT skull	47.34	0.9727	0.33	11.16:1
Sagittal T1 Weighted MRI brain	46.33	0.9822	0.34	12.55:1
Sagittal MRI Cerebral	43.73	0.9767	0.40	11.13:1
Coronal T1 Weighted MRI chest	46.25	0.9761	0.39	10.21:1
Average	**46.44**	**0.9751**	**0.38**	**11.72:1**

From Table 2.3 it is inferred that Radon transform based compression method result in significant PSNR with an average value of 46.44. Since the interpolation of radon projections secures the pixel strength, it results in high PSNR. Also an average SSIM value of 0.9751 infers that the structural information of compressed image closely matches to that of the input image. The average time utilized to compress the medical images is 0.38 s which is relatively low when compared with other transforms. Since radon transform eliminates the need for interpolation and calculates only the difference between projections, the algorithm requires very low computational time. It is also understood that Radon based compression yields better compression ratio with an average value of 11.72 and performs well on the compression of medical images.

5.4 Investigations on Contourlet Transform

The performance measures of Contourlet transform based image compression with SPIHT encoder is shown in Table 2.4.

From Table 2.4 it is inferred that Contourlet transform based compression method result in an average PSNR of 43.34 for a set of medical images. This is due to the fact that Contourlet transform effectively holds the scale-variability of intrinsic geometrical structure of images. The average value of SSIM achieved by Contourlet is 0.9145. A small number of clear directional features possessed by contourlet transform leads to artifacts in compression which further leads to relatively low similarity index. The average time utilized to compress the medical images is 0.55 s. Since this transform employs extended basis function with directional filtering, it does not require the conversion from continuous time-models. Therefore contourlet transform utilizes relatively low computational time. Besides, this transform provides an average compression ratio of 10.12 for the compression of medical images.

5.5 Investigations on Curvelet Transform

The performance measures of Curvelet transform based image compression with SPIHT encoder is shown in Table 2.5.

Curvelet transform efficiently represents boundary discontinuities in image structures except for C^2 singularities. This results in an average PSNR of 42.07 for a set of medical images. Curvelet transform enjoys the benefit of geometric regularity of image structures; however, its approximations are not accurate for irregular curves and edges. This results in relatively low similarity

TABLE 2.4
Performance Measures of Contourlet Based Medical Image Compression.

Images (256 × 256)	PSNR	SSIM	CT (s)	CR
Axial T2 Weighted MRI brain	44.11	0.9278	0.58	11.05:1
Axial CT abdomen	44.25	0.9278	0.64	10.33:1
Axial CT CVJ brain	43.18	0.9119	0.55	10.21:1
Axial CT skull	44.89	0.8978	0.51	9.84:1
Sagittal T1 Weighted MRI brain	42.64	0.9309	0.52	9.96:1
Sagittal MRI Cerebral	40.76	0.8778	0.59	9.73:1
Coronal T1 Weighted MRI chest	43.56	0.9278	0.49	9.74:1
Average	**43.34**	**0.9145**	**0.55**	**10.12:1**

TABLE 2.5
Performance Measures of Curvelet Based Medical Image Compression.

Images (256 × 256)	PSNR	SSIM	CT (s)	CR
Axial T2 Weighted MRI brain	43.553	0.9495	0.58	10.89:1
Axial CT abdomen	42.663	0.9095	0.62	9.78:1
Axial CT CVJ brain	43.582	0.9245	0.71	11.93:1
Axial CT skull	42.908	0.9368	0.69	9.61:1
Sagittal T1 Weighted MRI brain	41.754	0.9245	0.64	9.53:1
Sagittal MRI Cerebral	39.942	0.8995	0.62	9.24:1
Coronal T1 Weighted MRI chest	40.06	0.9295	0.63	10.31:1
Average	42.07	0.9248	0.64	10.18:1

TABLE 2.6 Performance Measures of DCT Based Medical Image Compression.				
Images (256 × 256)	**PSNR**	**SSIM**	**CT (s)**	**CR**
Axial T2 Weighted MRI brain	39.03	0.799	0.56	10.36:1
Axial CT abdomen	38.61	0.789	0.44	9.59:1
Axial CT CVJ brain	36.81	0.840	0.42	9.37:1
Axial CT skull	38.55	0.799	0.53	9.22:1
Sagittal T1 Weighted MRI brain	36.65	0.830	0.41	8.71:1
Sagittal MRI Cerebral	33.90	0.789	0.47	9.7:1
Coronal T1 Weighted MRI chest	38.74	0.789	0.51	9.83:1
Average	**37.47**	**0.8050**	**0.48**	**9.54:1**

TABLE 2.7 Performance Measures of Haar Transform Based Medical Image Compression.				
Images (256 × 256)	**PSNR**	**SSIM**	**CT (s)**	**CR**
Axial T2 Weighted MRI brain	45.90	0.9014	0.49	10.57:1
Axial CT abdomen	45.00	0.8554	0.38	10.21:1
Axial CT CVJ brain	41.38	0.9229	0.46	9.62:1
Axial CT skull	44.23	0.9054	0.42	9.37:1
Sagittal T1 Weighted MRI brain	42.95	0.9229	0.39	9.62:1
Sagittal MRI Cerebral	39.46	0.8954	0.43	10.22:1
Coronal T1 Weighted MRI chest	44.97	0.919	0.47	9.41:1
Average	**43.41**	**0.9032**	**0.44**	**9.86:1**

index of 0.9248. The average computational time utilized by this transform is 0.64 s. This is due to the redundancy of the curvelet frame. It is also inferred that curvelet transform yields better compression ratio with an average value of 10.18 owed to the anisotropy behavior and approximation with lesser coefficients.

5.6 Investigations on Discrete Cosine Transform

The performance measures of DCT based image compression with SPIHT encoder is shown in Table 2.6.

From Table 2.6 it is inferred that DCT based compression achieves an average PSNR of 37.47 for a set of medical images. In DCT, the quantization of sub block coefficients results in pseudo borders between the sub-blocks which further results in an average SSIM of 0.8050. The average time utilized by DCT to compress the medical images is 0.48 s. This is due to the fact that the compression by DCT is performed by absorbing the signal in the lower spatial frequencies. However, DCT experiences blocking artifacts produced by edge discontinuities which yields an average compression ratio of 9.54.

5.7 Investigations on Haar Wavelet Transform

The performance measures of Haar transform based image compression with SPIHT encoder is shown in Table 2.7.

Haar wavelets are capable of representing sharp point singularities. However, the weak orientation and directional selectivity of Haar transform results in inadequate depiction of higher dimensional singularities. Hence it has an average PSNR of 43.41 with an average SSIM of 0.9032. Since different spectral coefficients are produced in various decompositions, more exhaustive coefficients appear in less magnitude. Therefore it utilizes an average of 0.44 s to compress the images. Haar transform depends on averaging and differencing values stored in sparse matrix, which leads to reduced image sizes. Hence it achieves an average compression ratio of 9.86.

5.8 Investigations on RST (Geometric) Transform

Geometric transform uses scaling and rotation procedures to locate and analyze the region of interest at various sizes and angles with good diagnostic quality. The performance measures of RST (Geometric) transform based image compression with SPIHT encoder is shown in Table 2.8.

The RST transform preserves the quality of the compressed image by contemplating the association between neighboring pixels using bilinear interpolation. From Table 2.8 it is inferred that the RST based compression achieves an average PSNR of 46.20 with an average computational time of 0.39 s.

The performance evaluation is also performed based on the analysis of impact of scaling factors on compression ratio. Image scaling modifies the input image size without compromising the visual quality. Table 2.9 shows the compression ratios gained for various scaling factors.

It is inferred from Table 2.9 that the compression ratio is significantly high when the scaling factor is 1.1. As the scaling factor increases, the compression ratio is reduced and the visual quality is also compromised.

5.9 Investigations on Subjective Assessment

To facilitate the investigation of diagnostic quality of compressed images, an extensive subjective assessment is performed on image visual quality, edge representation, sharpness and contrast, observation of soft tissues like fat, muscle, nerves, fibrous tissues, blood vessels, or deep skin tissues etc. The emergence of artifacts and deformation of information signal are also assessed. The subjective assessment signifies that transform based compression methods maintain the diagnostic visual quality of medical images and offer better representation of edges and curves. Most of the test medical images are graded as excellent and no single image has failed the test of acceptability as shown in Table 2.10. The Appearance of artifacts, Vignetting and Visible distortion of information signal are also evaluated.

6 CONCLUSIONS

In this chapter, the emerging paradigms in transform based medical image compression methods has been presented and investigated. Among the analyzed transforms, it has been observed that compression based on Ripplet, Bandelet and Radon transforms provide better PSNR (0.46) and high compression ratio. High PSNR with greater Compression ratio is one of the major requirements in telemedicine based image compression. The anisotropy behavior of ripplet transform, decomposition of geometric regularized images in Bandelet transform and exact invertibility of radon transform lead to significantly high PSNR, SSIM and high compression ratio. Even though curvelet transform and contourlet transforms offer optimal sparse representation, they lack in exhibiting discontinuity down the curvature with bounded variation functions. Also if image edges are irregular curves, then curvelet

TABLE 2.8
Performance Measures of RST Transform Based Medical Image Compression.

Images (256 × 256)	PSNR	CT (s)
Axial T2 Weighted MRI brain	47.45	0.42
Axial CT abdomen	45.89	0.36
Axial CT CVJ brain	46.13	0.38
Axial CT skull	45.77	0.43
Sagittal T1 Weighted MRI brain	45.54	0.33
Sagittal MRI Cerebral	45.13	0.36
Coronal T1 Weighted MRI chest	47.46	0.45
Average	**46.20**	**0.39**

TABLE 2.9
Scaling Factors Versus Compression Ratios.

Scaling factors	Axial T2 weighted MRI brain	Axial CT abdomen	Axial CT CVJ brain	Axial CT skull	Sagittal T1 weighted brain	Sagittal MRI cerebral	Coronal T1 weighted MRI chest
1.1	14.36:1	14.42:1	13.07:1	12.84:1	13.59:1	11.72:1	13.45:1
1.3	13.47:1	12.62:1	11.90:1	11.43:1	11.86:1	10.36:1	11.81:1
1.5	11.31:1	10.65:1	8.62:1	8.32:1	10.03:1	8.19:1	9.36:1
1.7	8.22:1	7.37:1	7.46:1	6.42:1	7.83:1	6.59:1	7:01
1.9	7.35:1	6.84:1	5.97:1	5.21:1	6.37:1	5.03:1	5.34:1

TABLE 2.10
Sample Subjective Evaluations of Ripplet Transform Based Compression.

Evaluation parameters	MEDICAL IMAGES USED FOR EVALUATION				
	Axial T2 weighted MRI brain	Axial CT abdomen	Axial CT skull	Sagittal MRI cerebral	Coronal T1 weighted MRI chest
	EXCELLENT/GOOD/AVERAGE/POOR				
Image Visual quality	Excellent	Excellent	Excellent	Excellent	Excellent
Representation of edges in images	Excellent	Excellent	Good	Excellent	Excellent
Image Sharpness	Excellent	Excellent	Excellent	Excellent	Good
Image Contrast	Excellent	Excellent	Excellent	Excellent	Excellent
Ability to view bone and soft tissue	Excellent	Good	Excellent	Excellent	Excellent
Able to view air and water	Excellent	Excellent	Excellent	Excellent	Excellent
Image sharpness of soft tissue window	Excellent	Excellent	Excellent	Excellent	Good

approximations are not precise. This results in relatively low similarity index. DCT suffer from blocking artifacts which further leads to low SSIM.

Since Radon transform eliminates interpolation and requires only minimum number of projections for computation, the time utilized to compress the medical images is relatively low (0.37 s) when compared to the other transforms. However in Bandelet transform, a detailed and exhaustive search method is needed to identify the best directions of pixel scale, the computational time utilized by bandlet transform is relatively high with an average value of 1.43 s when compared with other compression methods. When the healthcare provider at rural hospital needs to locate the affected/diseased area and to converse to a specialist at a distance through telemedicine, Geometric transforms are used by varying the scaling factor. Experimental results on Geometric transform reveal that it has competing performance and achieves high PSNR with less computational time. In future, the research on telemedicine technology may be extended on the design of error protection techniques, knowledge based patient monitoring systems and Bigdata analytics in healthcare delivery systems.

REFERENCES

[1] World Population Review. http://worldpopulationreview.com/countries/india-population/.

[2] J.A. Villanueva, M.C. Suarez, O. Garmendia, V. Lugo, C. Ruiz, J.M. Montserrat, The Role of Telemedicine and mobile health in the monitoring of sleep-breathing disorders: improving patient outcomes, Smart Homecare Technol. TeleHealth 4 (2017) 1–11.

[3] A. David, K. Gregorij, Y.Y. Allen, B. Ruzena, Augmented telemedicine platform for real-time remote medical consultation, in: 23rd Int. Conf. On Multi Media Modeling, LNCS, 10132, 2017, pp. 77–89.

[4] P.N. Bideh, M. Mahdavi, S.E. Borujeni, S. Arasteh, Security analysis of a key based color image watermarking vs. a non-key based technique in telemedicine applications, Multimed. Tool. Appl. (2018) 1–23.

[5] N.K. Vinay, M. Kamesha, S. Michael, W. Eric, Telehealth in the delivery of home dialysis care: catching up with technology, Adv. Chron. Kidney Dis. 24 (1) (2017) 12–16.

[6] Y.L. Huang, The development of patient scheduling groups for an effective appointment system, Appl. Clin. Inf. 7 (1) (2016) 43–58.

[7] K.C. Ellen, E.S. Candjur, F.H. Marie, Shared electronic health record systems: key legal and security challenges, J. Diabetes Sci. Technol. 11 (6) (2017) 1234–1239.

[8] S.W. Laurence, R.S. Duncan, C. Patrick, Telehealth on advanced networks, Telemed J. E-health 16 (1) (2010) 69–79.

[9] Kankanhalli, J. Hahn, S. Tan, G. Gao, Big data and analytics in healthcare: introduction to the special section, Inf. Syst. Front. 18 (2016) 233–235.

[10] G. Manogaran, C. Thota, D. Lopez, V. Vijayakumar, K.M. Abbas, R. Sundarsekar, Big data knowledge system in healthcare, in: Internet of Things and Big Data Technologies for Next Generation Healthcare, Studies in Big Data, vol. 23, 2017, pp. 133–157.

[11] V.K. Bairagi, Big data analytics in telemedicine: a role of medical image compression, Big Data Manag. (2017) 123–160.

[12] S.J. Devaraj, K. Ezra, Current trends and future challenges in wireless telemedicine system, in: 3rd IEEE Int. Conf. On Electronics Computer Technology, vol. 4, 2011, pp. 417–421.

[13] W.-Y. Hsu, Clustering-based compression connected to cloud databases in telemedicine and long-term care applications, Telematics Inf. 34 (1) (2017) 299–310.

[14] J. Xu, L. Yang, D.O. Wu, Ripplet - a new transform for image processing, J. Vis. Commun. Image Represent. 21 (7) (2010) 627–639.

[15] E.L. Penneca, S. Mallat, Sparse geometric image representation with bandelets, IEEE Trans. Image Process. 14 (4) (2005) 423–438.

[16] S. Juliet, E.B. Rajsingh, K. Ezra, A novel image compression for medical images using geometric regularity of image structure, J. Signal Image Video Process. 9 (7) (2015) 1691–1703.

[17] I. Tali, I. Asaf, S. John, Z. Zeev, Optical realization of the radon transform, Opt. Express 22 (26) (2014).

[18] S. Juliet, E.B. Rajsingh, K. Ezra, Projection-based medical image compression for telemedicine applications, J. Digit. Imag. 28 (2) (2015) 146–159.

[19] M. Do, M. Vetterli, The contourlet transform: an efficient directional multiresolution image representation, IEEE Trans. Image Process. 14 (12) (2005) 2091–2106.

[20] A.N. Belbachir, P.M. Goebel, The contourlet transform for image compression, Phys. Signal Image Process. 4 (2005), 01.31-02.02, 251–256, (Toulouse, France).

[21] J.L. Starck, E.J. Candes, D.L. Donoho, Curvelets, multiresolution representation and scaling laws, IEEE Trans. Image Process. 11 (2000) 670–684.

[22] P. Piotr, L. Agnieszka, The haar wavelet transform in digital image processing: its status and achievements, Mach. Graph. Vis. 13 (1/2) (2004) 79–98.

[23] R. Rafael, B. Sebastian, K. Gitta, W. Thomas, A haar wavelet-based perceptual similarity index for image quality assessment, Signal Process. Image Commun. 61 (2018) 33–43.

[24] A. Hedi, K. Ali, G. Malek, S.B. Med, Medical image compression approach based on image resizing, digital watermarking and lossless compression, J. Signal Process. Syst. 87 (2) (2016) 203–214.

[25] D. Zheng, Y. Liu, J. Zhao, A.E. Saddik, A survey of RST invariant image watermarking algorithms, ACM Comput. Surv. 39 (2) (2007) 1–91.

[26] D.A. Huffman, A method for the construction of minimum redundancy codes, Proc. IRE 40 (9) (1952) 1098–1101.

[27] J.M. Shapiro, Embedded image coding using zerotrees of wavelet coefficients, IEEE Trans. Signal Process. 41 (1993) 3445–3463.

[28] A. Said, W. Pearlman, An image multiresolution representation for lossless and lossy compression, IEEE Trans. Image Process. 5 (1996) 1303–1310.

[29] X. Wu, N. Memon, CALIC: a context-based, adaptive, lossless image coding, IEEE Trans. Commun. 45 (4) (1997) 437–444.

[30] D. Taubman, High performance scalable image compression with EBCOT, IEEE Trans. Image Process. 9 (7) (2000) 1158–1170.

[31] W.A. Pearlman, A. Islam, N. Nagaraj, A. Said, Efficient, low complexity image coding with a set-partitioning embedded block coder, IEEE Trans. Circuits Syst. Video Technol. 14 (11) (2004) 1219–1235.

[32] S. Juliet, E.B. Rajsingh, K. Ezra, A novel medical image compression using Ripplet transform, J. Real-Time Image Process. 11 (2) (2016) 401–412.

CHAPTER 3

Adopting m-Health in Clinical Practice: A Boon or a Bane?

KURUBARAN GANASEGERAN[a] • SURAJUDEEN ABIOLA ABDULRAHMAN[b]
[a]Clinical Research Center, Seberang Jaya Hospital, Ministry of Health Malaysia, Seberang Jaya, Penang, Malaysia;
[b]Emergency Medicine Department, James Paget University Hospital, Great Yarmouth, Norfolk, United Kingdom

1 INTRODUCTION

Time and space constitute barriers to effective health communication! "Population avalanche" has caused a paradigm shift from disease to risk. National and international healthcare systems are challenged with the influx of patient admissions, disease outbreaks, population explosion, disasters, accidents, etc; all of which require fast and efficient interventions for timely survival. As a result, clinicians globally are burdened with higher demands of effective clinical communications, rapid information sharing and accessibility for successful patient management. Healthcare exigencies pose greater needs for continuous medical education, development of rigorous surveillance systems and strategies for widespread coverage and delivery of health services in physically remote areas. Achieving these necessities require rapid, accurate and constant communication platforms.

2 HISTORY AND EVOLUTION OF HEALTHCARE COMMUNICATIONS

The evolution of healthcare communication can be traced back to the 1870s. The telephone has been a means of communication between patients and clinicians and between healthcare providers [1]. Its potential grew when Einthoven transmitted the first ECG recordings over the telephone in 1906 [2]. Aronson succinctly transmitted heart and lung sounds via the telephone in late 1910's [3]. Around 1960's, advancements were seen with the prelude of "wired communications" — telemedicine was conceptualized to focus on the curative aspects of patients' ailments and constituted fields like teledermatology, teleradiology, telepharmacy, etc [4]. In 1978, Bennet and colleagues extended the scope of telemedicine by incorporating wider

functionalities [5]; the inclusion of the promotion, prevention and curative aspects of health to be accessible over a distance, collectively termed as telehealth. These communicative modes that promoted direct interactions in healthcare through telephone conversations and pager networks was used over decades. However, recent studies have alarmed potential adverse events and medical errors during patient care; due to incomplete, fragmented and disorganized communications of using telephony systems [6,7]. More precise, trustworthy, efficient and constant communication platforms were required during intense situations; such as to call a doctor urgently to patients' bedside or to report crucial laboratory investigations [7]. Owing to such necessities, "wired communications" was further revolutionized to "unwired communications" through the dawn of e-Health [8]. The terms e-Health and telehealth were often used interchangeably, but the distinctive concept between these two applications is that e-Health is not limited to healthcare over a distance [4].

3 M-HEALTH: DEFINITION, CONCEPTS AND MILESTONES

Mobile health (m-Health), a subset of e-Health, has gained extraordinary momentum within the medical arena since its inception in 2003 [9]. m-Health was conceptualized as medical practice supported by the use of ubiquitous gadgets like mobile phones, patient monitoring devices, personal digital assistants (PDAs) or other wireless devices [10]. Besides these mobile devices, smartphones and tablets have important implications in m-Health technology [9]. The m-Health revolution has undergone tremendous milestones since the introduction of the first mobile phone, the DynaTac in 1973 [11], followed by introduction of the PDA and

Telemedicine Technologies. https://doi.org/10.1016/B978-0-12-816948-3.00003-9

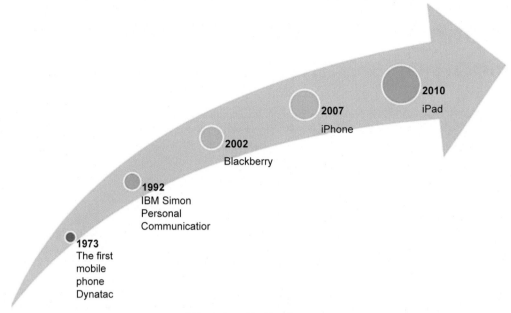

FIG. 3.1 m-Health milestones.

smartphone (the IBM Simon Personal Communicator) in 1992, Blackberry in 2002, iPhone in 2007 and iPad in 2010 [12−15] (Fig. 3.1). Expansions of these ultra-modern, versatile and ubiquitous devices with corresponding speedy internet connectivity overtime has enabled instantaneous access to loads of medical data, laboratory investigation outcomes, clinical practice guidelines and medication reference guides. By using these devices, health-related information dissemination and exchange is possible via text, audio, images, video or coded data in the form of short messaging services (SMSs), voice SMS, applications accessible via general packet radio service (GPRS), global positioning system (GPS), third and fourth generation mobile telecommunications and Bluetooth [16].

4 M-HEALTH FUNCTIONALITIES AND PROPERTIES

The ubiquitous availability and penetrance of mobile phones and devices in both developing and developed countries has not only expanded the horizon of health communication, but has also empowered patients, clinicians, policy makers and other stakeholders in healthcare to make fast, efficient and far-reaching decisions that impact patient care. The continuous evolution of mobile phones from basic (with only voice call capabilities) to feature phones (with text, voice,

video, multimedia, games and web capabilities), and now smartphones (with complex features such as capacity to run computer programs, connectivity with other health and non-health-related devices etc.) has expanded their capacity to support clinical practice in more targeted ways [17]. The exponential rise of smartphone use has saturated the healthcare market with almost 100,000 m-Health applications ("apps") in two major app stores (Apple and Android) that were specifically geared for health and well-being purposes. Dicianno and colleagues [18] have classified the range of m-Health app functionalities into three major categories (Table 3.1).

The unique feature of m-Health has supported individual expectations toward better doctor-patient interactions, self-health monitoring and disease management in healthcare. Fig. 3.2 exhibits m-Health properties.

5 CONTEXTS OF M-HEALTH APPLICATION IN CLINICAL PRACTICE

m-Health technologies have been adapted for a variety of health-related purposes, including but not limited to (1) enhancement of health service utilization and delivery (e.g., appointment reminders, mobile-based service delivery checklists), (2) encouraging better disease self-management practices (e.g., diabetes management,

TABLE 3.1
m-Health App Functionalities.

App type	Functionality	Examples
Patient-oriented apps	Advocates early self-identification of chronic diseases for prompt management. Such apps also promotes adherence to treatment.	1. i-Triage [19] allow individuals to detect particular disease symptom and signs, and to seek appropriate medical assistance. 2. HHeal [20], an app which integrates flu risk information and flu preventive behavior provides personal flu risk bar that rises when a user is near individuals with flu-like symptoms and drops when the user finishes one of the flu-preventive measures. 3. eMate [21] coaches patient with diabetes, HIV or cardiovascular disease to support their adherence therapy. 4. SMART [22] allows patients with sickle cell disease to record and assess disease symptoms and pain.
Clinician-oriented apps	Doctors are aided through quick reference guides, measurement tools and calculators to facilitate prompt clinical decisions.	1. Reference apps include encyclopedias and mini drug guides [23]. 2. Measurement tools and calculators designed include mobile accelerometers, inclinometers and goniometers [24,25].
Lifestyle-oriented apps	Promotes self-tracking of individual's healthy lifestyle programs like diet control, exercise and weight reduction activities.	Stand-alone apps like "Nexercise," [26] "Weight Watchers Mobile," [27] and "Lose It" [28].

medication adherence) and (3) delivering personalized health promotion interventions (e.g., smoking cessation, alcohol abstinence programs and weight loss programs) [29].

Many of these applications have sound theoretical basis; and regardless of the methodological diversity of the interventions, the overwhelming consensus from many clinical trials and systematic reviews favored beneficial effects of these interventions in improving clinical management, as well as promoting behavior change through improved knowledge, health promotion and disease prevention [30–32].

6 M-HEALTH BENEFITS IN CLINICAL PRACTICE

Informatics has been an integral part of medicine, allowing storage and access to "big data," which is built upon the foundations of e-Health. Despite the availability of vast information, the concept of e-Health was previously less than ideal in clinical practice, as healthcare professionals were tied down to immobile devices such as desktop computers. The e-Health revolution has reached the next stage of digital informatics — providing rapid access in storing, creating and retrieving material or information in a convenient manner with the emergence of smartphones as instruments for clinical communications. Studies have reported a commensurate increase of smartphone ownerships among healthcare professionals that ranged between 60 and 80% [6,33]. The rise of communication apps that is compatible with smartphones like WeChat, Tango, Line, Viber and WhatsApp have shown promising features for doctors to facilitate rapid interactions for effective clinical management [7]. As medical practice has advanced into a digitalized era, these gadgets have facilitated faster clinical communications, accurate diagnosis and better management and rehabilitation services.

6.1 Clinical Communications

The autonomous role enacted within a hierarchical structure in clinical practice frequently impede rapid decision making process as a consequence of fragmented collegial communications among multidisciplinary teams during patient care [7]. One common communication app that has been adopted rapidly across most local and international healthcare systems is WhatsApp. WhatsApp is an app compatible with smartphones which allows users across the globe to send, receive or share instant text messages and media contents via videos, voice messages or photographs with their contacts [6,34]. Its ability to create chat groups which allows multiple users to participate, monitor and reply

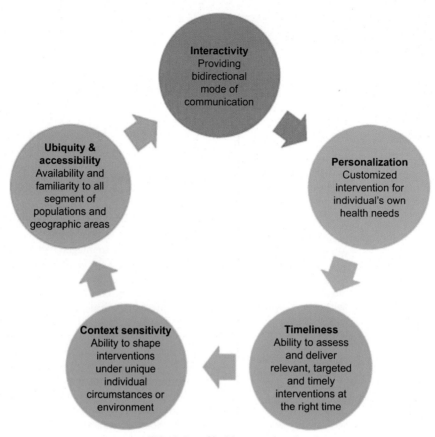

FIG. 3.2 m-Health properties.

conversations has added advantage to its usability across a range of clinical services, particularly to facilitate communication within intra-disciplinary groups. Two recently initiated investigations across medical and surgical teams from Malaysia and the United States have concluded that WhatsApp served as an excellent platform for the following communication events: answering clinical questions, providing information or instructions related to patient management and responding to administrative questions. WhatsApp was used during ward rounds, clinics, in procedure rooms and during on-calls frequently [6,7]. Such communication platform in clinical setting was vital to junior and senior doctors in conveying quick and accurate information for initiating rapid clinical actions [35]. The chat group functionality has helped suppress hierarchical dominance between multidisciplinary teams and the seniority between junior and senior doctors or consultants, thus reducing communication failures or role conflict [7]. Dedicated chat groups have enabled bidirectional benefits with the alleviation of novice doctors' hesitancy to

communicate directly with their seniors, while the senior clinicians could monitor and mentor their team's activities on patient management process [35,36]. Indeed, communication just got better with swifter and efficient handovers for continued patient care!

6.2 Diagnostics

As clinical practice anticipates zero tolerance to medical errors, diagnostic accuracy remains crucial for the execution of effective patient management. Recent studies have postulated the contribution of WhatsApp Messenger toward diagnostic accuracy.

6.2.1 Evaluating fractures in orthopedic setting
Giordanoa and colleagues [37] advocated WhatsApp as an adjunct "gold-standard" tool to diagnose tibial plateau fractures. They evaluated inter- and intra-observer reliability in the diagnosis and classification of tibial plateau fractures from plain radiographs and computer tomographic (CT) scan films that were photographed and sent via WhatsApp Messenger. Images were

obtained from thirteen subjects using iPhone 5 and were sent to six independent observers via WhatsApp. The observers determined the standard deviation, types of injury and the classification of fractures using the Schatzker and Luo scheme. They hypothesized that CT scan films changed the fracture classifications when evaluated across two different occasions, fifteen days apart. The Kappa statistics for both durations ranged from excellent to perfect ($0.75 < k < 1.0$). When queried if CT scan images would change their perceptions of X-ray classifications (Schatzker or Luo), they rated a perfect reliability ($k = 1$) on both assessment periods. Orthopedic surgeons were convinced that such results could enhance the best diagnostic procedures for rapid clinical management of fractures.

6.2.2 The evaluation of hematuria in surgical setting

In a recent study, Sener and colleagues [38] postulated that remote evaluation of hematuria was feasible using WhatsApp messenger. The investigation was undertaken by photographing two samples of voided urine in a sterile container and sent to two separate groups of urologists via WhatsApp to determine the seriousness of hematuria. Misdiagnosis of hematuria was around 6.5–7%. With the utilization of WhatsApp, the urologists were able to compare between normal urine and any form of hematuria precisely. The study proved that there was almost perfect agreement between the two groups of urologists with regards to the grades or seriousness of hematuria.

6.3 Rehabilitation

The use of m-Health technologies and applications for rehabilitation purposes has improved outcomes of aftercare in chronic disease states [18]. The following two real case scenarios (Boxes 3.1 & 3.2) have proven the advancement of rehabilitation services for patients of aftercare upon discharge from hospitals.

6.4 The Role of m-Health in Medication Non-Adherence

Non-adherence phenomenon afflicts 50–60% of all chronically ill patients [43]. Non-adherence negates treatment efficacy, increases hospital re-admissions and mortality, predisposes to complex medication prescribing or poly-pharmacy and escalates healthcare expenditures. Treatment plans are often challenged when patients blame accessibility, lack of knowledge about disease conditions, poor communication and rapport from healthcare providers and the lack of continuity of care as reasons for not wanting to adhere to prescriptions [44]. m-Health has created opportunities

---

BOX 3.1
The iMHere Project Experience in Spina Bifida Patients

CONCEPT

The Wellness Pilot Program was aimed to deliver coordinated care for spina bifida patients toward assisting them to manage skin integrity and neurogenic bladder, apart from tackling other health issues like medication management and addressing patients' emotional and mental well-being. In 2013, researchers' pilot tested iMHere [39], a web-based m-Health portal system used by wellness coordinators to be linked with smartphone apps used by spina bifida patients. The program delivered consistent reminders to patients related to self-care activities like catheterization, medication taking and self-skin examination for pressure ulcers that may have originated from the use of orthoses or limited weight shifting from wheelchair.

INTERVENTION

When patients encounter problems or difficulties, they can immediately report to the wellness coordinators. Current information regarding symptoms, photos and updates in relation to wound healing progress or complications are relayed rapidly to the wellness coordinator. Succinctly, appropriate information and instructions are provided to the patients via the portal's dashboard for urgent self-intervention.

OUTCOME

Wellness coordinators were able to identify patient's problems early and deliver rapid interventions to patients. The overall system was receptive to spina bifida patients, clinicians and caregivers.

---

to execute remote healthcare delivery via mobile phone systems to curb the non-adherence phenomenon. Rigorous randomized controlled trials (RCTs) that evaluated the application of m-Health technologies found leading solutions and opportunities to improve medication non-adherence. The bulk of these investigations were saturated across two diverse population groups: (1) HIV-infected patients; (2) Patients with chronic non-communicable illnesses [43,45,46].

6.4.1 Using m-Health to improve medication adherence in HIV-infected patients

Adherence to antiretroviral therapy (ART) is crucial to prevent disease progression. The use of text messaging services in HIV-infected patients was found to improve medication adherence rates to ART as compared to standard care. m-Health has supported novel approaches for medication adherence checking and intercessions using medicine stockpiling gadgets furnished with cell phone innovations. Sabin and colleagues [47] demonstrated real-time cell phone reminders in addition to counseling in a cohort of ART patients in China which resulted in significant improvements of adherent behaviors. A recent investigation from Malaysia randomized 242 adult HIV-patients into intervention and control groups to determine the efficacy of mobile phone reminder modules delivered through SMS and telephone call reminders, in addition to standard clinic counseling. This 24-week single-blinded, parallel group RCT found that patients in the intervention group that received mobile phone reminders (SMS and telephone calls) and peer counseling had improved adherence rates as compared to the controls [48].

6.4.2 Using m-Health to improve medication adherence in patients with hypertension
McGillicuddy and colleagues [49] applied the triadic interconnectedness of wireless GSM (Global System for Mobile communication) electronic medication tray, Bluetooth-enabled blood pressure monitoring device and a smartphone to improve treatment adherence in hypertensive patients. The system reminds patients through "blinks" from a particular drug compartment at a given time. Failure to respond within 30 minutes would trigger the device to ring for a subsequent 30 minutes, and if still not attended after that period, automated text message reminders or phone calls to patients will be executed. This study which was conducted over 3 months found that adherence rate at 3 months follow up significantly improved from baseline values in the m-Health arm as compared to the standard care arm, in addition to better clinical outcomes. Blood pressure improved from 138 to 122 mmHg (systolic) and from 88 to 81 mmHg (diastolic) in the m-Health arm; while in the standard care arm, it increased from 132 to 139 mmHg (systolic) and from 76 to 79 mmHg (diastolic) at post-3 months.

The SimCard Study [50], a one-year multi-center cluster RCT that recruited 2086 'high-risk' cardiovascular disease patients (those aged 40 years and above with confirmed diagnosis of ischemic heart disease, stroke, diabetes, or with a measured systolic blood pressure of 160 mmHg or more) were recruited to evaluate the efficacy of a smartphone-based electronic decision support system (EDSS) for cardiovascular disease management. The system was used by community healthcare workers (CHWs) in primary health practice across China, India and Tibet. The CHWs were given a one day orientation followed by delivery of the intervention monthly for a year using an Android app. The app consisted of the following functions: blood pressure parameter recording, current medication use, past medical history, contraindications to cardiac medications and lifestyle habits to aid decision support. In contrast, the control group received usual care without any additional interventions. Investigators observed that the intervention group had higher usage of anti-hypertensive medications, thus improving systolic blood pressure.

6.4.3 Using m-Health to improve medication adherence in ischemic heart disease patients
Vollmer and colleagues [51] examined the effects of treatment adherence through an interactive voice response call and electronic medical record (EMR) based feedback system among primary care doctors who handle diabetes and cardiac patients due for refills. The feedback system was delivered to the doctor if patients were ninety days overdue. The intervention delivered a positive effect with adherence to medications improving significantly, from 1.6% to almost 3.7% when compared to usual care.

6.4.4 Using m-Health to improve medication adherence in heart failure patients

Goldstein and colleagues [52] conducted an RCT among 60 heart failure subjects to compare the effectiveness of medication adherence using telehealth application (e-Pill box) with an m-Health system (using smartphone). The four armed study involved the use of one of the two devices (e-Pill box or smartphone), with or without active program reminders (active vs. passive medication-taking system). Participants rated the use of their device according to the following attributes: helpfulness, quality of life, willingness to recommend the device to a friend and satisfaction. Adherence rate overall was 78%. Across the four study arms, adherence rate was 80% for e-Pill box, 76% for smartphone use, 79% for active reminder and 78% for passive reminders. The findings concluded that patients preferred choice for motivation to adhere to agreed treatment plan was through the m-Health approach and rated the use of smartphone as a better devise compared to e-Pill box.

6.4.5 Using m-Health to improve medication adherence in stroke patients

The SMS4Stroke study [53] was a parallel-group, assessor-blinded, randomized controlled superiority trial that recruited 200 stroke subjects. Subjects in the intervention arm, apart from receiving the standard usual care, also received daily SMS prescription reminders for two consecutive months and bi-weekly health information updates. Adherence to medications was measured at baseline and at two months post-intervention in both arms. The trial concluded that improvement of medication adherence was greater in the intervention group as compared to the control group after adjusting for potential confounders.

6.4.6 Using m-Health to monitor glycemic control in diabetes patients

Istepanian and colleagues [46] conducted an RCT using m-Health technology in a sample of 137 diabetes patients. Patients in the intervention group (n = 72) were prepared to quantify their blood glucose level by using a sensor which transmits parameter readings to a cell phone by means of a Bluetooth remote connection. Clinicians were able to examine and respond to the readings through a web-based application. Patients in the control group (n = 65), however, only received standard care with their usual outpatient follow-up with a doctor. The study found that, patients in the intervention group had a significantly lower HbA1c reading than the control group.

7 IMPLICATIONS

There have been significant concerns with respect to potential negative effects of m-Health 'misuse' or 'overuse' with implications for privacy and confidentiality of patients' information, technological addiction and associated psycho-behavioral disturbances, oversimplification of health issues, possibility of developing hypochondria and questions of accountability.

7.1 Implications for Patients

A major concern for patients and users of m-Health relates to privacy and confidentiality considerations. This concern is of relevance given the possibility of unauthorized access to or sharing of patient information with or among members of the healthcare team who may have little or no role to play in the current management of the patient but have access by virtue of being members of the healthcare team. In addition, SMS-based interventions which convey information on sensitive health matters for which disclosure issues, stigma and discrimination may be rife (such as HIV/AIDS, STI, mental health etc), carry the potential risk of being circumstantially accessible to recipients' family and friends to which they may not be inclined to sharing such information with. Mobile data collection also raises similar concerns of privacy and confidentiality. Clinical pictures and patient particulars stored in mobile devices risk breaching patient's privacy and data protection laws when this information is shared in closed chat group discussions. Unauthorized access or contents forwarded to wrong recipients could alarm medico-legal consequences [7]. These concerns have become core areas of ethical considerations and approvals by several institutional and ethical review boards around the world, and have formed the basis of many guidelines to protect the privacy and confidentiality of participants volunteering to participate in m-Health intervention studies. The "Bitcoin" phenomenon that currently advocated cryptography; the science of information scrambling for ensuring privacy and security has been widely recommended to be adopted in healthcare. The concept which applies block-chain technology could be incorporated into m-Health applications, allowing exchange of data through secure connection by using crypto-keys managed exclusively and autonomously by the individual patient and the attending doctor. In addition, it preserves privacy and confidentiality of patient information, and may be exploited for a patient-controlled management of the consent process (for deciding participation in research, what data to be shared, stored and published) in clinical and public health practice [54,55].

Closely related to this consideration are negative behaviors that may potentially arise from m-Health misuse. These often include oversimplification of health issues and consequent poor health-seeking behavior which can arise when users overly rely on potentially diagnostic or prognostic m-Health apps. An example is the widely reported misuse of melanoma detection app. Quite easy to use, the app calculates the risk, from a snapshot taken by the user, that their skin mole may be cancerous [56]. While this appears as an easy and quick fix, what is best practice is to visit a doctor for consultation on such health issues, increasing the chances of early diagnosis and management, and better outcomes. Ready access to health information from mobile apps, SMS and internet is increasingly becoming the basis of 'do-it-yourself' approach, which unfortunately is beginning to worsen issues of self-diagnosis, self-medication and related harms being reported in many parts of the world, especially in countries with seemingly lax regulatory framework.

7.2 Implications for Healthcare Workers

A study conducted by Ganasegeran et al. [57] which examined WhatsApp addiction among Malaysian healthcare workers who often used WhatsApp messaging for clinical and social communication found a relatively high prevalence of sleep disturbances among young, female healthcare workers who initiate WhatsApp-ing activities immediately after sleep.

Concerns of interoperability of m-Health applications with existing Electronic Medical Record (EMR) systems, how to integrate m-Health into personalized management of patients, lack of strong and consistent evidence of effectiveness of m-Health from rigorous randomized studies and potential to bypass traditional primary care, remain some of the considerations that may explain why clinicians are slow to adopt m-Health in certain circumstances [58].

7.3 Implications for Stakeholders

Insurance companies, policy makers and regulators are all important stakeholders impacted by the m-Health revolution. While certain considerations may be of unique importance to particular stakeholders, the overall concerns revolve around issues of acceptance, value for money, safety and confidentiality, as well as sustainability. For example, health insurance companies are generally more concerned about implementation re-imbursements for m-Health-enabled care management programs, while also looking to maximize opportunities to enhance patient engagement, consumer satisfaction

and retention, and ultimately better outcomes. Additionally, issues relating to safety and confidentiality of patient information are of paramount importance to the insurance companies and regulators. For example, the 2015 reports of cyber attacks enabled by m-Health and related applications in the United States resulted in database breeches that compromised healthcare information of over 80 million subscribers and led to losses in millions of dollars [59,60]. Beyond this, regulators have a role to ensure that approved medical apps are reliable and used only as an accessory to regulated medical devices, while also enforcing compliance with data protection rules.

8 FUTURE DIRECTION

Future m-Health apps may facilitate the deployment of artificial intelligence (AI) into healthcare. Healthcare professionals may be offered novel decisional support tools that allow accessibility of AI powered apps to review patient's health records electronically or review clinical symptoms [61]. With increased number of smart-mobile devices, ubiquitous internet and cloud computing, the healthcare sector could see the rise of Wearable Internet of Things (WIoT) [62]. WIoT will increase capabilities of sensing, computing and communication allowing patients to be tracked by wearable sensors for personalized health and wellness information. However, the WIoT needs to overcome technical glitches of generating frameworks for networking, computing, storage and visualization while establishing its position toward crafting solutions that are clinically acceptable and functional [63].

m-Health is not a substitute for traditional care. Clinician engagement is ultimately crucial to maximize the benefits derived from m-Health systems. Nonetheless, data collected from m-Health technologies can be used to inform clinical manifestations rapidly to healthcare professionals, thus enabling optimal care to be provided at the right time for the right patient. As m-Health is still evolving, it is crucial for patients, providers, developers and other stakeholders to work closely together, to better understand the pressing problems in contemporary healthcare, and to enable better integration and interoperability of technological solutions with care delivery systems, hence maximizing benefit for all. Despite obvious benefits of block-chain in medical informatics, the proposed cryptographic technologies may still violate privacy in a patient controlled system when encryption keys are not efficiently managed. The autonomous direction of patient-doctor communication may hamper clinician workflow in the course of ensuring privacy, as urgent health data or medical history of the patient could not be decrypted for authorized access during emergencies in a multi-disciplinary healthcare team management plan [54]. In adopting block-chain technology for clinical purposes, developers must consider the need to overcome potential challenges of transparency and confidentiality, its' speed and scalability through appropriate designs and implementation of health application models [54].

9 CONCLUSIONS

In 2014, the European Commission estimated that by 2017, 3.4 billion people worldwide (roughly half of the world's population) will own a smartphone, and of these, about 50% will use health apps [64]. m-Health offers triple win by putting patients in control (driving positive behavior, and enabling them to take greater individual responsibility for their own health), making the health care system more efficient (allowing direct interaction between providers and clients while keeping face-to-face contacts at barest minimum and only when absolutely necessary, enhancing cost-savings and efficiency in service delivery and research), and creating massive opportunities for innovative services and linkages between patients, providers and app developers (thereby strengthening the value chain and improving client satisfaction and experiences). While the increasing access to m-Health apps by all strata of population offers significant benefits for clinical practice, it behoves on all stakeholders, patients and providers to work collaboratively to not only address its potential drawbacks, but also to establish concrete evidence of the safety and effectiveness of m-Health technology, in order to optimize the benefits for all.

ACKNOWLEDGMENTS

We thank the Director General of Health Malaysia for the support to publish this chapter.

REFERENCES

[1] M. Mars, R.E. Scott, WhatsApp in clinical practice: a literature review, in: A.J. Maeder, et al. (Eds.), The Promise of New Technologies in an Age of New Health Challenges, IOS Press, 2016, pp. 82–90, https://doi.org/10.3233/978-1-61499-712-2-82.

[2] W. Einthoven, Le telecardiogramme, Arch. Int. Physiol. 4 (1906) 132.

[3] S.H. Aronson, The Lancet on the telephone, Med. Hist. 21 (1977) 69–87.

[4] L. van Dyk, A review of telehealth service implementation frameworks, Int. J. Environ. Res. Public Health 11 (2014) 1279–1298, https://doi.org/10.3390/ijerph110201279.

[5] A.M. Bennet, W.H. Rappaport, E.L. Skinner, Telehealth Handbook, US Department of Health, Education and Welfare, Washington, DC, 1978. PHS Publication No. 79-3210.

[6] M.J. Johnston, D. King, S. Arora, N. Behar, T. Athanasiou, N. Sevdalis, A. Darzi, Smartphones let surgeons know WhatsApp: an analysis of communication in emergency surgical teams, Am. J. Surg. 209 (1) (2015) 45–51, https://doi.org/10.1016/j.amjsurg.2014.08.030.

[7] K. Ganasegeran, P. Renganathan, A. Rashid, S.A.R. Al-Dubai, The m-Health revolution: exploring perceived benefits of WhatsApp use in clinical practice, Int. J. Med. Inform. 97 (2017) 145–151, https://doi.org/10.1016/j.ijmedinf.2016.10.013.

[8] S. Tachakra, X.H. Wang, R.S.H. Istepanian, Y.H. Song, Mobile e-Health: the unwired evolution of telemedicine, Telemed. J. E Health 9 (3) (2003) 247–257.

[9] E.E. Ali, L. Chew, K.Y.-L. Yap, Evolution and current status of m-Health research: a systematic review, BMJ Innov. 2 (2016) 33–40, https://doi.org/10.1136/bmjinnov-2015-000096.

[10] WHO, mHealth: New Horizons for Health through Mobile Technologies: Second Global Survey on eHealth, Global Observatory for eHealth, Switzerland, 2011.

[11] Handheld Wireless Telephone. IEEE Global History Network [Internet]. Available from: http://www.ieeeghn.org/wiki/index.php/Handheld_Wireless_Telephone.

[12] R.H. Wiggins, Personal digital assistants, J. Digit. Imag. 17 (2004) 5–17.

[13] J.H. Yoo, The meaning of information technology (IT) mobile devices to me, the infectious disease physician, Infect Chemother. 45 (2) (2013) 244–251.

[14] C.L. Ventola, Mobile devices and apps for health care professionals: uses and benefits, P T 39 (2014) 356–364.

[15] D. Malvey, D.J. Slovensky, mHealth: Transforming Healthcare, Springer, New York, 2014.

[16] S.F.V. Sondaal, J.L. Browne, M. Amoakoh-Coleman, A. Borgstein, A.S. Miltenburg, M. Verwijs, K. Klipstein-Grobusch, Assessing the effect of m-Health interventions in improving maternal and neonatal care in low- and middle-income countries: a systematic review, PLoS One 11 (5) (2016) e0154664.

[17] L.C. Abroms, N. Padmanabhan, W. Evans, Mobile phones for health communication to promote behavior change, in: e-Health Applications: Promising Strategies for Behavior Change, first ed., Routledge, New York, NY, 2012, pp. 147–166.

[18] B.E. Dicianno, B. Parmanto, A.D. Fairman, T.M. Crytzer, D.X. Yu, G. Pramana, D. Coughenour, A.A. Petrazzi, Perspectives on the evolution of mobile (m-Health) technologies and application to rehabilitation, Phys. Ther. 95 (3) (2015) 397–405.

[19] iTriage, LLC. Available at: https://www.itriagehealth.com/.

[20] N. Li, C. Zhao, E.K. Choe, F.E. Ritter, HHeal: a personalized health app for flu tracking and prevention, in: Proceedings of the 33rd Annual ACM Conference Extended Abstracts on Human Factors in Computing Systems, April 2015, ISBN 978-1-4503-3146-3, pp. 1415–1420, https://doi.org/10.1145/2702613.2732804.

[21] M. Klein, N. Mogles, A. van Wissen, Intelligent mobile support for therapy adherence and behavior change, J. Biomed. Inf. 51 (2014) 137–151.

[22] N. Shah, J. Jomassaint, L. De Castro, Patients welcome the sickle cell disease mobile application to record symptoms via technology (SMART), Hemoglobin 38 (2) (2014) 99–103.

[23] Epocrates. Available at: http://www.epocrates.com/.

[24] G. Ferriero, S. Vercelli, F. Sartorio, S. Munoz Lasa, E. Ilieva, E. Brigatti, C. Ruella, C. Foti, Reliability of a smartphone-based goniometer for knee joint goniometry, Int. J. Rehabil. Res. 36 (2013) 146–151, https://doi.org/10.1097/MRR.0b013e32835b8269.

[25] P.A. Salamh, M. Kolber, The reliability, minimal detectable change and concurrent validity of a gravity-based bubble inclinometer and iphone application for measuring standing lumbar lordosis, Physiother. Theory Pract. 30 (2014) 62–70, https://doi.org/10.3109/09593985.2013.800174.

[26] Nexercise. Available at: http://www.nexercise.com/.

[27] Weight Watchers Mobile. Available at: https://itunes.apple.com/us/app/weight-watchers-mobile/id331308914?mt=8.

[28] FitNow. Lose It! Available at: http://www.loseit.com/.

[29] B.S. Fjeldsoe, Y.D. Miller, A.L. Marshall, Text messaging interventions for chronic disease management and health promotion, in: S.M. Noar, N.G. Harrington (Eds.), e-Health Applications: Promising Strategies for Behavior Change, Routledge, New York, 2012, pp. 167–186.

[30] B.S. Fjeldsoe, A.L. Marshall, Y.D. Miller, Behavior change interventions delivered by mobile telephone short-message service, Am. J. Prev. Med. 36 (2) (2009) 165–173, https://doi.org/10.1016/j.amepre.2008.09.040.

[31] H. Cole-Lewis, T. Kershaw, Text messaging as a tool for behavior change in disease prevention and management, Epidemiol. Rev. 32 (1) (2010) 56–69, https://doi.org/10.1093/epirev/mxq004.

[32] K.J. Head, S.M. Noar, N.T. Iannarino, H.N. Grant, Efficacy of text messaging-based interventions for health promotion: a meta-analysis, Soc. Sci. Med. 97 (2013) 41–48, https://doi.org/10.1016/j.socscimed.2013.08.003.

[33] K.B. Payne, H. Wharrad, K. Watts, Smartphone and medical related app use among medical students and junior doctors in the United Kingdom (UK): a regional survey, BMC Med. Inf. Decis. Mak. 12 (2012) 121, https://doi.org/10.1186/1472-6947-12-121.

[34] C. Montag, K. Blaszkiewicz, R. Sariyska, B. Lachmann, I. Andone, B. Trendafilov, M. Eibes, A. Markowetz, Smartphone usage in the 21st century: who is active on WhatsApp? BMC Res. Notes 8 (2015) 331, https://doi.org/10.1186/s13104-015-1280-z.

[35] V. Khanna, S.N. Sambandam, A. Gul, V. Mounasamy, "WhatsApp"ening in orthopedic care: a concise report from a 300-bedded tertiary care teaching center, Eur. J. Orthop. Surg. Traumatol. 25 (2015) 821–826.

[36] C. Nguyen, L.M. McElroy, M.M. Abecassis, J.L. Holl, D.P. Ladner, The use of technology for urgent clinician to clinician communications: a systematic review of the literature, Int. J. Med. Inform. 84 (2015) 101–110.

[37] V. Giordanoa, H.A. Koch, C.H. Mendes, A. Bergamin, F.S. de Souza, N.P. do Amaral, WhatsApp messenger is useful and reproducible in the assessment of tibial plateau fractures: inter-and intra-observer agreement study, Int. J. Med. Inform. 84 (2015) 141–148.

[38] T.E. Sener, S. Buttice, B. Sahin, C. Netsch, L. Dragos, R. Pappalardo, C. Magno, WhatsApp use in the evaluation of hematuria, Int. J. Med. Inform. 111 (2018) 17–23, https://doi.org/10.1016/j.ijmedinf.2017.12.011.

[39] B. Parmanto, G. Pramada, D.X. Yu, A.D. Fairman, B.E. Dicianno, M.P. McCue, iMHere: a novel m-Health system for supporting self-care in management of complex and chronic conditions, JMIR Mhealth Uhealth 1 (2013) e10, https://doi.org/10.2196/mhealth.2391.

[40] S. Chatterjee, A. Price, Healthy living with persuasive technologies: framework, issues, and challenges, J. Am. Med. Inf. Assoc. 16 (2009) 171–178, https://doi.org/10.1197/jamia.M2859.

[41] B.E. Dicianno, J. Arva, J.M. Lieberman, M.R. Schmeler, A. Souza, K. Phillips, M. Lange, R. Cooper, K. Davis, K.L. Betz, RESNA position on the application of tilt, recline, and elevating leg rests for wheelchairs, Assist. Technol. 21 (2009) 13–22, https://doi.org/10.1080/10400430902945769, quiz 24.

[42] D. Ding, H.Y. Liu, R. Cooper, R.A. Cooper, A. Smailagic, D. Siewiorek, Virtual coach technology for supporting self-care, Phys. Med. Rehabil. Clin 21 (2010) 179–194, https://doi.org/10.1016/j.pmr.2009.07.012.

[43] H. Anglada-Martinez, G. Riu-Viladoms, M. Martin-Conde, M. Rovira-Illamola, J.M. Sotoca-Momblona, C. Codina-Jane, Does m-Health increase adherence to medication? Results of a systematic review, Int. J. Clin. Pract. 69 (1) (2015) 9–32, https://doi.org/10.1111/ijcp.12582.

[44] K. Ganasegeran, A. Rashid, The prevalence of medication non-adherence in post-myocardial infarction survivors and its perceived barriers and psychological correlates: a cross-sectional study in a cardiac health facility in Malaysia, Patient Prefer. Adherence 11 (2017) 1975–1985, https://doi.org/10.2147/PPA.S151053.

[45] Y. Gandapur, S. Kianoush, H.M. Kelli, S. Misra, B. Urrea, M.J. Blaha, G. Graham, F.A. Marvel, S.S. Martin, The role of m-Health for improving medication adherence in patients with cardiovascular disease: a systematic review, Eur. Heart J. 2 (2016) 237–244, https://doi.org/10.1093/ehjqcco/qcx030.

[46] R.S.H. Istepanian, K. Zitouni, D. Harry, N. Moutosammy, A. Sungoor, B. Tang, K.A. Earle, Evaluation of a mobile phone tele-monitoring system for glycemic control in patients with diabetes, J. Telemed. Telecare 15 (3) (2009) 125–128, https://doi.org/10.1258/jtt.2009.003006.

[47] L.L. Sabin, D.M. Bachman, C.J. Gill, L. Zhong, T. Vian, W. Xie, F. Cheng, K. Xu, G. Lan, J.E. Haberer, D.R. Bangsberg, Y. Li, H. Lu, A.L. Gifford, Improving adherence to antiretroviral therapy with triggered real-time text message reminders: the China adherence through technology study, J. Acquir. Immune Defic. Syndr. 69 (5) (2015) 551–559, https://doi.org/10.1097/QAI.00000000 00000651.

[48] S.A. Abdulrahman, L. Rampal, F. Ibrahim, A.P. Radhakrishnan, H.K. Shahar, N. Othman, Mobile phone reminders and peer counseling improve adherence and treatment outcomes of patients on ART in Malaysia: a randomized clinical trial, PLoS One 12 (5) (2017) e0177698, https://doi.org/10.1371/journal.pone.0177698.

[49] J.W. McGillicuddy, M.J. Gregoski, A.K. Weiland, R.A. Rock, B.M. Brunner-Jackson, S.K. Patel, B.S. Thomas, D.J. Taber, K.D. Chavin, P.K. Baliga, F.A. Treiber, Mobile health medication adherence and blood pressure control in renal transplant recipients: a proof-of-concept randomized controlled trial, JMIR Res. Protoc. 2 (2013) e32, https://doi.org/10.2196/resprot.2633.

[50] M. Tian, V. Ajay, D. Dunzhu, S. Hameed, X. Li, Z. Liu, C. Li, H. Chen, K. Cho, R. Li, X. Zhao, D. Jindal, I. Rawal, M.K. Ali, E.D. Peterson, J. Ji, R. Amarchand, A. Krishnan, N. Tandon, L.Q. Xu, Y. Wu, D. Prabhakaran, L.L. Yan, A cluster-randomized controlled trial of a simplified multifaceted management program for individuals at high cardiovascular risk (SimCard trial) in rural Tibet, China, and Haryana, India, Circulation 132 (2015) 815–824, https://doi.org/10.1161/CIRCULATIONAHA.115.015373.

[51] W.M. Vollmer, A.A. Owen-Smith, J.O. Tom, R. Laws, D.G. Ditmer, D.H. Smith, A.C. Waterbury, J.L. Schneider, C.H. Yonehara, A. Williams, Improving adherence to cardiovascular disease medications with information technology, Am. J. Manag. Care 20 (SP17) (2014) SP502–SP510.

[52] C.M. Goldstein, E.C. Gathright, M.A. Dolansky, J. Gunstad, A. Sterns, J.D. Redle, R. Josephson, J.W. Hughes, Randomized controlled feasibility trial of two telemedicine medication reminder systems for older adults with heart failure, J. Telemed. Telecare 20 (2014) 293–299, https://doi.org/10.1177/1357633X14541039.

[53] A.K. Kamal, Q. Shaikh, O. Pasha, I. Azam, M. Islam, A.A. Memon, H. Rehman, M.A. Akram, M. Affan, S. Nazir, A randomized controlled behavioral intervention trial to improve medication adherence in adult stroke patients with prescription tailored Short Messaging Service (SMS)-SMS4Stroke study, BMC Neurol. 15 (2015) 1, https://doi.org/10.1186/s12883-015-0471-5.

[54] T. Kuo, H. Kim, L. Ohno-Machado, Blockchain distributed ledger technologies for biomedical and health care applications, J. Am. Med. Inf. Assoc. 24 (6) (2017) 1211–1220, https://doi.org/10.1093/jamia/ocx068.

[55] A. Omotosho, J. Emuoyibofarhe, C. Meinel, Ensuring patients' privacy in a cryptographic-based electronic health records using bio-cryptography, Int. J. Electron. Healthc. 9 (4) (2017) 227–254, https://doi.org/10.1504/IJEH.2017. 10003030.

[56] L. Kobayashi, The Ethics of Mobile Health Technology. Public Health Perspectives, 2014. Available at: http://blogs.plos.org/publichealth/2014/10/16/health-apps/.

[57] K. Ganasegeran, S.A. Abdulrahman, S.A. Al-Dubai, A. Rashid, M. Perumal, P. Renganathan, Identifying factors associated with sleep disturbances among health workers using WhatsApp in Malaysia, J. Hosp. Manag. Health Policy 1 (2017) 2. https://doi.org/10.21037/jhmhp.2017.08.01.

[58] Z.J. Eapen, M.P. Turakhia, M.V. McConnell, G. Graham, P. Dunn, C. Tiner, C. Rich, R.A. Harrington, E.D. Peterson, P. Wayte, Defining a mobile health roadmap for cardiovascular health and disease, J. Am. Heart Assoc. 5 (7) (2016) e003119.

[59] Reuters, Premera Blue Cross says data breach exposed medical data, The New York Times (March 17, 2015). Available at: http://www.nytimes.com/2015/03/18/business/premera-blue-cross-says-data-breach-exposed-medical-data.html?_r=1.

[60] R. Abelson, M. Goldstein, Millions of anthem customers targeted in cyberattack, The New York Times (February 5, 2015). Available at: http://www.nytimes.com/2015/02/05/business/hackers-breached-data-of-millions-insurer-says.html.

[61] Mobile Business Insight, 2018. Available at: https://mobilebusinessinsights.com/2018/03/mobile-apps-for-healthcare-professionals-current-and-future-trends/.

[62] R.K. Kher, Mobile and E-healthcare: recent trends and future directions, J. Health Med. Econ. 2 (3) (2016) 10.

[63] S. Hiremath, G. Yang, K. Mankodiya, Wearable Internet of Things: concept, architectural components and promises for person-centered healthcare, in: 2014 EAI 4th International Conference on Wireless Mobile Communication and Healthcare (Mobihealth), Athens, Greece, 2014, pp. 304–307.

[64] European Commission, Healthcare in Your Pocket: Unlocking the Potential of M-Health. Press Release, 2014. Available at: http://europa.eu/rapid/press-release_IP-14-394_en.htm.

CHAPTER 4

Investigation of Telecardiology System to Detect Cardiac Abnormalities

G. GOPU[a] • T. ANITHA[b] • M. NAGARAJAPANDIAN[b] • M. PRABHAKARAN[b]

[a]Department of Bio Medical Engineering, Sri Ramakrishna Engineering College, Coimbatore, India; [b]Department of Electronics and Instrumentation Engineering, Sri Ramakrishna Engineering College, Coimbatore, India

1 THEORETICAL BACKGROUND

In conventional medical diagnosis, Electrocardiography plays a vital role which records the patient activity as an electrical output. The heart consists of specialized organ muscles which have been self-excited by periodic pulses. As per the human anatomy principle, the excitation is spread from right atrium to left atrium and left ventricle then right ventricle.

Fig. 4.1 illustrates that the pattern of ECG waveform. Every heart pulses of ECG. In every heart pulses, the following elements were namely P, Q, R, S, T and U These are represented as ECG Waves where P denotes auricular depolarization and QRS referred as ventricular depolarization and also known as resting position of were. For the purpose to complete the ECG waveform "U Element" is considered and its origin is unknown.

2 OVERVIEW OF CARDIAC DISEASE

In general cardiovascular disease comprises heart and blood vessel disease. The World Health Organization (WHO) listed the possible cardiovascular diseases which are given as atherosclerosis, coronary heart disease, stroke, hypertensive heart disease, inflammatory heart disease, rheumatic heart disease. From the above-mentioned list, atherosclerosis is one type of diseases which might be caused by plaque occurred in the arteries outer wells. Due to this plaque sediments, arteries are shrunken and blood flow will be restricted which leads to heart attack. If this condition prolongs affected patient life may be at a risk.

But the proper diagnosis of the first heart attack may save the patient's life from critical conditions abnormalities in cardiac rhythms.

Basically abnormal cardiac disorders are identified as
1. Sinus tachycardia
2. Sinus Bradycardia
3. Wide QRS complex
4. Atrial fibrillation
5. Cardiac asystole

Fig. 4.2 shows sinus tachycardia abnormal rhythms start from the Sino arterial node during this time heartbeat reaches nearly 100 beats/min. It increases depolarization. There is a possibility of getting inappropriate tachycardia fairly.

Fig. 4.3 shows the Sinus Bradycardia abnormal rhythm due to fall down of blood rate pulse. In this condition, the heartbeat will be less than 60 beats/min. Generally, it affects the patient during the deep sleep. It has symptoms like fainting, fatigue, heart failure.

Fig. 4.4 shows the Wide QRS complex abnormal rhythm due to slow activation of the ventricular and Purkinje system.

Heartbeat will be 120 beats/min. It is due to the depolarization in the ventricles. It has the symptoms like hypotension, confusion, angina.

Fig. 4.5 shows the Atrial fibrillation as similar to wide QRS but mostly affects the older people. Due to these abnormality P elements of the ECG wave may not be identified.

It has the symptoms like palpitations, weakness, shortness of breath. It creates abnormal signals in the atria. Cardiac asystole is an absence of rhythm which is equivalent to heart dead condition. Table 4.1 shows the various methods used for telemonitoring.

3 METHODOLOGY

Fig. 4.6 shows that the methodology of the system. Remote patients are monitored by ECG devices where the information are transmitted to the wireless medium and the patient monitoring system with a diagnosis unit provides the prescriptions to the concerned patient.

Telemedicine Technologies. https://doi.org/10.1016/B978-0-12-816948-3.00004-0

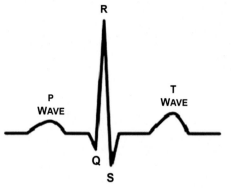

FIG. 4.1 ECG waveform. (Wikipedia electrocardiography.)

Diagnosis unit has an expert doctor whose job is to analyze the patient conditions regularly.

Once we diagnose the abnormal rhythms properly the patient life-saving activity will be improved. Similarly, modern telemetry systems offer the ways to rectify the problems quickly.

4 TELECOMMUNICATIONS

Telecommunication is the field of engineering which transfers the data from one point another through the wire or wireless communication protocol. Possibly, the communication can occur from short distance to several miles.

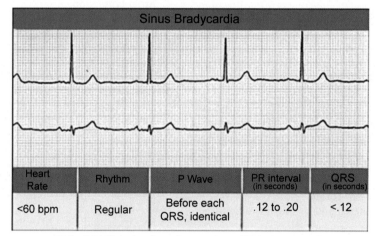

FIG. 4.2 Sinus bradycardia waveform. (Slideplayer.com.)

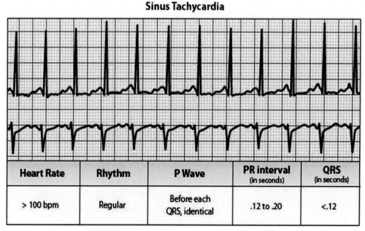

FIG. 4.3 Sinus tachycardia waveform. (CEUFast.com.)

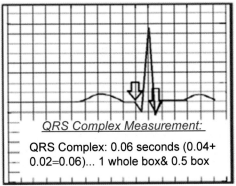

QRS Complex Measurement:

QRS Complex: 0.06 seconds (0.04+ 0.02=0.06)... 1 whole box& 0.5 box

FIG. 4.4 QRS complex waveform. (registerednursern.com.)

homes etc. IoT also provides the option to communicate among the devices hence the technology may be termed as Intelligent of Things. Notably, it focuses on ZigBee, 6LoWPAN, Bluetooth Low Energy, Lora, and also the completely distinct variations of Wi-Fi in addition to the latest IEEE 802.11ah protocol. The studies appraise the abilities and behaviors of these technologies concerning several metrics in addition to the data vary and rate, community size, RF Channels, and information degree and electricity intake. It's entirely that there may be a choice to broaden a multifarious era technique to alternate realistic and secure communications inside the IoT.

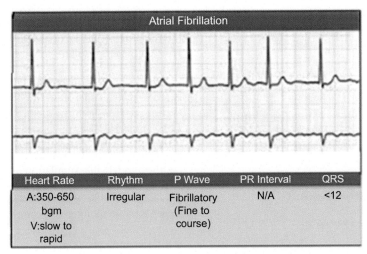

Heart Rate	Rhythm	P Wave	PR Interval	QRS
A:350-650 bgm V:slow to rapid	Irregular	Fibrillatory (Fine to course)	N/A	<12

FIG. 4.5 Atrial fibrillation waveform. (pinterest.com.)

5 WI-FI

Wi-Fi is a portable wireless network that can communicate to the internet. Wi-Fi gives the same speed same as the wired Ethernet.

6 BLUETOOTH

Bluetooth can be connected with more than the other seven devices. The data is transferred according to the master-slave technique. The one which transfers the data is called master and the other receives it.

The Internet of Things (IoT) carries a couple of lengthy-range, short-variety, and personal area wireless networks and technologies into the forms of IoT packages. It consists of numerous enterprise possibilities in fields such as e-health, smart cities, sensible

7 WEARABLE DEVICE FOR ECG MONITORING

In the modern scenario, wearable devices are most commonly used for human healthcare monitoring. Wearable devices are basically System on Chip (SoC) which provide the compact environment to the patient. The main function of the wearable device is to monitor the human health parameters and passes those values into wireless medicine. A Commonly wireless medium may be Bluetooth, Zigbee depending upon the distance we used. Fig. 4.7 illustrates the Remote Monitoring and Diagnosis.

Recent days low pass VLSI based chips are available to communicate with the cloud server where the server sends the data to the remote station. Remote station has experts in doctor team whose objective is to monitor the patient parameters continuously.

TABLE 4.1
Existing Method and Drawbacks of Telecardiology.

S. No	Existing methodology used for telecardiology	Drawbacks
1	Wavelet threshold based ECG	It can be used for band limited telephone networks
2	Representative Signals	Transmission of ECG Signal is delayed
3	Automatic patient authentication with mobile-based remote monitoring	ECG Biometric pertinence of ectopic beats
4	Wireless Technology	Estimation is used for HR and QRS Complex
5	RFID/sensor based telecardiology	RFID is used for observing medical facility in the nursing home
6	Real-time ischemia	It is used in a home monitoring system

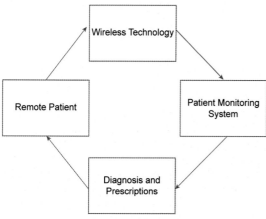

FIG. 4.6 Methodology of the system.

The wearable devices simply act as two-way data acquisition systems send the data from a patient to a remote location and similarly receive the prescription information from the remote stations. Apart from those smart wearable fixed end ECG units are available. In that case, electrodes are covered in the bedspread and again in contact with the patient. This entire unit also used to diagnose and send the data to the remote monitoring system.

Fig. 4.8 illustrates the methodology of ECG signal reception. Electrodes are fixed with the patient and the information is received. Then this information is feeble hence it has been amplified by the preamplifier. High-frequency noise signals are rejected through a bandpass filter and further amplified. SOC may communicate with portable devices through Bluetooth module.

8 RESULTS AND DISCUSSION

Based on the previous chapter studies and from standard literature this work aims to develop the different rhythmic pattern of ECG signals for remote diagnosis. The simulated results shown below explained the abnormal rhythms of ECG.

Most of the industry obey the principle "Model-based system Engineering" [MBSE] otherwise it is known as 3Ds namely design, Develop and Deploy. The concept of Design and Develop involves the preliminary construction of the model which is based on the data received from the real-time (i.e) patient data. Based on the model, the simulation will be performed to attain the desired responses. If the responses are not obtained properly means the model modifications have been carried out to meet the desired specifications. Deployment is the final stage which means that real-time implementation of concern work. All these terminologies must be fulfilled during the time of fabrication. This is achieved by MATLAB software which will provide the solution by the simulated environment.

9 SIMULATION OF THE PROPOSED IDEA

Fig. 4.9 shows the Block Diagram of ECG Detection of Various Cardiac Rhythms The simulations are done using MATLAB software. For simulation of the ECG signals of two different patients are taken and amplified by means of an amplifier. The amplified signals are given to the bandpass filter to select the particular band. The signals of the previous stage are given to the notch filter to get power line interference. Those signals are again amplified and the numbers of pulses are counted by means of a counter. Finally, the signals are displayed.

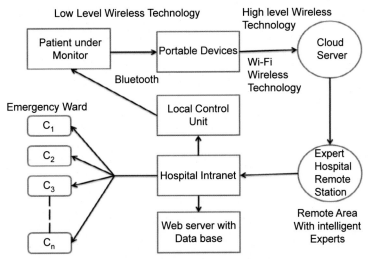

FIG. 4.7 Remote monitoring and diagnosis.

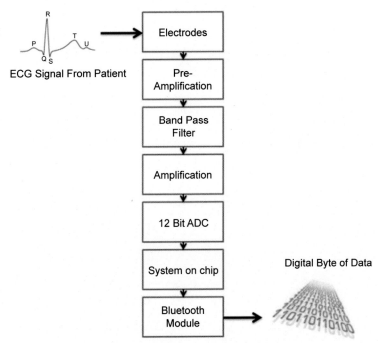

FIG. 4.8 Methodology to adopt the ECG Signal.

Fig. 4.10 shows the subsystem layout. In case of the up counter is used to generate the signal and its combined with the relational operator. The pulse generator creates the square wave pulses of these signal. The signal from the relational operator and pulse generator are combined and passed through the multi-port switch. The switch act as a multiplexer to produce a single output signal from several signals. Then the signal is passed through the up counter and compared with the constant value to produce the desired ECG signal.

Fig. 4.11 illustrates the information received by Bluetooth is assumed as constant value over the time. Hence it is considered as the gain value in MATLAB. This value will be compared with various abnormalities standard values for diagnosis.

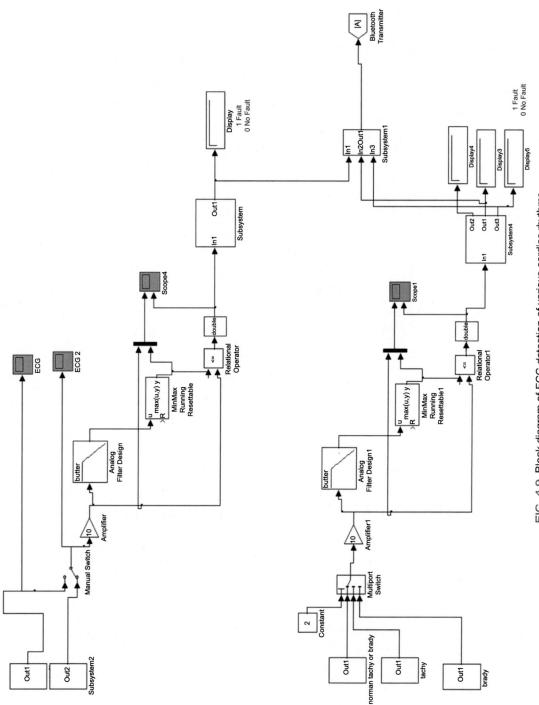

FIG. 4.9 Block diagram of ECG detection of various cardiac rhythms.

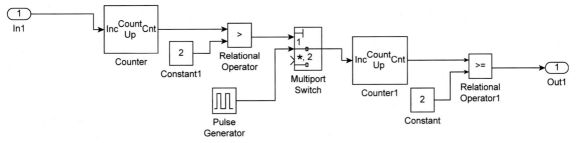

FIG. 4.10 Sub-system layout.

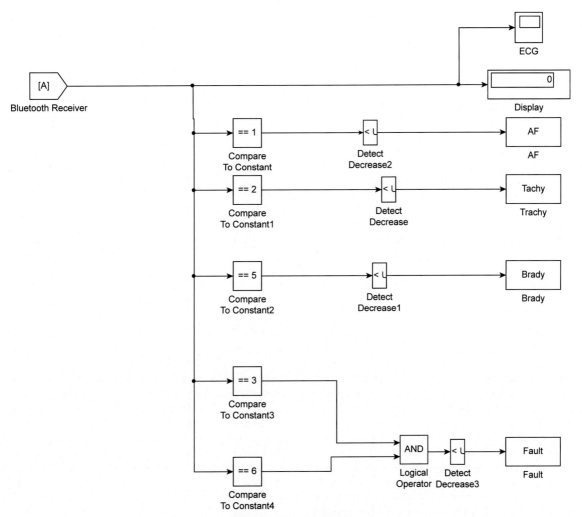

FIG. 4.11 Portable device.

Fig. 4.12 shows the abnormality of the Atrial Fibrillation affects the atrial contract. This may occur when the heart beats from 100 to 175 beats a minute. It needs certain treatment to control and rhythm of their heart.

Fig. 4.13 shows the condition of Atrial Fibrillation it commonly affects the older people. The person may have symptoms like palpitations, weakness, confusion, chest pain.

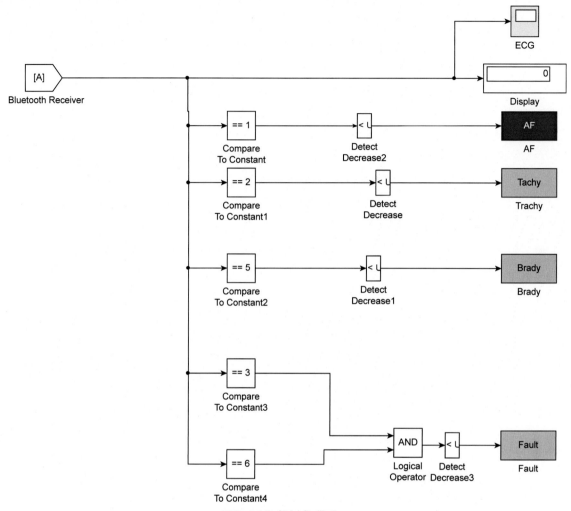

FIG. 4.12 Atrial fibrillation.

Fig. 4.14 illustrates the abnormality of the sinus tachycardia. Treatment is needed if the condition is prolonged and often occurs. A Treatment plan can be done along with the expert system.

Fig. 4.15 illustrates the condition of cardiac fault and it resembles the severe cardiac disorder.

Fig. 4.16 shows the changes in the sinoatrial node and having the heartbeat range greater than 100 beats/min. The person may have symptoms like stress, fever, chest pain, shortness of breath, congestive heart failure, lung disease.

Fig. 4.17 shows the condition occurs due to due to a drop in blood rate pulse. It shows the electrical impulse slower than the normal. The person may have the

symptoms like high blood pressure, psychological stress, the inability of the heart to pump enough blood.

Fig. 4.18 shows the abnormality of sinus bradycardia in which the heartbeat rate occurs less than 60 beats/min. It shows the discharge of an electric impulse is blocked due to a contract. The expert system is designed to monitor the activity of the heart when the problem is involved.

10 CONCLUSION
In this work, the abnormal rhythmic pattern of the cardiac vascular system has been discussed. Initially, a pattern of ECG wave was discussed and several diseases

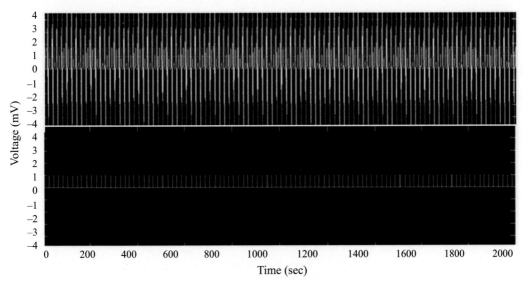

FIG. 4.13 Output of atrial fibrillation.

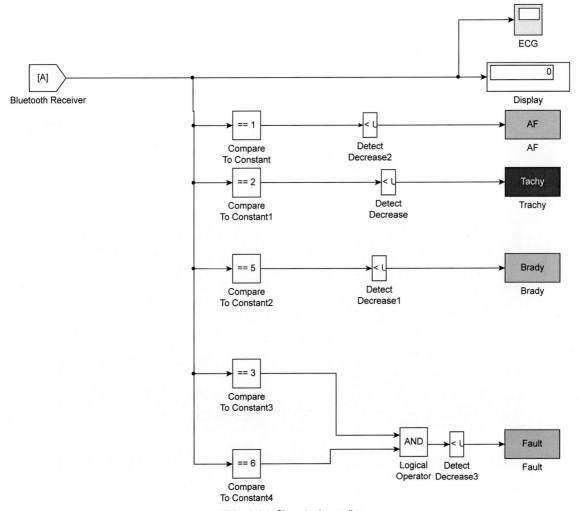

FIG. 4.14 Sinus tachycardia.

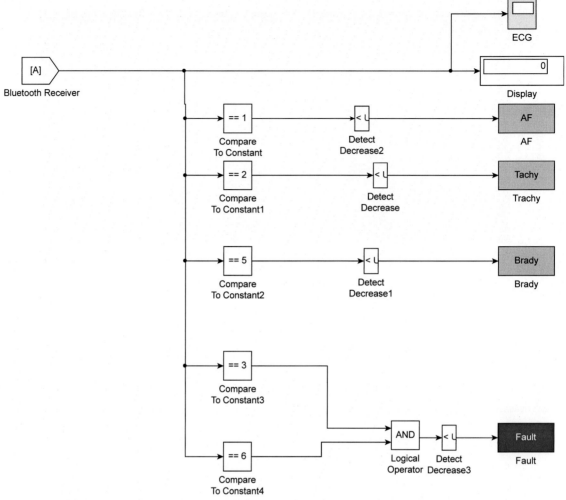

FIG. 4.15 Cardiac disorder.

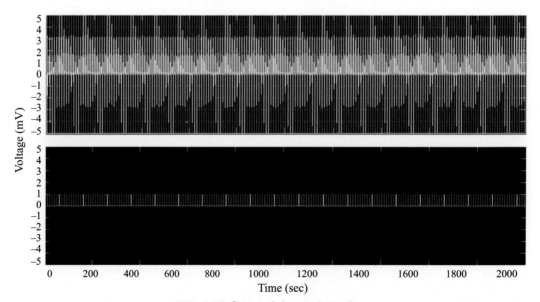

FIG. 4.16 Output of sinus tachycardia.

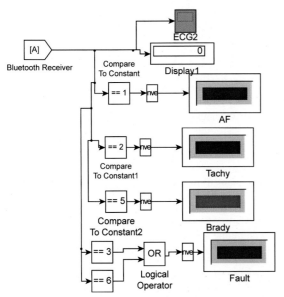

FIG. 4.17 Sinus bradycardia.

transmit the data over the cloud server and send the information to the emergency system, in parallel data are stored in the database for future reference. In order to investigate abnormal rhythmic patterns, the signals were simulated on MATLAB environment for better visualization and results were discussed. This study will improve the understandings of ECG abnormal rhythmic patterns and its teletransmission techniques. Based on these survey one can fabricate the real-time system which may help the people of rural areas.

11 FUTURE SCOPE

The concept discussed in this work under the telecommunication system is fixed with the Wi-Fi-based protocol where distance is the major problem for practical implementation. In the future, telecommunication will evolve the dimensions of the internet of things where distance is not an issue. The IoT server such as Amazon Web Server (AWS), ThingSpeak will provide the virtual storage of data through a cloud server. Hence

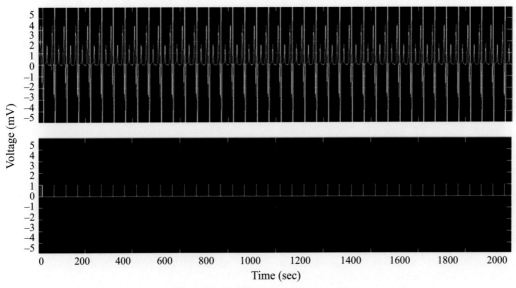

FIG. 4.18 Output of sinus bradycardia.

relevant to the cardiac vascular system were also analyzed. Based on abnormal rhythms and it is beat pattern one can classify the cardiovascular diseases as sinus tachycardia, sinus Bradycardia, wide QRS complex, Atrial fibrillation, and cardiac asystole. This work discussed the telecommunication system included to

big data analytics will play the major role in the medical field. This may help to implement an expert system with the developed data model. New data are continuously trained through the neural network for enhancing the expert systems.

ACKNOWLEDGMENT

The authors acknowledge their indebtedness to the management, the director and the Principal of Sri Ramakrishna Engineering College, Coimbatore for their support and for permitting us to use the facilities to conduct this research work.

FURTHER READING

[1] World Health Organization, The World Health Report 2008, World Health Organization, Geneva, 2008. Available: http://www.who.int/whr/2008/whr08_en.pdf.

[2] S. Koch, Home telehealth—current state and future trends, Int. J. Med. Inform. 75 (2006) 565–576.

[3] R.J. Romanow, Building on Values: The Future of Healthcare in Canada—Final Report, Commission of the Future of Health Care, Ottawa, Canada, November 2002 [Online]. Available: http://www.cbc.ca/healthcare/final_report.pdf.

[4] C.-T. Lin, K.-C. Chang, C.-L. Lin, C.-C. Chiang, An intelligent telecardiology system using a wearable and wireless ECG to detect atrial fibrillation, IEEE Trans. Inf. Technol. Biomed. 14 (3) (2010) 726–733.

[5] J. Hurst, Naming of the waves in the ECG, with a brief account of their Genesis, Circulation 98 (1998) 1937–1942.

[6] W.B. Kannel, D.L. McGee, Diabetes and cardiovascular disease. The Framingha study, J. Am. Med. Assoc. 241 (19) (1979) 2035–2038. ACC/AHA/ESC 2006 guidelines for the management of patients with atrial fibrillation, J. Am. Coll. Cardiol. 48 (2006) 149–246.

[7] G. Vanagas, R. Zaliunas, R. Benetis, R. Slapikas, Factors affecting the relevance of tele-ECG systems application to high risk for future ischemic heart disease events patients group, Telemed. J. e-Health 14 (4) (2008) 345–349.

[8] C.T. Lin, L.W. Ko, C.J. Chang, Y.T. Wang, C.H. Chung, F.S. Yang, J.R. Duann, T.P. Jung, J.C. Chiou, Wearable and wireless brain computer interface and its applications, in: Presented at the 13th Int. Conf. Human-Computer Interface, San Diego, CA, 2009.

[9] H. Strubb, K. Johnson, A. Allen, V. Bellotti, T. Starner, Privacy, wearable computers, and recording technology, Panel Discussion, the Second International Symposium on Wearable Computers, October 19–20, 1998, Pittsburgh, PA. G. Bell, The body electric, communications of the ACM, 40(2) (February 1997) 31–32.

[10] D. Hoyer, B. Pompe, H. Herzel, U. Zwiener, Nonlinear Coordination of cardiovascular autonomic control, IEEE Eng. Med. Biol. 17 (6) (1998) 17–21.

[11] M. Akay (Ed.), Time Frequency and Wavelets in Biomedical Signal Processing, IEEE Press, 1999.

[12] M. Vaage-Nilsen, V. Rasmussen, C. Sorum, G. Jensen, ST-segment deviation during 24-hour ambulatory electrocardiographic monitoring and exercise stress test in healthy male subjects 5 1 to 75 years of age: the Copenhagen city heart study, Am. Heart J. 137 (6) (1999) 1070–1074.

[13] J. Pan, W.J. Tompkins, A real time QRS detection algorithm, IEEE Trans. Biomed. Eng. BME-32 (3) (March 1985) 230–236.

[14] H.H. So, Development of QRS detection method for real-time ambulatory cardiac monitor, in: Proceedings of the Annual International Conference of the IEEE Engineering in Medicine and Biology Society, 1997, pp. 282–292.

[15] L. Pang, I. Techoudovski, M. Braecklein, K. Egorouchkina, W. Kellermann, A. Bolz, Real time ischemia detection in the smart home care system, in: Proc. 27th Annu. Conf. IEEE Eng. Med. Biol., Shanghai, China, Sep. 1–4, 2005, pp. 3703–3706.

[16] E. Jovanov, A. O'Donnell Lords, D. Raskovic, P.G. Cox, R. Adhami, F. Andrasik, Stress monitoring using a distributed wireless intelligent sensor system, IEEE Eng. Med. Biol. Mag. 22 (Iss. 3) (2003) 49–55.

[17] T. Martin, E. Jovanov, D. Raskovic, Issues in wearable computing for medical monitoring applications: a case study of a wearable ECG monitoring device, in: Proc. 4th Int. Symposium on Wearable Computers, 2000, pp. 43–49.

[18] P.F. Binkley, Predicting the potential of wearable technology, IEEE Eng. Med. Biol. Mag. 22 (Iss. 3) (2003) 23–27.

[19] P. Bonato, Wearable sensors/systems and their impact on biomedical engineering, IEEE Eng. Med. Biol. Mag. 22 (Iss. 3) (2003) 18–20.

[20] O. Boric-Lubeke, V.M. Lubecke, Wireless house calls: using communications technology for health care and monitoring, IEEE Microwave Mag. 3 (Iss. 3) (2002) 43–48.

[21] K.Y. Kong, C.Y. Ng, K. Ong, Web-based monitoring of real-time ECG data, Comput. Cardiol. 27 (2000) 189–192.

[22] R. Fensli, E. Gunnarson, O. Hejlesen, A wireless ECG system for continuous event recording and communication to a clinical alarm station, in: 26th Annual InternationalConference of the IEEE Engineering in Medicine and Biology Society, San Francisco, USA, 2004, pp. 2208–2211.

[23] R.J. Myerburg, A. Interian Jr., R.M. Mitrani, K.M. Kessler, A. Castellanos, Frequency of sudden cardiac death and profiles of risk, Am. J. Cardiol. 80 (1997) 10F–19F.

[24] P. Corabian, Accuracy and Reliability of Using Computerized Interpretation of Electrocardiograms for Routine Examination, Alberta Heritage Foundation for Medical Research HTA 25: Series A Health Technology Assessment, 2002.

[25] U. Rajendra Acharya, N. Kannathal, S.M. Krishnan, Comprehensive analysis of cardiac health using heart rate signals, Physiol. Meas. 25 (2004) 1139–1151.

[26] Wikipedia, the Free Encyclopedia. http://www.wikipedia.com.

[27] T.R.F. Fulford-Jones, G. Wei, M. Welsh, A portable, low-power, wireless two-lead EKG system, in: 26th Annual International Conference of the IEEE EMBS, September 2004.

[28] U. Anliker, et al., AMON: a wearable multi parameter medical monitoring and alert system, IEEE Trans. Inf. Technol. Biomed. 8 (5) (December 2004) 415–427.

[29] E. Jovanov, A. Milenkovic, C. Otto, P.C. de Groen, A wireless body area network of intelligent motion sensors for computer assisted physical rehabilitation, J. Neuro Eng. Rehabil. 2 (6) (March 2005).

[30] J.E.P. Waktare, Atrial fibrillation, Circulation 106 (2002) 14–16.

[31] https://en.wikipedia.org/wiki/Electrocardiography.

[32] https://slideplayer.com/slide/7760035/.

[33] https://ceufast.com/course/ecg-interpretation.

[34] http://www.registerednursern.com/how-to-measure-a-qrs-complex-on-an-ekg-strip-qrs-complex-measurement-quiz/.

[35] https://www.pinterest.com/edavis0516/ekg/.

CHAPTER 5

Microaneurysms Detection for Early Diagnosis of Diabetic Retinopathy Using Shape and Steerable Gaussian Features

G. INDUMATHI • V. SATHANANTHAVATHI
Mepco Schlenk Engineering College, Sivakasi, India

1 INTRODUCTION

Diabetic persons may have more possibility of eye disease which affects vision. Retina, a light sensitive tissue which lines the rear end of eye is affected by Diabetic retinopathy. All types of diabetics may lead to Diabetic retinopathy and the risk of Diabetic retinopathy increases with respect to longer the days, the person has diabetics. It is more common among the working age people and good blood glucose control helps to reduce the risks of diabetes retinopathy. At the initial stages, diabetic retinopathy usually doesn't have any noticeable symptoms. Even the disease progression is unnoticed by the diabetic patients until it affects vision. At this stage only periodic retinal fundus imaging helps to identify the disease progression.

Diabetic retinopathy is initiated by uncontrolled high blood glucose levels of diabetic patients for a longer time. Due to this high sugar glucose levels, tiny blood vessels within the retina will get weakened and damaged. It leads to cause bleeding from abnormal retinal blood vessels known as hemorrhages, leakage of some nutrient fluids named as exudates and even it may cause swelling of the retina. Due to this bleeding of fluids and blood, the retinal region couldn't get sufficient oxygen and hence abnormal weak vessels may grow. These new blood vessels leak and form scar tissue that can cause vision loss. The earliest symptom of Diabetic retinopathy is presence of microaneurysms appears as a small red spot in fundus images. It is a sphere-shaped swellings of the capillaries due to the weakening of the vascular walls by high level of diabetics. At this stage no symptoms are sensible to the diabetic persons and identified through fundus imaging. Doctors diagnose the disease progression only with the presence of microaneurysms and hemorrhages. But clinical diagnosis will be a tedious process due to the small size and intensity similarity of microaneurysms with other regions of fundus images. Also the factors such as uneven image illumination, resolution and contrast makes the diagnosis complex and time consuming process. Edges are also not clearly defined for the microaneurysms with respect to the background; hence it is still a tedious task to identify those regions.

Blood vessels intensity is also have similar intensity to that of microaneurysm, hence possibility of considering it as microaneurysm is more. Also detecting microaneurysms which are located close to the blood vessel is difficult. Even in a high resolution condition, due to the non uniform illumination effect it is difficult to detect microaneurysms in the fundus image. Hence it is the time consuming process in the clinical diagnosis process and there is a need of automated system.

2 DIABETIC RETINOPATHY

Diabetic Retinopathy (DR) is possible to develop anyone who has affected by diabetes even if they maintain their sugar level in a controlled manner with the help of medicines. A person is possible to be affected by DR if he/she has diabetics for a prolonged period of years. DR is also become incurable if a person has very high blood pressure and cholesterol level. It is also possible to develop Diabetic Retinopathy during the pregnancy period.

Telemedicine Technologies. https://doi.org/10.1016/B978-0-12-816948-3.00005-2

Diabetic retinopathy leads to the abnormal growth of blood vessels in the retina. Complications can lead to serious vision problems like Vitreous hemorrhage, Retinal detachment, Glaucoma and Blindness. Due to Diabetic retinopathy, blood vessels may perhaps bleed into the center of the eye. If the quantity of bleeding is less it look like a dark spots called as floaters. If the disease severity increases, bleeding block the vitreous cavity hence the vision. The development of abnormal blood vessels related with diabetic retinopathy excite the development of scar tissue may leads to the formation of spots floating in the vision or may cause severe vision loss. The new blood vessel formed in the front region of eye may block the normal flow of fluid leads to increase in pressure causes Glaucoma. This increase in pressure will impair the nerve which carries vision information from the eye to the brain. Due to the above factors gradually diabetic retinopathy leads to blindness.

2.1 Introduction to Microaneurysms

A microaneurysm is a tiny aneurysm, or swelling, in the retinal region or in the side of a blood vessel. Some research point out that these microaneurysms can predict the progress of diabetic retinopathy. These miniature aneurysms can break and leak blood. A sample micro aneurysm from DiaretdB01 database is shown in Fig. 5.1.

In addition to diabetic retinopathy, any form of vascular disease formation of high blood pressure in the retinal region may cause the formation of retinal microaneurysm. There are no initial signs or symptoms like reddishness, pain or irritation for microaneurysm. It can be detected only by the careful examination of the retina, frequently through a dilated pupil. A dilated retinal examination will expose the microaneurysm, and additional investigations through fluorescein

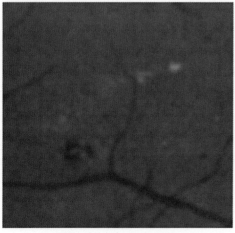

FIG. 5.1 Microaneurysm.

angiogram will support the doctors to find the origin and severity level. A microaneurysm does not necessitate any treatment on it's own. The disease or disorders underlying it to cause needs to be treated. The prevention of microaneurysms includes the maintenance of healthy lifestyle includes diet, exercise, periodic checkup and controlling of diabetes and high blood pressure with medicine will reduce the retinal microaneurysms.

3 LITERATURE SURVEY

Numerous methods have been proposed to detect the Mas present in the retina and the severity of the disease of the diabetic retinopathy affected patients as mentioned here. Saiprasad Ravishankar, Arpit Jain and Anurag Mittal used geometric features and correlations to distinguish between Mas and blood vessels. The authors treat Mas as holes (i.e. small dark blobs surrounded by brighter regions) and morphological filling is performed on the green channel to identify them [1]. The unfilled green channel image is then subtracted from the filled one and thresholded in intensity to yield an image with microaneurysm patches. The threshold is chosen based on the mean intensity of the retinal image in the red channel. Blood vessels can also appear as noise in the microaneurysm and hemorrhage detection as they have similar color and contrast to the clots. In order to remove this additional noise, the full blood vessel network skeleton is first obtained. The resulting blood vessel network is dilated and subtracted from the original image to remove the noise due to vessels. The remaining patches are further classified using intensity properties and a color model based on the detected blood vessels. The authors found that many features of the candidates have common intensity and so geometric features, morphological features and correlation were formulated so as to distinguish them. However, it was found that adding few more learning techniques apart from geometric and morphological features would produce high accuracy.

T. P. Karnowski S, Aykac, L. Giancardo, Y. Li, T. Nichols, K.W. Tobin and E. Chaum proposed a segmentation algorithm [2] which uses a vascular tree segmentation. The vessel density, orientation, and thickness are measured and combined to compose a four dimensional feature vector for each pixel. Two main lesions or anomalies were detected with the system: exudates (small lipid deposits with a bright yellow color and distinctive spatial distribution) and microaneurysms (dilations of a small retinal capillary vessel, producing a small round dark spot in the vicinity of a capillary vessel.) The authors proposed a system which uses specific detectors designed for these defects.

The detectors produce candidate regions of interest which are likely positive examples of each lesion. While these detectors were designed to provide good lesion detection without post-processing, they use supervised learning to separate the candidates into "true lesions" and "nuisance blobs". Both detectors generate candidate blobs which are filtered into "true lesions" and "nuisance blobs" using neural network classifiers trained on a set of ground-truth lesions. The detected, filtered lesions create an overall fundus description that consists of a measurement of features related to the lesions detected (such as the number detected, histograms of the sharpness of the lesion edges, shape properties, etc.), the vascular density within the lesion population, population moments, and textural features of the macula region. This vector, which they refer to as the "population vector", was used to determine an index that can be used to locate similar images in an archive which then generates an automatic detection of disease by analysis of the ground-truth disease states of the similar images. They use PCA to reduce the dimensionality of the features space to a more efficient and effective search space, followed by a classification using a K-NN classifier. However it was found that only severe DR can be detected. So this method cannot be adopted if the image quality is low.

Salim Lahmiri, Christian Gargour and Marcel Gabrea used the empirical mode decomposition (EMD) to analyze retina digital images in the frequency domain and statistical features are extracted from high frequency components of the analyzed images [3]. The purpose was to classify normal versus abnormal images. Three different pathologies are considered including, circinates, drusens, and microaneurysms (MAs). The most informative and non-redundant features are ranked and selected by use of statistical features selection techniques; namely Adaptive Histogram Equalization, t-statistic, entropy, Battacharrayia statistic, and the area between the receiver operating characteristic (ROC), and the principal component analysis (PCA). Also six statistical textural features are extracted from the retina processed images. The statistical features are the mean, standard deviation, smoothness, third moment, uniformity, and entropy. Finally, support vector machines (SVMs) with polynomial and radial basis function (RBF) kernels are used to classify retina digital images based on the selected features. It was inferred that Feature selection based on Battacharrayia statistic provided accurate results.

Bo Wu, Hongying Lilian Tang, Lutfiah Ismail Al turk, Yin Hu and Saeid Sanei proposed a method in which the normalization of illumination and contrast limited adaptive histogram equalization (CLAHE) enhancement methods have been applied in the preprocessing step to enhance the input image for the better candidate extraction and feature extraction [4]. It was found that this method is more robust preprocessing method. At first, the preprocessing steps are applied to enhance the input images for candidate extraction and feature extraction. A step of CLAHE enhancement is applied. It was found that the enhancement method of CLAHE had a good performance in making MA more visible while suppressing noises. The preliminary candidate pixels are extracted by applying peak detection on each profile. The region growing is adopted to grow preliminary candidate pixels back to original pathology shape, in which, the dynamic transformation is applied to obtain the optimal threshold. For feature extraction, 27 features are used which contain not only profile features but also local features. For classification, three classifiers of KNN, NB and Adaboost are compared and tested to find a suitable classifier for the feature set. In the feature extraction step, authors have used much more features. A total of 27 features which contain not only profile features but also local features are used for classification. Peak detection and the candidates profiles is used to extract a set of statistical features that can discriminate true MA from non-MA. It was found that there was a better classification of affected regions, various regions with different resolutions, quality, and of different ethnic origins.

Su Wang, Hongying Lilian Tang, Lutfiah Ismail Al turk, Yin Hu, Saeid Sanei, George Michael Saleh and Tunde Peto proposed a method in which separation of MAs and its background was performed by increasing the difference in profiles of candidate and non candidate MAs [5]. Candidate extraction scheme was adopted to extract more MA candidates including those close to vessels. Then, for every candidate, its cross-section profiles along 12 directions were obtained. Singular spectrum analysis (SSA) was then used to decompose each profile and reconstruct a new one that is of a slow varying trend. Finally, each filtered profile is scaled using the correlation coefficients (CCs) between itself and an ideal Gaussian shape assuming this candidate is a true MA. This will enable an enhancement of the profiles in all directions for true MA candidates while decrease the similarity among profiles in all directions for non-MA candidates. Features are then extracted from the scaled profiles of each candidate for MA/non-MA classification. This will help in achieving more discriminative features for MAs and non-MA candidates. But it was observed that MAs were labeled as normal because of low contrast or blurry outline. So they were missed during candidate extraction thus producing less accuracy.

Wei Zhou, Chengdong Wu, Dali Chen, Zhenzhu Wang, Yugen Yi and Wenyou Du extracted diverse features and a dictionary was made to maximize the sensitivity and specificity [6]. MA detection algorithm named joint dynamic sparse representation based on multiple features (JDSRMF), which integrates joint dynamic sparse representation and multiple channels multiple features into a framework. This method not only promotes the multiple features sparse representation vectors belonging to the same test sample shared at class-level, but also allows varying sparsity patterns within each class to facilitate flexible representation at atom-level. However the drawback here is that the degree of DR severity is not considered.

M.M. Habib, R.A. Welikala, A. Hoppe, C.G. Owen, A.R. Rudnicka and S.A. Barman presented a new technique for MA detection based on an ensemble classifier for classification [7]. The authors introduced 70 of the most common features like Gaussian matched filter, Shape and moment invariants, morphological, Gaussian, Flemings and intensity features are extracted and feature ranking was performed in order to identify the features that are most important for discriminating MA candidates from spurious objects. They have also introduced a new groundtruth dataset for MA detection based on the MESSIDOR dataset and combined using ensemble classifier. It was inferred that this approach eliminates false detection to a high extent however the feature selection is not complete. Identification of major features and using those in classification would have provided better results. In addition to this intensity profile based features [8] are considered to be the discriminant features in MA detection in many existing methods.

Although the above proposed methods of detection of candidate region have many advantages, there are problems related to elimination of the false positives on blood vessels and locating Mas that are located near to veins which has to be addressed. So this paper propose a method to detect a lesion namely, MAs using multiple essential features and aggregating the results finally in an ensemble classifier which combines the result of multiple weak regression learning trees and this helps in obtaining a high accuracy.

4 PROPOSED METHODOLOGY

The general flow diagram of the proposed method in detection of MA candidates are as given in Fig. 5.2.

The input retinal image of varying illumination is obtained from the publicly available Messidor database. Then the input retinal image is preprocessed to correct its nonuniform Illumination and to enhance the contrast of the same so as to easily identify the region of interest required for diagnosis of disease. Thus, the preprocessing steps involves Illumination Equalization, contrast enhancement using CLAHE, Histogram Equalization and then Gaussian smoothening filtering operation. The preprocessed image is then used to detect the initial set of candidate regions where MAs are likely to exist. For this purpose, Morphological closing operation is done to eliminate the major blood vessels followed by Binarization and noise removal by specifying a threshold to eliminate the isolated pixels.

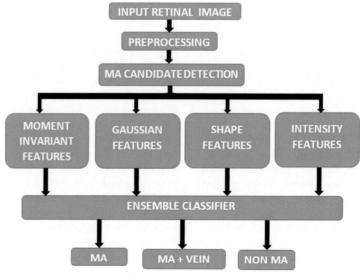

FIG. 5.2 Proposed methodology.

The resultant images from Illumination equalization, CLAHE, Gaussian smoothening, isolated pixels removed binarized image are stored as patches of equal size (56×81) and used for feature extraction. Various features like Gaussian, Shape, Moment Invariant, intensity Features are extracted from the patches and the results are compared based on the feature vectors. These patches are used as training images by assigning a label to every patches.

Ensemble classifier is then trained using the extracted features and based on the three categories - Non MA, Only MA, MA with vein the input patches are identified. The classification model is obtained which can be used to test the new patches. Then the test images are processed, their features are extracted and accuracy in detecting MA is determined.

4.1 Input Retinal Image
The input retinal image used for experimentation is obtained from publicly available MESSIDOR and Diaretdb01 database. A database of infected images with varying illumination and varying number and size of MAs were collected for training and testing.

4.2 Preprocessing
Preprocessing [9] is the step where the initial enhancement of the retinal images are done so as to distinguish the MAs from blood vessels. It is done for two main purposes.
- To correct the non-uniform illumination of retinal images
- To enhance the contrast of retinal images

When examined the input retinal images that were collected from the Messidor database, it was found that each of the retinal image is non-uniformly illuminated. There were shade variations in different regions which was corrected initially using illumination equalization algorithm. Then the green channel is alone extracted and contrast enhancement is done to distinguish MAs from veins using CLAHE algorithm. In order to further increase the contrast, histogram equalization is performed by modifying the intensity distribution of the histogram of the image. Thus, a highly contrast image is obtained from which the further preprocessing steps are carried out.

4.3 Illumination Equalization
The retinal images collected are found to have varying illumination which is made uniform by subtracting the background estimated image that is obtained by applying median filtering and then adding the mean of the original image in order to maintain the same illumination as that of the original image. This process is represented by the equation given below

$$I_{ie} = I - I_{bg} + u \qquad (4.3.1)$$

where I_{ie} represents the illumination equalized image of individual RGB channels, I_{bg} denotes the background estimated image which is obtained by applying a 51×51 median filtering and u denotes the mean of the individual RGB channels.

4.4 Contrast Limited Adaptive Histogram Equalization (CLAHE)
Contrast Limited Adaptive Histogram equalization (CLAHE) [10] is a computer image processing technique used to improve contrast in images. Contrast Limited Adaptive Histogram Equalization (CLAHE) differs from ordinary adaptive histogram equalization in its contrast limiting. It is therefore suitable for improving the local contrast and enhancing the definitions of edges in each region of an image. Contrast Adaptive Histogram Equalization (CLAHE) is applied to the Illumination equalized Green channel with standard deviation of 0.01.

4.5 Histogram Equalization
Histogram equalization [11] is applied in order to adjust the contrast of the image by modifying the intensity distribution of the histogram. This method usually increases the global contrast of images, especially when the usable data of the image is represented by close contrast values. This adjustment can better distribute the intensities on the histogram. This allows for areas of lower local contrast to gain a higher contrast. Histogram equalization accomplishes this by effectively spreading out the most frequent intensity values. This method is useful in images with backgrounds and foregrounds that are both bright or both dark.

4.6 Gaussian Smoothening Filtration
In this work, Gaussian smoothening filter with size 3 and sigma 1 to smoothen the histogram equalized image. Thus the preprocessing helps in correcting the non-uniform illumination and improving the contrast of the image.

4.7 MA Candidate Detection
MA candidate detection phase is performed so as to detect the initial set of candidate regions where MAs are likely to exist. Morphological processing is carried out in order to eliminate the major blood vessels present in the preprocessed output. Then, Binarization

and noise removal is done to detect the initial regions where MAs are present.

4.7.1 Morphological processing

Morphological image processing is a collection of non-linear operations related to the shape or morphology of features in an image. Morphological techniques probe an image with a small shape or template called a structuring element. The structuring element is positioned at all possible locations in the image and it is compared with the corresponding neighborhood of pixels. Here, morphological processing is adopted to eliminate the blood vessels Minimum of 9 Closing operated image using a linear (line) structuring element with length $= 20$ and degree $= 1$ as given by the equation below

$$I_{bothat} = \min_{i=0\ldots9} \ Z_i$$

$$Z_i = I_{hist}.strel(\pi i/n) - I_{hist} \qquad (4.7.1.1)$$

where $strel(x.n)$ represents the linear structuring element at an angle $X(X = \pi i)$ and of length n, I_{hist} represents the histogram equalized output and I_{bothat} represents the result of bottom hat filtration. Bottom hat filtration technique used here is adopted since the candidate regions appear with high contrast in compared with the background. Performing closing operation using a linear structuring element followed by bottom hat filtration enables the removal of veins in the image.

4.7.2 Binarization

Binarization is applied so as to represent the blood vessels and candidate regions with binary 1 and other background regions with binary 0. By removing the isolated pixels with neighborhood connected value less than 25, the noise components which appears similar to MAs can be removed efficiently from the binarized image. The preprocessed outputs such as individual channels of illumination equalized RGB image, the individual channels of illumination equalized HSV image, CLAHE output, Gaussian filtered output and the MA candidate detected output obtained after the removal of isolated pixels are divided into patches of size 56×81. These equally sized patches from various outputs are used to extract the features that are essential to discriminate the true MAs and false positives.

4.8 Feature Extraction

4.8.1 Intensity features

Intensity Features are calculated directly from the intensity in the image at multiple bands: the red (R), blue (B), green (G) band in the RGB color space.

The Hue (H), Saturation (S) and Value (V) bands of the HSV space. The 35 intensity features are listed in Table 5.1.

Sum of the candidate intensity of a RGB color space is given in the equation below

$$Intensity = (R + G + B)/3 \qquad (4.8.1.1)$$

TABLE 5.1
Set of Features Considered in the Proposed Method.

Features	Feature count	Properties	Images
Intensity Features	7	Sum of Candidate Intensity	R,G,B,H,S,V, Preprocessed Channels
	7	Mean of Candidate Intensity	
	7	Std. dev of Candidate Intensity	
	7	Range (Max-Min Candidate Value)	
	7	Candidate contrast	
Moment Invariants	7	7 Moment Invariants	Binarized Image
Shape Features	2	Major Axis Length	Binarized Image
	2	Minor Axis Length	
	2	Aspect Ratio	
	2	Perimeter	
	2	Area	
	2	Eccentricity	
Gaussian Features	1	Max 1D Gaussian Response	Gaussian Image
	1	Min 1D Gaussian Response	
	1	Std. Dev of 1D Gaussian Response	
	1	Mean of 1D Gaussian Response	

Range of candidate intensity is the total number of distinctive pixel values that occur in the image. Range is given by the equation below

$$Range = Max - Min \qquad (4.8.1.2)$$

where *Max* denotes the maximum value of candidate intensity and *Min* denotes the minimum value of candidate intensity. Candidate contrast is the amount to which the candidates are visually distinguished from the background. The equation of candidate contrast is

$$Contrast = (\max(I) - \min(I))/(\max(I) + \min(I)) \qquad (4.8.1.3)$$

where *I* denotes the intensity of the pixels in the image. The mean and standard deviation values are also calculated in order to provide accurate results.

4.8.2 Shape features

The shape features [12] describe various shape properties of the detected candidates. The maximum and minimum response are used for feature extraction to distinguish MAs and vessels. Area is the actual number of pixels in the region i.e. count of the number of pixels with intensity 1 in a binarized image given by the following equation.

$$Area = \sum 1_{r,c \in R} \qquad (4.8.2.1)$$

where *r,c* is the size of region *R*, i.e. *r* represents the row and *c* represents the column. Major axis length is the length (in pixels) of the major axis of the ellipse that has the same central moments as the region. Similarly, Minor axis length is the length (in pixels) of the minor axis of the ellipse that has the same central moments as the region. Thus, the Aspect Ratio is defined by,

$$Aspect\ Ratio = Major\ axis\ length/Minor\ axis\ length \qquad (4.8.2.2)$$

Eccentricity is a measure of Aspect ratio. Perimeter accounts to the distance around the boundary of the region.

4.8.3 Moment invariants

Moment Invariants [13] are 7 features that represent various shape properties of an object. Here, I, II, III, IV, V, VI, and VII order moments are calculated in additional to the shape features. General moment $M_{pq}(f)$ of an image $f(x,y)$, where p, q are non-negative and $r = p + q$ is called the order of the moment defined as

$$M_{pq}(f) = \iint P_{pq}(x,y)f(x,y)dxdy \qquad (4.8.3.1)$$

where $P_{pq}(x,y)$ is the polynomial basis function defined in the region covered by the integral.

4.8.4 Gaussian features

A Gaussian steerable filter is an orientation-selective convolution kernel used for image enhancement and feature extraction that can be expressed via a linear combination of a small set of rotated versions of itself. The oriented first derivative of a 2D Gaussian is a steerable filter. The oriented first order derivative can be obtained by taking the dot product of a unit vector oriented in a specific direction with the gradient. The process by which the oriented filter is synthesized at any given angle is known as steering. The various Gaussian features extracted from the Gaussian output in the preprocessing stage is as given in Table 5.1.

4.9 Ensemble Classification

A collection of many learning algorithms is known as Ensemble methods [14] which can achieve better predictive performance compared to any of the principal learning algorithms alone. It is a supervised learning algorithm, because it can be trained and the model can be used to make predictions.

The reason for choosing Ensemble classifier in the proposed method is

- It is a multiple classifier system
- It can rank features giving insight about important features
- It is able to manage with small training sets

New testing sets are then fed as input to classifier to test the accuracy of the classifier. As a result, the output falls under three categories as listed below

- Class 0: Non-MA region
- Class 1: Only MA region
- Class 2: MA along with blood vessels

Thus the classification results are obtained by performing all the above processes and the accuracy for the proposed method is calculated for every features separately. Then all the features are combined to check for the improved accuracy.

5 RESULTS AND DISCUSSIONS

The detection of Microaneurysms proposed in this work involves the steps of preprocessing, candidate detection and they were followed by dividing each of the preprocessed image into patches. From each of the patches, various features were extracted and these were given as input to the ensemble classifier which is a multiple classifier. The classifier classifies whether the patch contains only MA or only vein or both MA and vein. The proposed methodology is validated by the performance measures like accuracy, specificity and sensitivity. The work was implemented in Matlab R2017a-64 bit.

5.1 Database

The retinal images are obtained from the public databases namely MESSIDOR (Methods to evaluate segmentation and indexing techniques in the field of retinal ophthalmology) and DIARETDB01. (Standard Diabetic Retinopathy Database-Calibration level 1). The Messidor database has been established to facilitate studies on computer-assisted diagnoses of diabetic retinopathy. The data included in this database can be used, free of charge, for research and educational purposes.

Within the scope of diabetic retinopathy, the primary purposes of the Messidor project is to compare and evaluate both

- The various segmentation algorithms developed for the detection of lesions present in color retinal images
- A tool to index and manage the image databases.

DIARETDB01 is a public database for benchmarking diabetic retinopathy detection from digital images. It can be used as a testing protocol to benchmark diabetic retinopathy detection methods. By using this database and the defined testing protocol, the results between different methods can be compared.

The MESSIDOR database contains both the ground-truth and the screened retinal images. The images were classified as close to vessel, obvious, regular and subtle. In each set, few images were taken and divided into patches and processed as mentioned above. The retinal images from MESSIDOR were used for both training and testing. Few patches from these images were trained and few other were used for testing. Out of them, many patches were mutually exclusive. Patches from the DIARETDB01 were used only for testing purpose. The images from MESSIDOR and DIARETDB01 are shown in Fig. 5.3 respectively.

5.2 Preprocessing

In preprocessing, first the green channel of the image is extracted. The retinal images are usually low contrast images. Microaneurysms are clearly visible in green channel due to high contrast. Preprocessing is performed in order to enhance the contrast of green channel. In Fig. 5.4, the extracted channels from input image, namely, red, green and blue channels are shown.

Each of the individual channels of RGB and HSV is then illuminated equally and concatenated finally to obtain the illumination equalized image which is shown in Fig. 5.5.

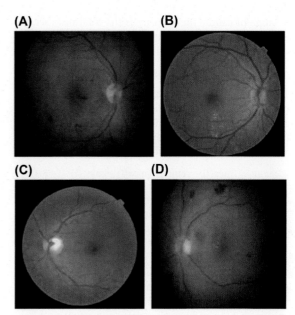

FIG. 5.3 Database images. (A) and (B) Messidor Database; (C) and (D) Diaretdb01 database.

Green channel is alone extracted from this concatenated RGB and CLAHE (Contrast Limited Adaptive Histogram Equalization) is done on that in order to improve the contrast on tile level. Adapthisteq calculates the contrast transform function for each tile individually. CLAHE is followed by histogram equalization where the images are equalized in an overall basis. This is done in addition to CLAHE in order to enhance the contrast further. The presence of other noises is removed by Gaussian filtering. Fig. 5.6 shows the CLAHE, histogram equalized and Gaussian filtered output. Thus at the end of the preprocessing the images are equally illuminated and contrast is enhanced. The presence of noise is also removed to obtain a smooth image suitable to detect the presence of MAs.

5.3 MA Candidate Detecton

Candidate detection is performed in order to detect the first set of MAs where they are likely to be present.

5.3.1 Morphological operation

The output of morphological operation where the closing operation is performed by choosing a length of 20 and an angle of 1 (minimum of 9 times) through trial and error method is shown in Fig. 5.7.

The bottom-hat filtered image is then binarized making easy for the detection of MAs. The clots and veins are represented by one and the background is zero.

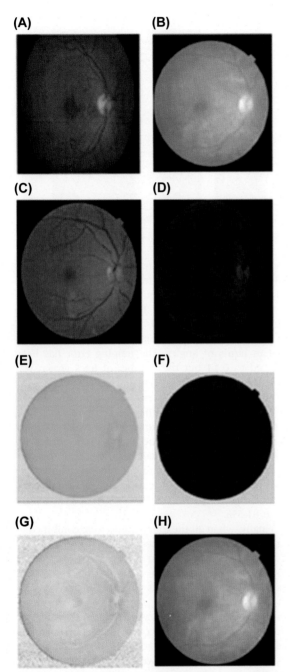

FIG. 5.4 Channel separation. (A–D) RGB and individual channel; (E–H) HSV and individual channel.

Since noise is also present in the bottom hat filtered image, it has to be removed for accurate detection of MAs. So there arises a necessity to remove them, which can be achieved by removing the isolated pixels. The connected

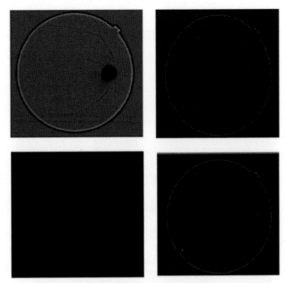

FIG. 5.5 Illuminated images.

pixels from the binarized image are removed by applying a threshold value of less than 25. This removes the noise present in the binarized image which is shown in Fig. 5.7.

5.4 Feature Extraction

In order to eliminate the false positives, various features are extracted by dividing the image into patches. Totally 86 patches were used for training and 27 patches for testing. About four features were considered, each having its own significance. The four features include Intensity, Shape, Gaussian and 7 Moment invariants. The retinal images that are given as input are divided into smaller patches. All the original input images were equally resized to the dimension of 450×650 and then were divided to bring them to the dimension of about 56×81 so that the presence of MAs can be easily detected. The time consumption to process is also significantly reduced. The patches divided are shown in Fig. 5.8.

In order to differentiate the MAs from background, intensity variations were taken into account. Since the intensity of MA will be greater than the background, it is easy to distinguish between the both. Totally 35 features were considered which includes candidate intensities' sum, mean standard deviation, range and contrast. From Fig. 5.9, it is inferred that the error rate got saturated at a point about 0.15 and the number of trees grown was about 800. Also some of the features were of less significance and were neglected i.e., feature 21, 27, 34. The testing accuracy was about 66.66%.

(A)   **(B)**   **(C)**

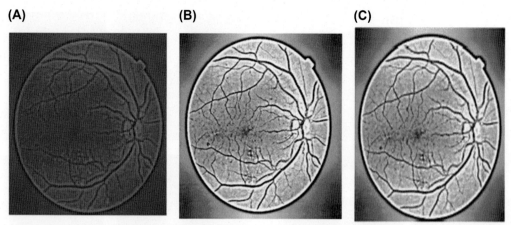

FIG. 5.6 Preprocessing stages. (A) CLAHE output; (B) Histogram equalized image; (C) Smoothened image.

Removal of isolated pixels

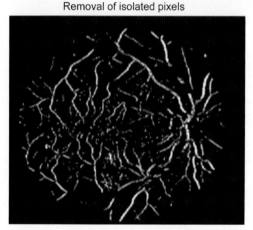

FIG. 5.7 Binarized image.

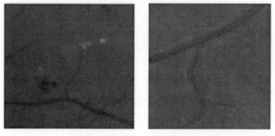

FIG. 5.8 Sample patches.

Shape features were considered in order to find the variations between veins and MAs. This feature was found to be more significant than other features. In shape features, six features were extracted from the binarized image patches. From the six features, maximum and minimum of each were taken into account for processing. The training error rate was about 0.162 and the number of trees developed was within 100 which is an efficient training. Moreover, the accuracy of shape feature testing was about 74.3% which is considerably good.

Steerable Gaussian feature output considers variation in terms of angles. The training error has been reduced to 0.14 but the number of trees grown was 300. The four features of Gaussian namely maximum, minimum, standard deviation and mean were applied to the Gaussian filtered image. The accuracy in this feature has been increased up to 80.2% which is higher than above two features. The 7 moment invariants did not prove to be an efficient feature. The 7 features were extracted here. The training error was about 0.15 and the number of trees grown was 900. Though the number of trees grown was 900, the accuracy was built up to 81.5%.

Finally, when all the four features were combined to train, the error rate was increased to 0.17 and saturated at about 0.157; the number of trees grown was about 800. The main reason to combine the entire four features was to find out if it improved the accuracy by any means. Fortunately, there was a significant improvement in the accuracy rate. Many of the feature indices were insignificant as compared to the individual features considered. The accuracy of all combined feature was about 85.2% which is the highest of all. Fig. 5.9 shows the error rate and feature importance for intensity, shape and steerable Gaussian feature based classification. Fig. 5.10 shows the error rate and Feature importance of Moment Invariant and all feature based classification.

In the proposed work, we randomly select the regions from the different retinal images having different.

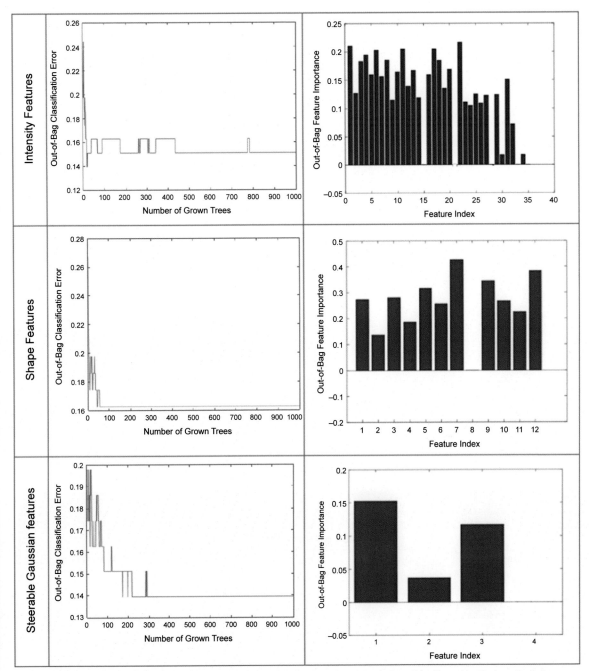

FIG. 5.9 Error rate and Feature importance of Intensity, shape and steerable Gaussian feature based classification.

The selected regions include regions having only veins, veins with micro aneurysms and only micro aneurysms. Those regions are used for testing by the classifier model based on intensity features, shape features, Gaussian features, moment invariant features and combined features. It is observed by our experimentation that when the classifier model considers all the four categories of features then it is possible to achieve minimum error

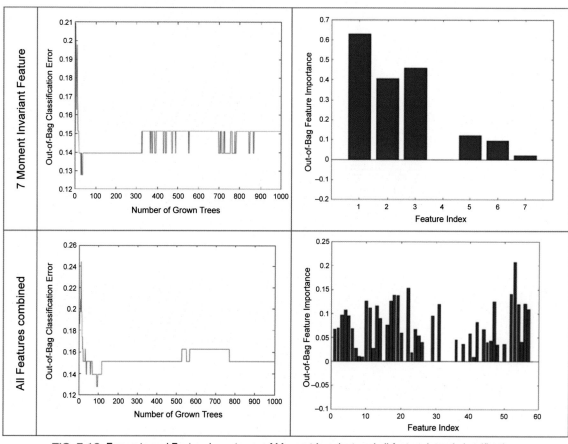

FIG. 5.10 Error rate and Feature importance of Moment Invariant and all feature based classification.

TABLE 5.2
Accuracy Comparison – Ensemble Classifier.

	CLASS WISE ACCURACY (%)			
Classifier model	**Only vein**	**Only MA**	**MA + vein**	**Overall accuracy (%)**
Intensity feature	80	16.66	84	66.66
Shape feature	93.33	33.33	66.56	74.3
Steerable Gaussian feature	100	33.33	83.33	80.2
Moment invariant feature	86.66	50	100	81.5
Combined	100	66.66	66.66	85.2

of 0.126. We also observed the importance of feature in each classifier model. Accuracy comparison for the proposed work is given Table.

From the above table, it can be absorbed that when all the individual categories of features were combined, the accuracy of detecting MAs were increased and it ranged up to 66.66% which is the highest of each of individual categories of features. The overall accuracy for each of the feature is compared. It is found out that when all the features were combined, the accuracy

TABLE 5.3
Accuracy Comparison — K NN and Ensemble Classifier.

Classifier Model	K NN	Ensemble
Intensity feature	66.66	66.66
Shape feature	70.37	74.3
Steerable Gaussian feature	66.66	80.2
Moment invariant feature	74.07	81.5
Combined	81.48	85.2

rate increased considerably which is shown in Table 5.2. The overall accuracy of each of the categories of features hiked up to 85.2% which indicates that the detection of each class was nearly correctly done when ensemble classifier was used. Ensemble classifier performs better compared to K NN classifier and the result is shown in Table 5.3.

6 CONCLUSION

An automatic detection of microaneurysms present in retinal images which leads to diabetic retinopathy is proposed in this work. A robust preprocessing step is carried out in this work to enhance the contrast of the MA candidates from its background. Also, various features like intensity, shape, moment, Gaussian features are extracted to improve the accuracy of the classification result. The feature importance plot is obtained for all the features used in this work. Classification is done on patch level using Ensemble classifier. It is inferred that using 58 features, a high degree of accuracy was achieved since the final classification is based on the aggregation of individual regression trees. The feature selection step has to be improved in this method in future to achieve even better results. This proposed methodology doesn't compute the severity of the DR which is also the future work to be carried out.

REFERENCES

[1] S. Ravishankar, A. Jain, A. Mittal, Automated feature extraction for early detection of diabetic retinopathy in fundus images, in: IEEE Conference on Computer Vision and Pattern Recognition, 2009. CVPR 2009.

[2] T.P. Karnowski, D. Aykac, Giancardo, Y. Li, T. Nichols, K.W. Tobin, E. Chaum, Automatic detection of retina disease: robustness to image quality and localization of anatomy structure, in: Engineering in Medicine and Biology Society, EMBC, 2011 Annual International Conference of the IEEE, 2011.

[3] S. Lahmiri, C. Gargour, M. Gabrea, Statistical features selection from intrinsic mode functions for pathologies detection in retina digital images, in: IECON 2012 - 38th Annual Conference on IEEE Industrial Electronics Society, 2012.

[4] B. Wu, L. Tang, L.I. Al turk, Y. Hu, S. Sanei, Automatic detection of microaneurysms in retinal fundus images, Comput. Med. Imag. Graph. 55 (January 2017).

[5] S. Wang, H.L. Tang, L.I. Al turk, Y. Hu, S. Sanei, Localizing microaneurysms in fundus images through singular spectrum analysis, IEEE Trans. Biomed. Eng. 64 (5) (May 2017).

[6] W. Zhou, C. Wu, D. Chen, Z. Wang, Y. Yi, W. Du, Automatic microaneurysm detection of diabetic retinopathy in fundus images, in: IEEE Control and Decision Conference (CCDC), 2017 29th Chinese, 2017.

[7] M.M. Habib, R.A. Welikala, A. Hoppe, C.G. Owen, A.R. Rudnicka, S.A. Barman, Detection of microaneurysms in retinal images using an ensemble classifier, Inf. Med. Unlocked 9 (2017).

[8] I. Lazar, A. Hajdu, Retinal microaneurysm detection through local rotating crosssection profile analysis, IEEE Trans. Med. Imaging 32 (2) (February 2013) 400—407.

[9] M. Niemeijer, B. van Ginneken, M. Cree, A. Mizutani, Retinopathy online challenge: automatic detection of microaneurysms in digital color fundus photographs, IEEE Trans. Med. Imaging 29 (2010).

[10] K. Zuiderveld, Contrast Limited Adaptive Histogram Equalization, 1994, pp. 474—485.

[11] A.D. Fleming, S. Philip, K.A. Goatman, J.A. Olson, P.F. Sharp, Automated microaneurysm detection using local contrast normalization and local vessel detection, IEEE Trans. Med. Imaging 25 (2006) 1223—1232.

[12] L. Seoud, T. Hurtut, J. Chelbi, F. Cheriet, J.M. Pierre Langlois, Red lesion detection using dynamic shape features for diabetic retinopathy screening, IEEE Trans. Med. Imaging 35 (4) (April 2016).

[13] M.K. Hu, Visual pattern recognition by moment invariants, IEEE Trans. Inf. Theory 8 (2) (1962) 179—187.

[14] M.M. Fraz, et al., An ensemble classification-based approach applied to retinal blood vessel segmentation, IEEE Trans. Biomed. Eng. Sep. 59 (9) (2012) 2538—2548.

CHAPTER 6

Telemetry System for Early Detection of Hyperbilirubinemia in Neonates

K. UMA[a] • J. REVATHI[a] • J. ANITHA[b]

[a]Department of Biomedical Instrumentation Engineering, Avinashilingam Institute for Home Science and Higher Education for Women, Coimbatore, India; [b]Department of Electronics and Communication Engineering, Karunya Institute of Technology and Sciences, Coimbatore, India

1 INTRODUCTION

1.1 Hyperbilirubinemia

Hyperbilirubinemia means an increased level of bilirubin content in the blood. Bilirubin is a yellow pigment obtained in the liver during the breakdown of red blood cells. When bilirubin content increased, skin and the white region of the eyes become yellow in color and it is called as jaundice [3]. The word jaundice is acquired from the french word jaune [23]. Hyperbilirubinemia is classified into two types: Conjugated bilirubin (indirect) and unconjugated bilirubin (direct). High conjugated bilirubin could be caused by liver diseases such as hepatitis infections and bile duct blockage. High unconjugated bilirubin could be due to the breakdown of excess red blood cells and genetic conditions. Usually, bilirubin levels are higher for the newborn babies for the first few days of life [11]. Normal indirect bilirubin level for a newborn would be under 8 mg/dL within the first 24 h of birth, 13 mg/dL for 48 h old babies, 16 mg/dL for 72 h old babies and 17 mg/dL for 96 h old babies.

1.1.1 Formation of bilirubin process

This section explains the process of bilirubin formation [24]. The hemoglobin is liberated from the dead cell and it is disintegrated into heme and globin. Heme is oxidized and it is converted into biliverdin and globin is converted to amino acids. The biliverdin is reductive to form bilirubin in two forms. The indirect jaundice is formed at the initial stage. Then albumin transports the indirect bilirubin into the liver. After reacting with gluconic acids the indirect bilirubin becomes direct bilirubin. Bilirubin mingles with bile and it is disintegrated into stercobilin and urobilinogen. The stercobilin is combined with wastes from the intestine. The urobilinogen is the waste and excreted in the form of urine.

The concentration of serum bilirubin when greater than 12.9 mg/dL is called as severe hyperbilirubinemia [25]. In newborns, high level of bilirubin content in the blood may cause damage to the brain [10]. It may cause bilirubin encephalopathy or kernicterus. Continuous monitoring of bilirubin content is essential to prevent the neonates at high risk [10].

2 EXISTING METHOD AND TECHNOLOGIES

Several methods are available for the estimation of bilirubin levels which include the standard clinical test with blood or urine sample and the advanced technologies that uses image processing techniques.

2.1 Clinical Methods

In this subsection, the various conventional methods like Diazo method, High-Performance Liquid Chromatography (HPLC) and Direct Spectrophotometry method were reported [8]. The Total Serum Bilirubin Test and Transcutaneous Bilirubinometers were also described [26].

2.1.1 Diazo method

This method measures bilirubin content in newborns. It depends on the various properties of solubility of conjugated bilirubin and unconjugated bilirubin. The chemical reaction takes place between bilirubin and diazotized sulphanilic acid. The azobilirubin compound is formed. It shows the fraction of conjugated bilirubin. Then, it reacts with alcohol or sodium-benzoate to form an unconjugated bilirubin.

Telemedicine Technologies. https://doi.org/10.1016/B978-0-12-816948-3.00006-4

The intensity of the color of the bilirubin is measured using spectrophotometry.

2.1.2 High-performance liquid chromatography (HPLC)

HPLC method is applied to separate and quantify the fraction of bilirubin such as unconjugated bilirubin, monoconjugated bilirubin and diconjugated bilirubin. This method was widely used in research to get the information about bilirubin species available in blood. The disadvantage of this method is that it doesn't provide any additional clinical information and this method is more expensive.

2.1.3 Direct spectrophotometry

Direct Spectrophotometry is a simple method to measure bilirubin content in newborns. The preparation of blood sample is required to acquire serum in it. This method depends upon the bilirubin absorbance at 454 nm and hemoglobin absorbance at 454 and 528 nm. The difference between the absorbance between 528 nm and 454 nm is obtained to remove the hemolysis effects. The resultant value is proportional to bilirubin content. The disadvantages of this method are that it absorbs the other pigments like carotenoids at 454 nm. Hence, this method is limited to 2–3 weeks age of newborns.

2.1.4 Total Serum Bilirubin test

Total Serum bilirubin (TSB) measures the bilirubin content from the blood sample directly. It is an invasive technique and also more precise method. It meets medical gold standard. The healthcare provider refers a specialized graph to measure this bilirubin content called bhutani nomogram. It specifies age of newborns, weeks of gestation and level of bilirubin. The Risk zone of nomogram is shown in Fig. 6.1.

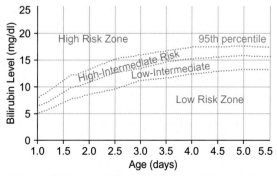

FIG. 6.1 Risk zone of Bhutani nomogram (Lilian de Greef, 2014).

2.1.5 Transcutaneous bilirubinometers

The transcutaneous bilirubinometers (TcB) is used to measure bilirubin content of neonate. It is a non-invasive technique that indirectly measures bilirubin content. Healthcare providers place the device into the neonates' forehead [21]. It emits light and measures the resultant reflectance and absorbance of the skin to measure the bilirubin content [14]. This method is unpredictable when the value of the bilirubin is above 14.5 mg/dL. This method requires regular calibration and its cost also high. The TcB method provides good correlation with Total Serum Bilirubin (TSB). But TcB method cannot correlated to TSB nomogram and it requires some adjustment to eliminate error measurement of bilirubin content.

2.1.6 ETCOc analyzer

This paragraph describes the measurement of bilirubin content by measuring End-tidal Carbon Monoxide (ETCOc) [20]. The heme molecules breakdowns and lead to the production of carbon monoxide. The bilirubin level can be determined by measuring carbon monoxide in the exhaled breath. The CO-Stat End Tidal Breath Analyzer was used to analyze the infant's breath for the measurement of End-tidal Carbon Monoxide (ETCOc). The analyzer consists of the nasal sampler which draws air continuously to determine End-tidal Carbon Monoxide (ETCOc). The hyperbilirubinemia in neonates can be determined by investigating the value of End-tidal Carbon Monoxide. This method achieves 6.4% of positive predictive value and 99% of negative predictive value.

2.1.7 Estimation of metalloenzymes by atomic absorption spectrophotometer

This section was reported to estimate certain metalloenzymes for the bilirubin measurement [3]. Few minerals are necessary for the development and maintenance of healthy tissues. Zinc present in a human body consists of metalloenzymes which include alkaline phosphates, RNA and DNA polymerases, thymidine kinase, carbonic anhydrase. The structures of proteins, nucleic acids are stabilized by zinc. Copper is also another metalloenzyme present in the human body which includes dopamine −β− hydroxylase, ceruloplasmin, superoxide dismutase, lysyl oxidase, ascorbate oxidase and tyrosinase. It includes oxidation and reduction process. Manganese present in a body is responsible for cell respiration, glycolysis. It is mainly involved in the development of connective and bony tissues. It includes enzymes such as manganese superoxide, arginase and pyruvate carboxylase.

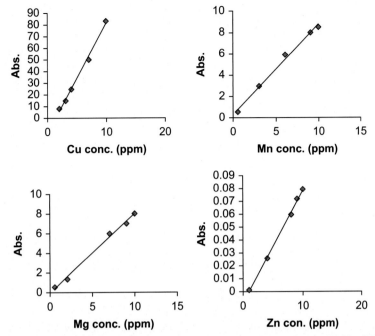

FIG. 6.2 Calibration curve (A) Copper (B) Manganese (C) Magnesium (D) Zinc [3].

The blood samples of neonates were collected using the syringe and it is placed in a plain tube at a room temperature for nearly 10 min for clotting. The clots were removed by a wooden applicator stick. Then it was centrifuged at 3000 rpm for 10 min. Then the serum was placed in another tube and maintained at a temperature of 20 °C. The Atomic Absorption Spectrophotometer (AAS) is used for the analysis of copper, Zinc and Manganese.

A calibration curve was acquired from copper, zinc, and manganese shown in Fig. 6.2. The level of zinc was less for the neonates who are affected by jaundice when compared to healthy neonates. The levels of copper, manganese and magnesium levels were higher for the neonates who are affected by jaundice when compared to healthy neonates. The elevated level of zinc affects the bilirubin metabolism. The elevated level of copper leads to liver diseases such as obstructive jaundice, cholestasis and cirrhosis. The abnormal level of manganese and magnesium leads to hemolysis.

2.1.8 Enzymatic method

This section describes the bilirubin measurement based on enzymatic method [5]. This method is based on oxidation of bilirubin with bilirubin oxidase to biliverdin with molecular oxygen. This method is highly sensitive to pH changes and thus prone to errors.

It measures the level of bilirubin content based on the flow of current through the bilirubin oxidase electrode. The cell consists of a platinum disk with 2 mm inner diameter and 12 mm outer diameter. The silver ring with 1 mm diameter acts as a reference electrode. Both the silver and platinum electrodes were placed in a sealed thermostatic incubation chamber.

1. Preparation of solution

    Bilirubin oxidase solution is prepared with lyophilized bilirubin oxidase. It was again mixed up with phosphate buffer with the pH value of 7.5. The bilirubin solution is prepared by dissolving 60 mg of bilirubin in 3 mL of 0.05 M phosphate buffer with the pH value of 7.5 Then, NaOH is diluted to phosphate buffer with the pH value of 7.5 and it is stored at 20 °C.

2. Assembling of electrode

    The bilirubin oxidase membrane was composed by crosslinking the enzyme with glutaraldehyde and BSA. It is bonding with preactivated membrane. The 10 mg BSA was dissipated in 300 mL of 0.1 M phosphate buffer with pH value of 7.5. The 10 mL of this prepared solution were added with the 15 mL of bilirubin oxidase and 5 mL of 9% aqueous glutaraldehyde solution were mixed. It was placed in Immunodyne ABC membrane. Then it was dried by using air for 30 min. After that it was washed in phosphate buffer to eliminate the unbounded protein. The half saturated potassium chloride solution is filled in the electrolyte chamber. The bilirubin oxidase membrane was laid over the oxygen membrane. The voltage of −800 mV is applied between the electrodes to measure the bilirubin content.

3. Procedure to measure bilirubin content using enzymatic method

   To measure the bilirubin content using enzymatic method, $(1.0 - x)$ mL of 0.1 M glycylglycine with pH value of 9.0 was inserted in the chamber. The parameter x represents unknown bilirubin solution. It allows the oxygen pressure to achieve equilibrium state with respect to the atmospheric pressure by varying the temperature of solution to 34 °C. Then the current flowing through the electrodes were measured to obtain the bilirubin level.

2.2 Advanced Technologies

Advancement in latest technologies incorporates the use of smart phones and image processing techniques for the early detection of jaundice in neonates.

2.2.1 Biliscan method

This section describes the measure of bilirubin level in neonates using smart phone [16]. Biliscan application is installed in iPhone smart phone which is used to record the details of gestational age, gender, date of birth, photographs of chest and abdomen. The color calibration card is placed on the chest and abdomen of newborn. The color card is aligned within the frame which is provided by the biliscan application. The color calibration card is utilized for color balancing due to the variation of skin color with the different lighting conditions. This color card consist of 12 squares with various colors such as black, gray, green, blue, yellow, pink and white. The photograph of color calibration card with different lighting conditions was captured.

The advanced technique for the estimation of bilirubin is shown in the Fig. 6.3. The image segmentation

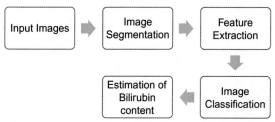

FIG. 6.3 Block diagram of the advanced technique for the estimation of bilirubin.

process was carried out to identify the different color patches depends on the values of hue and saturation. The red, green and blue values of card are adjusted to the red, green and blue values of skin. This process is called color balancing. The feature extraction techniques were applied to obtain various features of the skin patches. And the estimation of bilirubin content can be obtained by applying various regression techniques. To analyze the performance and the accuracy of this method, the blood samples were collected from 35 newborn and it was analyzed by using diazo-coupling method.

The correlation between the serum bilirubin level and Biliscan bilirubin level were measured by using Pearson's correlation method. The correlation value of biliscan sternum value was 0.6 and biliscan abdomen value was 0.55 which is shown in Figs. 6.4 and 6.5. This value is high for sternum value than abdomen value.

3 COMPUTATIONAL IMAGE PROCESSING TECHNIQUES

This subsection describes image processing techniques to measure bilirubin content [11]. The calibration card with 8 different colors was placed on the abdominal position of newborn (Fig. 6.6). Photos of the whole body of the baby were taken using Smartphone camera without flash. Some of the sample pictures were shown in Fig. 6.7. These images were analyzed using MATLAB environment.

The first stage of the image processing is color balancing method followed by feature extraction. The color based balancing method includes the various stages like image segmentation method, pixel similarity method and white balancing method. Segmentation of these images was carried out to preserve region of interest and other areas were changed into the black color tone. Next, in the pixel similarity method, few of the missing colors on the color card are brought back again. And finally, the white balancing method is used to build the image with more independent of light, reflections, and shadows.

Second, the feature extraction provides information about the skin of babies and the colors of calibration card. Later, the K nearest neighbor and support vector regression were used to estimate the bilirubin content present in the blood.

3.1 Image Segmentation

Image segmentation is the process of dividing an image into multiple segments. The captured images were divided into various segments to analyze the color variation between calibration card and the jaundice

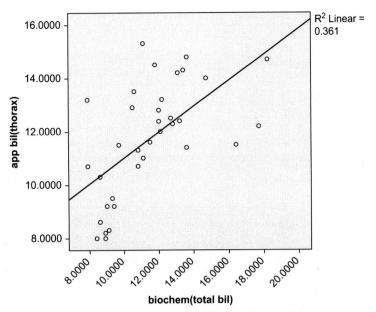

FIG. 6.4 Correlation graph between serum bilirubin and biliscan bilirubin over sternum [16].

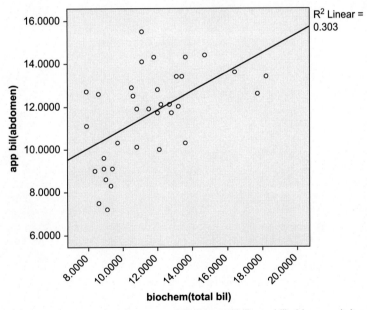

FIG. 6.5 Correlation graph between serum bilirubin and biliscan bilirubin over abdomen [16].

region of the newborn. The point-based segmentation and region based segmentation techniques were applied to segment jaundice region and other parts were eliminated. Gauss filter was applied to eliminate the noises from environment.

The objective of the image segmentation is that to represent an image into more meaningful information. In this process, labels are assigned to each pixel and the pixels with the similar label consist of certain characteristics.

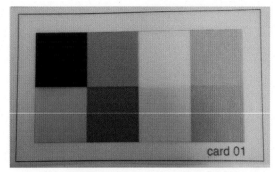

FIG. 6.6 Sample color calibration card [11].

technique, the gray scale of a particular point will be different from the gray level of neighboring points.

3.1.2 Region-based segmentation technique

Region-based segmentation is a technique where the images are divided into regions by grouping adjacent pixels together with similar properties. It can be applied in four different ways. They are region growing, region splitting, region merging, split and merge. In the region growing technique, pixels are clustered into a larger region depending on confined condition. In region merging operation, false boundaries are removed by

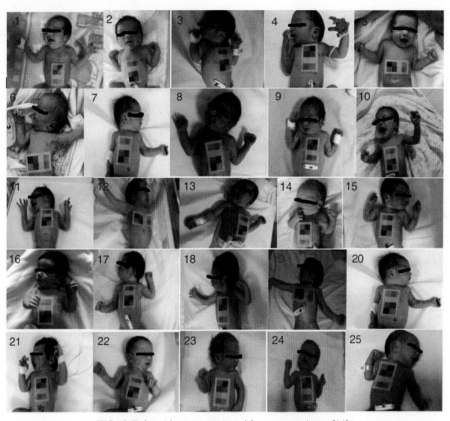

FIG. 6.7 Input images captured from smartphone [11].

3.1.1 Point-based segmentation technique

In this process of image segmentation, a particular point is detected at a location p(i,j). The mask is centered if $|R| > T$, where R is the mask response at any point (x, y) and T is a non-negative threshold value. In this

combining adjacent regions of similar characteristics. In region splitting operation, missing boundaries are added by portioning regions of different characteristics. In region splitting and merging operation, first, region splitting splits the entire image into four quadrants.

This process continues until all the sub-region satisfies the property of homogeneity. Then, each pixel is considered as a small region. Each small region is combined to form a larger region based on the property of homogeneity. If the region with similar characteristics is small, region merging technique is preferred otherwise region splitting is preferred (Fig. 6.8).

3.2 Pixel Similarity Method
After segmentation, few of the colors present in the calibration card may be lost. So, except the region of interest, all the other parts were changed into black color tone. By pixel similarity method, missing colors of calibration card were identified and again brought back on the card (Fig. 6.9).

3.3 White Balancing Method
The white balancing method makes the original images independent of the luminous intensity. While processing the image some difference were obtained to determine the color of the skin. To avoid these differences, normalized red, blue and green values were calculated (Fig 6.10).

3.4 Feature Extraction
Color map transformation was carried out to detect the yellow region present in the body parts of a baby. The images were changed from RGB values to YCbCr color space and the image size was also reduced. With the use of this color map transformation, all the colors on the calibration card were compared with the newborn skin color.

3.4.1 Feature calculation
The mean value of YcbCr and lab color spaces were estimated and totally 9 features were obtained. By using linear color gradient, the difference in color values was estimated. These calculations were changed by Sobel filter. The three features were obtained at the end of the processing.

FIG. 6.8 Results of image segmentation [11].

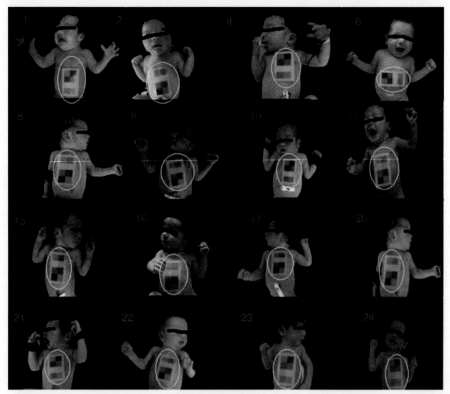

FIG. 6.9 Images after applying Pixel similarity method [11].

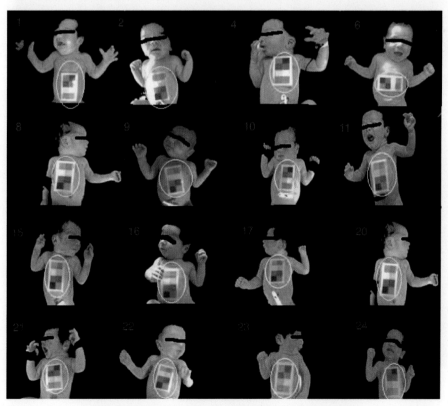

FIG. 6.10 Output image of white balancing method [11].

Finally, 12 features were acquired from the calculation and these features were applied to machine learning algorithm such as The KNN and SVR to compare feature data at different angles.

3.4.2 Results
In this study, input dataset was obtained from 40 healthy and 40 jaundiced neonates. For eighty input images and 9 channels the value of dataset is $440 \times 440 \times 9 = 17, 42,400$ black areas to be avoided in each dataset. So, the value of a dataset is 9, 00,000. The calculated value is $110 \times 180 \times 3 = 59,400$. By applying the KNN and SVR methods, 9, 59,400 new features were obtained. By trial and error method threshold value is selected as 1.91. The difference between the maximum and minimum output value was determined. If it was less than the threshold value then, the mean of these values was calculated to determine the bilirubin level.

The f-statistical test was applied to determine the dataset accuracy and system performance. The result of bilirubin estimation level was compared with blood test result. The similarity of the results based on both the methods was investigated by using p value. The p-value is greater than 0.06. The results of both the test were found to be correlated. The 85% of results were obtained by this system.

The Receiving Operating Characteristics (ROC) analysis was implemented to investigate the dataset (Fig. 6.11). The threshold value was 205 μmol/L. It was considered as true positive value (TPV) and False Positive rate values (FPV) when the values were greater than the threshold value. The sensitivity and specificity values were 0.83 and 0.195 respectively.

3.5 Homomorphic Filtering
In this section, the applications of homomorphic filtering in the diagnosis of jaundice were reported [15]. Homomorphic filtering is a method to enhance image which was captured under various illuminous conditions. In some situation, images are acquired under poor lighting conditions. Hence, some part of the image appears as bright and some part appears as dark which makes the process difficult to diagnose the disease. It is one of the best methods to enhance the images.

The image consists of two parts. The first part refers to the number of illuminations from the image and it is considered as i(x,y), where i(x,y) refers to low-frequency components. The second part is considered

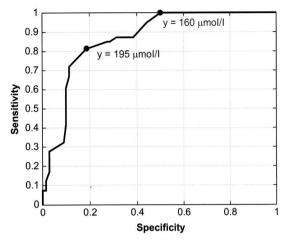

FIG. 6.11 ROC curve for TSB (205 μmol/L) [11].

as the quantity of reflectance from the image r(x,y). The image f(x,y) is represented as i(x,y)r(x,y). Homomorphic filtering is applied to minimize the impact of i(x,y) by decreasing the low-frequency components present in the image. This process is carried out by converting the image into the frequency domain by using Fourier transform.

3.5.1 Bilirubin detection based on homomorphic filtering
The bilirubin level is detected by collecting the images of the strips which is made up of different bilirubin concentration levels. Images of these strips were captured under various light effects. The bilirubin level is directly proportional to yellow color and the yellow color is the reciprocal of the blue color. Hence the intensity of yellow color can be calculated by determining the intensity of blue color. The level of bilirubin content is high when the intensity of the blue color is low. To eliminate the effect of luminance, homomorphic filter is implemented.

In this work, 10 strips were prepared which contain bilirubin solution with different concentration ranges from 0 to 30 mg/L. Then it is diluted with blood serum. Images of these strips were captured under different illumination conditions. The images consist of 3 components such as Red, Blue, and Green (RGB). From each image blue components alone extracted to detect the bilirubin content. The correlation between bilirubin concentrations is evaluated by taking images of the strips with and without a homomorphic filter.

The correlation coefficient is between $+1$ and -1. The coefficient lies between these values then the degree of correlation between two variables is perfect. If the correlation coefficient is nearer to zero, then the degree of correlation between two variables is weaker.

In this case, the blue color component is extracted and its intensity is evaluated with and without homomorphic filtering. Both the data sets are compared with bilirubin levels. The correlation coefficient between the bilirubin levels and blue intensity values with and without filtering is calculated and it was concluded that correlation coefficient is increased for the images with homomorphic filtering than without homomorphic filtering.

4 INTRODUCTION TO MOBILE COMMUNICATION

Mobile health uses information and communication technologies that are accessible to the individual through mobile devices such as mobile phones, patient monitoring devices, personal digital assistants and other wireless device [27]. The smartphone is a handheld device that plays a vital role in remote healthcare service to reduce health risks. It allows data analysis, monitoring, reporting and transmitting data to healthcare practitioners and other individuals (Lamia Ben Amor, 2017). Effective transmission of medical images from remote locations for consultation is a challenging task in the healthcare sector. The major issue in the transmission of the medical image is that it requires very high bandwidth [13]. To reduce the bandwidth many image compression techniques have been proposed by experts to reduce the bandwidth required for the transmission [13]. The need for the image compression is to increase the storage space and reduce the transmission time [13]. It reduces the size of the image without compromising the image quality. It is measured in terms of compression ratio. It is defined as a ratio of original medical image size to compressed image size [18].

The compression techniques broadly classified into two namely Lossy and Lossless compression techniques [2]. Image compression plays a vital role in transmitting images through a wireless network.

4.1 Lossy Compression Technique

Lossy compression technique permanently deletes some of the information and hence compressed image is not similar to the original image [7]. This technique is usually used in videos and sounds. It yields high compression ratio than the lossless compression technique [9].

The various Lossy compression techniques are Block truncation coding, Code Vector quantization, Fractal coding, Transform coding and Sub-band coding [1].

4.1.1 Block truncation coding (BTC)

In this section, Block Truncation Coding (BTC) method was described [1]. It is based on dividing an image into different blocks of the same size. It changes the original intensity value of each pixel in a block either as low mean intensity value or high mean intensity value depends on the mean intensity of the pixels in the block. Then a bitmap of that matrix is achieved by changing all pixels values which are greater than or equal to the threshold value by 1. In each segment, a pixel value is determined based on the mean value of the corresponding pixels in the original code. The advantages of BTC are that it provides less complexity and preserve edges.

4.1.2 Code vector quantization

In this section, Code Vector quantization method were reported [1]. Vector quantization develops a look-up table of fixed size vectors are known as code vectors. An input image is further split up into non-overlapping blocks known as image vectors. Then for each image vector, the nearest mapping vector in the look-up table is identified and its index in the look-up table is utilized for encoding the original image vector.

4.1.3 Fractal compression

This paragraph explains Fractal Compression technique [1]. The image is split up into various segments based on color differences, texture and edges. There is a lookup table named as fractal segments. This lookup table consists of codes which consist of a packed set of values. Executing the algorithm fractals are executed and it encodes the image.

4.1.4 Transform coding

This section explains transform coding method [1]. This transform coding's used to alter the pixel specifications from the spatial domain into frequency domain. In the image transform process, a correlation between the pixels is minimized because correlation between one pixel and another are the same. Once the correlation is reduced, statistical characteristics can be analyzed. Thus this information is packed into few numbers of coefficients. Then it is quantized to reduce the number of output values to smaller bits and it is further minimized to the desired number of bits by entropy coding scheme.

4.1.5 Sub-band coding

In this technique, an input image is disintegrated into its subbands using low pass and high pass digital filters. In this scheme, quantization technique and coding process are applied to each of the sub-bands. This coding is very useful because quantization and coding is more accurately applied to the sub-bands to compress the image.

4.2 Lossless Compression Techniques

In this technique, the input image is exactly retrieved from the compressed image [17]. Few of the Lossless compression techniques are Huffman encoding, Run length encoding, Arithmetic coding [9]. The widely used algorithm for the transmission of medical images is wavelet-based algorithms and Set Partitioning in Hierarchical Tree (SPHIT) algorithms [8]. The combination of these two algorithms provides a better result. The Long Term Evolution (LTE) network is one of the fourth generation networks. It uses Carrier Aggregation (CA) for effective resource allocation [8]. The SPHIT algorithms were applied to compress medical images and transmitted to a remote location over a LTE network [8].

Some of the lossless color image compression algorithms that are used to transmit through mobile networks are Modified Hierarchical Prediction and Context Adaptive model, Modified Hierarchical Predictive and Block-Based Lossless Image Coding (MHPBLI), Coefficient Density Adaptive Quantization (CDAQ) Approach Based Lossless Image Compression. The image formats widely used in lossless compression techniques are JPEG, JPEG-LS, LOCO-I, CALIC and JPEG2000.

4.2.1 Huffman coding

In this section, the Huffman Coding algorithm was reported [9]. Huffman Coding algorithm assigns short code word for the most frequently presented symbols and the long code word for the least frequently presented symbols to the input. The Huffman code is constructed by the process of combining lowest possible symbols repeatedly until one symbol remains. A code tree is constructed based on this process. It is optimal prefix code obtained from a set of combining probable symbols. It reduces the length of the code.

4.2.2 Run length coding

This technique is useful when the data consists of repeated symbols or characters. Run length coding assigns a data in terms of length and value. The values are repeated character or symbols. The length is the number of repetitions of a character or symbols. The most effective format for the run length encoded data is TGA, PCX, and ILBM. It is most useful when the data contains graphic images.

For example,

1111111111111100000000000001111 = 31 bits

(14,1), (13,0), (4,0) Since $2^4 = 16$; (1010,1)(1101,0) (0100,0)

Compression ratio = 15/31 Or 1:2.06

4.2.3 Arithmetic coding

It is a variable-length entropy encoding. Instead of splitting the input data into symbols and replaced into codes, it encodes the whole information into a single number. It consists of a probability line. For each character, a range is assigned based on the probability.

4.2.4 Set partitioning in hierarchical tree (SPHIT) algorithm

This subsection describes the Set Partitioning in Hierarchical Tree (SPHIT) algorithm for the effective transmission of data through the mobile network [13]. The Set Partitioning in Hierarchical Tree (SPHIT) algorithm is one of the lossless image compression techniques. It was developed by Said and Pearlman. The medical images were converted into a set of transform coefficients by applying a wavelet transform with the desired sub bands. SPHIT algorithm was applied to this coefficient. The encoded data were sent to the LTE network.

The mapping of coefficients in SPHIT algorithm is achieved by two stages such as sorting pass and refinement pass. In this process, three lists of coefficients were acquired. These lists are List of Insignificant Sets (LIS), List of Insignificant Pixels (LIP) and List of Significant Pixels (LSP) [4]. The List of Insignificant Sets (LIS) consists of wavelet coefficients that are described by a tree structure which have magnitude lesser than the threshold value. The set of pixels which have a magnitude above the threshold value of significant pixels are called List of Significant Pixels (LSP). The singular pixel which has a magnitude lesser than the threshold value is called List of Insignificant Sets (LIS). The following steps were implemented in encoding.

Step 1: Initialize target bit rate = n

$$n = \log_2(\max/\text{coeff}/)$$

Step 2: Sorting process
   i. Examine the List of Insignificant Pixels
   ii. Examine the List of Insignificant Sets

Step 3: Refinement process

Examine the List of Significant Pixels (LSP)

Step 4: Update coefficients and go to next iteration by varying n deserve

The SPHIT algorithm is applied to medical images. The quality of compressed images was measured using various parameters such as Compression Ratio, PSNR and MSE are calculated for different bit rates. The PSNR is calculated by the equation given below.

$$PSNR = 10\log_{10}\frac{L^2}{MSE} \qquad (6.1)$$

where L = maximum pixel in the image

$$MSE = \frac{1}{MN}\sum_{m=1}^{M}\sum_{n=1}^{N}\left((x(m,n) - x'(m,n))^2\right) \qquad (6.2)$$

where x(m,n) = original image; $x'(m,n)$ = reconstructed medical image.

Compression Ratio is the ratio of original medical image size to compressed image size.

4.2.5 Modified hierarchical prediction and context adaptive model

This section describes the lossless compression algorithm [12].

This model predicts the pixels to be encoded based on the vertical, horizontal and diagonal pixels. In the initial stage, RGB is transformed to YC_uC_v by Residual Color transform method (RCT). They channel is encoded by the conventional image compression algorithm. The C_u and C_y are encoded by MHPCA algorithm. In this, chrominance images disintegrated in to two subimages. They are even numbered row subimage and odd numbered row subimage. First, even numbered row subimage are encoded then it is used to identify the pixels of odd numbered row subimage.

4.2.6 Modified hierarchical predictive and block-based lossless image coding (MHPBLI) technique

This paragraph describes the Modified Hierarchical Predictive and Block-Based Lossless Image Coding [12]. The MPHCA encodes the chrominance images. The luminance image gradients were computed and it is converted into nonoverlapping blocks. Then the block regions are classified into black, gray and white blocks based on the variation of colors. Then quantization schemes were applied to this block. The vector quantization scheme is employed for the black blocks, the side Match Vector Quantization is employed for the

gray blocks and the in-painting method is applied for the white blocks which maintain the image sharpness. The process of quantization is carried out to encode image blocks. Thus images are compressed without deteriorating the quality of the image.

4.2.7 Coefficient Density Adaptive Quantization (CDAQ) approach based lossless image compression technique

This section explains Coefficient Density Adaptive Quantization (CDAQ) Approach Based Lossless Image Compression [12]. In this method, the image blocks are quantized by determining the density of the DCT coefficients. The luminance images were disintegrated into 8×8 non-overlapping blocks. Each block is analyzed for the edge content. Then the block regions were categorized into structural, grated, featured and nonfeatured regions. By using an in-painting method, the white blocks are quantized and the other two blocks are adaptively quantized.

Discrete Cosine Transform is implemented in the black and gray blocks. The image is transformed from the spatial domain to the frequency domain. It minimizes the decorrelation of the adjacent pixels. Then by using the probability density function, the global and local density functions of the DCT coefficients for the blocks were determined. The global and local arithmetic mean values are evaluated. It is the magnitude of the DCT coefficients. Hence the image blocks can be quantized effectively and it is encoded by context adaptive arithmetic coding.

5 HYPERBILIRUBINEMIA IMAGE TRANSMISSION SYSTEM

This section clearly describes various methods of image transmission through wireless network.

5.1 Image Transmission System on Mobile Devices

The architecture diagram given in Fig. 6.12 describes the image transmission system on mobile devices [28]. We require a technique that sends images of hyperbilirubinemia immediately to the doctor that reduces the time through portable system. The Personal Digital Assistant (PDA) is one of the popular system but it's computing capacity is large. Power consumption is also high. The various communication techniques with PDA are Bluetooth, infrared data access (IRDA), Wireless Local Area Network (WLAN), Code Division Multiple Access (CDMA). Infrared Data Access (IRDA) is a short

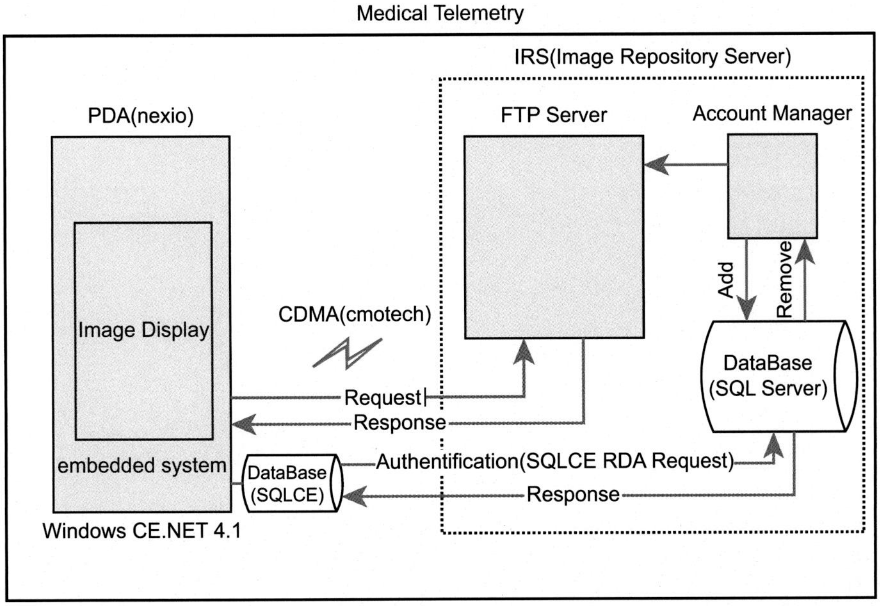

Medical Telemetry

FIG. 6.12 Architecture of medical image transmission system on mobile devices (SangBock Lee, 2008).

distance and low speed communication technique. Bluetooth is a short distance communication between systems. WLAN is a wireless technique which requires nearby access point and also it utilizes routers so it cannot be used without such facilities. Code Division Multiple Access (CDMA) utilizes a network which built for EV-DO (evolution-data optimized) modem. It was developed by Qualcom and implemented on chips with 3GPP2 standards. But it's not suitable for voice communication. Nowadays, CDMA based mobile networks have been employed mostly so that it can be utilized for voice communication. This study utilizes a CDMA communication technique to access the images of hyperbilirubinemia by the doctors through PDA.

This proposed system is composed of server and client system. The server system is utilized for the user and image management. The client system is employed for receiving images in a portable terminal. It was implemented with Nexio XP30 PDA. Since PDA has a memory capacity of 32−64 MB it was not sufficient for the storage of large volume of data. So that each DICOM image is transformed in to jpg file. The image is compressed and provided to PDA.

The server system stores the image which is to be sent to PDA. The server system also named as image repository server. It controls PDA and updates information to the doctors. It also provides authentication to the users and act as a file server. The PDA connection employs file transfer protocol. The image repository server system can be separated in two components such as FTP server which takes care of client connection and account manager which governs the images. Account manager creates an user account and utilized by PDA at any time. The image repository server system maintains image file and database separately. The database contains patient and doctors information.

CCU-550 is a CDMA modem which is USB type and its used in Nexio XP30. Its frequency is 800 MHz and the data transmitted at 2.4 Mbps. It can be used as modem similar to WLAN. The USB port present in XP30 is utilized after the driver is registered in the system so that it can be recognized.

The account for the doctor and the patient is created based on the predefined rules present in the directory of FTP. The program in account manager implements the menu and receives the input from the user. The portable equipment employs a CDMA modem. The user has to provide user id and password to access the image information present in the portable system. The PDA access the server database. The server employs Microsoft SQL 2000 SP4 as a database Management system. The SQLCE 2.0 database system was constructed to access the database through PDA.

The server environment consists of internet information server which accomplishes synchronization with SQL server by using Internet server application programming interface (ISAPI) (Fig. 6.13). SQLCE implements Client request to the SQL server through SQL server

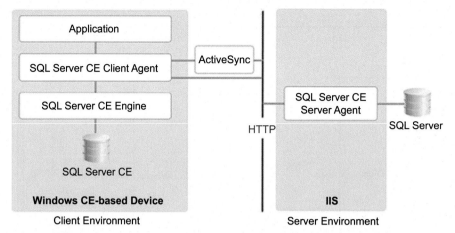

FIG. 6.13 Interconnection between server and PDA (SangBock Lee, 2008).

CE server Agent. Based on the utilization the server interacts with the clients either through RDA (Remote Data Access) or MR (merge and replication). The DBMS based merge and replication function synchronizes MR with SQLCE. This technique reproduces a data in the server and places it in a SQLCE database in PDA. Through network connection SQL server is accessed in order to perform a certain task by Remote Data Access. PDA synchronous database through RDA, Image files can be downloaded by using FTP. FTP is an efficient file transmission and reception protocol. Certain FTP file functions were constructed in PDA and its employed in directory listing. The functions of FTP are separated into connection formation, initialization, command transmission, data reception. This protocol supports in binary and ASCII mode. The code is executed by using. NET framework. The image transmission through this technique achieves very high functionality. It allows physicians to access the image and diagnose the diseases at any place at any time.

5.2 Transmission of Image With High Quality Through Erroneous Wireless Network

The image transmission model shown in Fig. 6.14 describes the transmission and reception of images over erroneous wireless network [22]. This proposed method utilizes unequal error protection and median filtering to transmits the images through wireless communication (mostly cellular mobile network). The images were applied to source encoder. The source encoder implements the suitable image compression method to the source image. The compressed image is applied to channel encoder where the suitable channel encoding method is implemented. The modulator

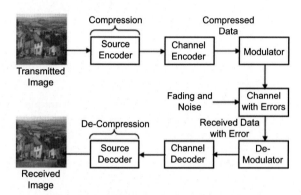

FIG. 6.14 Image transmission model [22].

transmits the data through wireless communication. The channel generates noise to the transmitted data. The demodulator accepts the image data with error. The image data are decompressed followed by channel decoding technique.

5.2.1 Modulation technique

The Unequal Error Protection (UEP) techniques are employed to transmit high quality images through wireless communication. The UEP is mainly split up in two parts which includes data partitioning and applying UEP.

The data partitioning classifies important and less important data based on various levels of importance of sources of images. The level of error protection will be allocated based on the importance of image data. This is implemented in the channel coding phase. The important image data is secured by using channel coding level at high rate. The less important image data is secured by channel coding level at low rate.

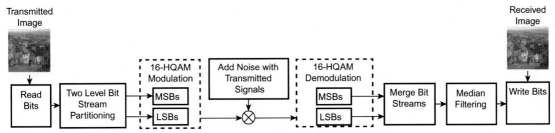

FIG. 6.15 Transmission and reception of images using HQAM technique [22].

This method uses a Hierarchical Quadrature Amplitude Modulation (HQAM) technique for the image transmission through wireless channel (Fig. 6.15). This is an asymmetric modulation technique. Hierarchical Quadrature Amplitude Modulation (HQAM) employs various levels of protection to the data bits which is to be transmitted. The most important data bits are depicted to the Most Significant Bits (MSB) and the less important data bits are depicted to Least Significant Bits (LSB) of modulation constellation points. Hence HQAM will provides better quality of image. The median filtering is employed to reduce the noise in the image.

This technique provides more efficient transmission and reception of image through erroneous wireless network.

5.3 Context-Aware Wireless Networks for Image Transmission

This subdivision describes the Image Transmission through Context-Aware Wireless Networks [12]. The wireless networks such as IEEE 802.11e, WiMAX and UMTS are utilized for the transmission of compressed images based on adaptive compressed schemes. An efficient transmission of images cannot be performed using traditional networks because of large volumes of data, limited network bandwidth. In the networks, there may be some failure due to the congestion of network and limited network resources. To overcome these difficulties a new technique is introduced which is called as Context-Aware Scheme.

The context awareness point identifies the system capability that responds to the concerned environment. The devices and network components consist of information about their environment, operating protocols and its responses to the activity imposed on it. In the area of image processing, the context-aware scheme identifies the pixel position in the images and appropriate adaptation method is applied. This method minimizes the loss when the image is transmitted over wireless networks.

The wireless networks such as UMTS, IEEE 802.11e and WiMAX are considered for transmission of the image. The adaptive compression technique provides efficient transmission of the data through a wireless network. It was proved that Packet loss, Bit Error Rate and Packet delay is very less for WiMAX than UMTS, IEEE 802.11e. It was concluded that the WiMAX network provides more efficient transmission than other networks.

6 MOBILE COMMUNICATION SECURITY

This subsection describes the secured data transmission through mobile network [6]. In our day today life mobile phones play a vital role in data accessing and sharing through a wireless network. The maintenance of these data with confidentiality and reliability is a major concern in our society. The various error correcting codes has been identified to maintain the reliability of the data. Turbo Code (TC) is an excellent encoder for the reliable data transmission but it is no guarantee for the secured transmission of data. Vidya Sawant et al. proposed a technique called Cryptographic Turbo Code (CTC) for the reliable and secured data transmission [19]. Data Encryption is obtained by the new method called Elliptic Curve Cryptographic Interleaver (ECCI) of CTC [19]. This method rearranges the sequence depends on Elliptic Curve (EC) arithmetic and a private key [19]. This bit shuffling sequences enhance the performance of error correction code. This method helps the effective transmission of image over a mobile network.

A two similar rate half convolutional encoder is connected in parallel to form a CTC encoder. Each encoder has a memory M and constraint length K. where $M = K - 1$. These two encoders are distinct by Elliptic Curve Cryptographic Interleaver (ECCI) [19]. The input data bit sequence was received by encoder 1 and interleaved bit sequences were received by encoder 2. Encoder 1 was protected by its code word and it improves the secure transmission of data through a mobile network. This method rearranges the bit sequences applied to the two encoders [19]. The authenticated persons can retrieve the original data sequence by using correct key.

The turbo decoder is performed by using Bahl-Cocke- Jelinek-Raviv (BCJR) with Max-Log-Maximum A Posteriori (MAP) algorithm.

6.1 Public Key Generation Procedure

There are two private keys K1 and K2. K1 is for the encoder and K2 is for the decoder. The public keys are generated by the Elliptic Curve Diffie-Hellman Key Exchange (ECDHKE) algorithm. These public keys are interchanged via a wireless channel. It is assigned as input to the process of interleaved position generation.

The input is that G is the base point on elliptic curve E_p (a, b). The keys K_1, K_2 and lookup table with each bit position mapped to a point on EC. The public key for the encoder is generated by $P_a = K_1 \times G$ and for the decoder is generated by $P_b = K_2 \times G$.

The ECCI is mapped to each bit position to P_m which is a point on the Elliptic Curve (EC). It attained coding gain of 6.5 dB with PSNR of 29 dB.

7 CONCLUSION

This chapter provides insight knowledge about the various invasive and non-invasive methods to detect hyperbilirubinemia in neonates. It also discussed various image compression techniques to transmit images through mobile network and secured transmission of data through mobile network. The feasibility of determining hyperbilirubinemia in neonates using image processing is not limited. In summary, the figure shows the generalized block diagram which represents how mobile network can assist a healthcare provider to diagnose hyperbilirubinemia in neonates using various image processing and image compression techniques. This system really helps the people in rural area for the early diagnosis of hyperbilirubinemia in neonates.

REFERENCES

[1] D.V. Rojatkar, N.D. Borkar, B.R. Naik, R.N. Peddiwar, Image compression techniques: lossy and lossless, Int. J. Eng. Res. Gen. Sci. 3 (2) (2015) 912–917.

[2] Dinu, R. Ganesan, A.A. Kebede, B. Veerasamy, Performance analysis and comparison of medical image compression techniques, in: Control, Instrumentation, Communication and Computational Technologies (IC-CICCT), 2016 International Conference on, IEEE, 2016, pp. 738–745.

[3] E. Hasan, Evaluation of copper, zinc, manganese, and magnesium levels in newborn jaundice in baghdad, Ibn Al-Haitham J. Pure Appl. Sci. 24 (3) (2011).

[4] J. Wang, F. Zhang, Study of the image compression based on SPHIT algorithm, in: Intelligent Computing and Cognitive Informatics (ICICCI), 2010 International Conference on, IEEE, 2010, pp. 130–133.

[5] J. Klemm, M.I. Prodromidis, M.I. Karayannis, An enzymic method for the determination of bilirubin using an oxygen electrode, Electroanalysis 12 (4) (2000) 292–295.

[6] K. Vidhya, G. Karthikeyan, P. Divakar, S. Ezhumalai, A review of lossless and lossy image compression techniques, Int. Res. J. Eng. Technol. 3 (4) (2016) 616–617.

[7] P. Khobragade, S. Thakare, Image compression techniques-a review, Int. J. Comput. Sci. Inf. Technol. 5 (1) (2014) 272–275.

[8] S. Kudavelly, P. Keswarpu, S. Balakrishnan, A simple and accurate method for estimating bilirubin from blood, in: Instrumentation and Measurement Technology Conference, IEEE, 2011, pp. 1–4.

[9] M. Singh, S. Kumar, S. Singh, Various image compression techniques: lossy and lossless, Int. J. Comput. Appl. 142 (6) (2016).

[10] M. Mansor, S. Yaacob, M. Hariharan, S. Basah, S.A. Jamil, M.M. Khidir, M. Rejab, K.K. Ibrahim, A.A. Jamil, A. Junoh, et al., Jaundice in newborn monitoring using color detection method, Proc. Eng. 29 (2012) 1631–1635.

[11] M. Aydn, F. Hardala, B. Ural, S. Karap, Neonatal jaundice detection system, J. Med. Syst. 40 (7) (2016) 166.

[12] P.S. Babu, S. Sathappan, Effective compressed image transmission through context aware wireless networks, in: Advances in Computer Applications (ICACA), IEEE International Conference on, IEEE, 2016, pp. 82–86.

[13] B. Panjavamam, P. Bhuvaneswari, Performance analysis of SPHIT algorithm for biomedical image transmission, in: Signal Processing, Communication and Networking (ICSCN), 2017 Fourth International Conference on, IEEE, 2017, pp. 1–5.

[14] G. Bertini, F.F. Rubaltelli, Non-invasive bilirubinometry in neonatal jaundice, Semin. Neonatol. 7 (2002) 129–133. Elsevier.

[15] R. Singla, S. Singh, A framework for detection of jaundice in new born babies using homomorphic filtering based image processing, in: International Conference on Inventive Computation Technologies, vol. 3, IEEE, 2016, pp. 1–5.

[16] S. Swarna, S. Pasupathy, B. Chinnasami, B. Ramraj, et al., The smart phone study: assessing the reliability and accuracy of neonatal jaundice measurement using smart phone application, Int. J. Contemp. Pediatr. 5 (2) (2018) 285–289.

[17] N. Surabhi, S.N. Unnithan, Image compression techniques: a review, IJEDR 5 (2017).

[18] T. Brahimi, A. Melit, F. Kheli, An improved SPIHT algorithm for lossless image coding, Digit. Signal Process. 19 (2) (2009) 220–228.

[19] V. Sawant, A. Bhise, Cryptographic turbo code for image transmission over mobile networks, in: Advances in Computing, Communications and Informatics (ICACCI), 2016 International Conference on, IEEE, 2016, pp. 844–850.

[20] V.K. Bhutani, L.H. Johnson, Jaundice technologies: prediction of hyperbilirubinemia in term and near-term newborns, J. Perinatol. 21 (S1) (2001) S76.

[21] Z.A. El-Kabbany, N.N. Toaima, A.M. Shedid, Implementation and validating transcutaneous bilirubinometry for neonates, Egypt. Pediatr. Assoc. Gaz. 65 (2) (2017) 38–42.

[22] M.A. Kader, F. Ghani, R.B. Ahmad, Image transmission over noisy wireless channels using HQAM and median filter, Int. J. Inf. Electron. Eng. 3 (5) (2013) 529.

[23] Vidyasagar Ramappa, Guruprasad P. Aithal, Jaundice: applying lessons from physiology, Surgery (Oxford) 27 (1) (2014) 11–18.

[24] Ankan Gupta, Ashok Kumar, Preeti Khera, Jaundice Prediction through Non-Invasive Techniques: Issues and Challenges, *Annual IEEE India Conference (INDICON)*, IEEE 1–5 (2015).

[25] Suresh K. Alla, Adam Huddle, Joseph F. Clark, Fred R. Beyette, Signal processing system to quantify bilirubin in the jaundice clinical model spectra, 32nd Annual International Conference of the IEEE EMBS Buenos Aires, Argentina, 2010, pp. 1356–1359.

[26] Lilian de Greef, Mayank Goel, Min Joon Seo, Eric C. Larson, James W. Stout, James A. Taylor, Shwetak N. Patel, BiliCam: Using Mobile Phones to Monitor Newborn Jaundice, Proceedings of the 2014 ACM International Joint Conference on Pervasive and Ubiquitous Computing, ACM (2014) 331–342.

[27] Lamia Ben Amor, Imene Lahyani, Mohamed Jmaiel, PCA-based Multivariate Anomaly Detection in Mobile Healthcare Applications, Proceedings of the 21st International Symposium on Distributed Simulation and Real Time Applications, IEEE Press, 2017, pp. 172–179.

[28] SangBock Lee, Taesoo Lee and Gyehwan Jin, An Implementation of Wireless Medical Image Transmission System on Mobile Devices, Journal of medical systems 32 (6), 471–480.

FURTHER READING

[1] A.K. Chowdhary, S. Dutta, R. Ghosh, Neonatal jaundice detection using colour detection method, Int. Adv. Res. J. Sci. Eng. Technol. 4 (7) (2017) 197–203.

[2] Mariakakis, M.A. Banks, L. Phillipi, L. Yu, J. Taylor, S.N. Patel, Biliscreen: smartphone-based scleral jaundice monitoring for liver and pancreatic disorders, Proc. ACM Interact. Mobile Wearable Ubiquitous Technol. 1 (2) (2017) 20.

[3] P. Puppalwar, K. Goswami, A. Dhok, Review on evolution of methods of bilirubin estimation, IOSR-JDMS 1 (3) (2012) 17–18.

[4] G.A. KC, A. Shriyan, B. Raj, Cord blood albumin as a predictor of neonatal hyperbilirubinemia in healthy neonates, Int. J. Contemp. Pediatr. 4 (2) (2017) 503–506.

[5] B. Pooja, P. Kulkarni, J. Sneha, K. Madhuri, G. Indumathi, Comparison of Algorithms for Image Compression, 2011.

[6] M. Shahian, P. Rashtian, M. Kalani, Unexplained neonatal jaundice as an early diagnostic sign of urinary tract infection, Int. J. Infect. Dis. 16 (7) (2012) e487–e490.

[7] S. Lee, T. Lee, G. Jin, J. Hong, An implementation of wireless medical image transmission system on mobile devices, J. Med. Syst. 32 (6) (2008) 471–480.

[8] S.K. Alla, A. Huddle, J.D. Butler, P.S. Bowman, J.F. Clark, F.R. Beyette, Point-of-care device for quantification of bilirubin in skin tissue, IEEE Trans. Biomed. Eng. 58 (3) (2011) 777–780.

[9] T.S. Leung, K. Kapur, A. Guilliam, J. Okell, B. Lim, L.W. MacDonald, J. Meek, Screening neonatal jaundice based on the sclera color of the eye using digital photography, Biomed. Opt. Express 6 (11) (2015) 4529–4538.

[10] W. Liu, Research on image compression algorithm based on SPHIT, in: International Conference on Intelligent Networks and Intelligent Systems (ICINIS), 2010 IEEE, 2010, pp. 104–107.

[11] Z. Xiong, K. Ramchandran, M.T. Orchard, Y.Q. Zhang, A comparative study of DCT-and wavelet-based image coding, IEEE Trans. Circuits Syst. Video Technol. 9 (5) (1999) 692–695.

[12] Z. Osman, A. Ahmad, A. Muharam, Rapid prototyping of neonatal jaundice detector using skin optics theory, in: Biomedical Engineering and Sciences (IECBES), 2014 IEEE Conference on, IEEE, 2014, pp. 328–331.

[13] V. Ramappa, G.P. Aithal, Jaundice: applying lessons from physiology, Surgery 27 (1) (2009) 11–18.

CHAPTER 7

WBAN: Driving e-healthcare Beyond Telemedicine to Remote Health Monitoring
Architecture and Protocols

PIJUSH KANTI DUTTA PRAMANIK[a] • ANAND NAYYAR[b] • GAURAV PAREEK[c]
[a]National Institute of Technology, Durgapur, India; [b]Duy Tan University, Da Nang, Vietnam; [c]National Institute of Technology, Goa, India

1 INTRODUCTION

Twenty-first century has been marked as the trendsetter for recent advancements and developments in the area of wireless communications, micro-electromechanical systems (MEMS) technology, and integrated circuits. This has led to the development of low-power, small-sized, cost-efficient, intelligent, invasive/non-invasive micro and nano-technology derived sensor nodes that can be placed strategically inside or outside the human body to be utilized for various applications like health monitoring, etc. This has revolutionized the way healthcare services are implemented and utilized in the real world. The primary focus of modern healthcare services is early detection and prevention of diseases in patients and to deliver best results in almost all situations as compared to traditional healthcare equipment's in terms of treating illness.

The increasing growth of the aging population has triggered the demand for health care and cure, and in all possibility, the trend will continue for the next decades also. According to the latest US Census Bureau data, the number of old people between 65 and 84 years is going to double by 2025 to 70 million [1]. The population could touch 761 million by 2023. During the next 40 years, one-fourth of the population will be over 60. This will put an intense financial burden on the governments. The overall expenditure by the U.S. on healthcare was $9 trillion in 2015 which is expected to triple and touch $27 trillion by 2020. Along with that, the gap between the total population and the number of healthcare professionals will be increasing at an alarming rate. The shortage of doctors and medical personnel will be severed in some particular regions. The increase in the number of doctors has been stagnant for the last few years in many of the regions of some specific countries. The decreased ratio between specialist doctors and general practitioners has made the situation worse. The number of general physicians, in comparison to the specialists, is getting downsized. This has acutely hampered basic healthcare, especially for age-related and chronic diseases.

Traditional healthcare systems suffer from by many shortcomings which can be overcome through new ICT supported healthcare services such as remote and pervasive healthcare [2], healthcare data analytics [3], etc. Advanced ICT-based healthcare systems are able to deliver prompt and efficient services to patients not only in medical intensive care units in hospitals but also in their homes and even workplaces, which in turn is cost-effective and improvises the patient's life quality. The impending health crisis attracts researchers, organizations, and scientists to search for optimal and best health solutions. A term called "eHealth" evolved where healthcare was supported via electronic processes, and now healthcare is extended to becoming mobile known as *mHealth*. In order to fully utilize and optimize wireless technologies, a new type of network has evolved termed as *Wireless Body Area Network* (WBAN) [4].

A WBAN comprises a number of health sensors that collects the information of vital parameters of a human body. The WBAN data sensed by these sensors are

Telemedicine Technologies. https://doi.org/10.1016/B978-0-12-816948-3.00007-6

collected by the remote monitoring application and sent to the appropriate destination for processing and analysis on the basis of which treatment decisions are taken. This enables doctors to continuously monitor patients remotely.

The concept of wireless sensor-based patient monitoring systems via WBAN has evolved and has brought out the revolutionary change in healthcare systems. WBAN technology has been implemented in clinical research laboratories as well as medical test centers. A WBAN system can be deployed at medical care units for old-aged people or at home to act as a source for monitoring without affecting their day to day life activities.

The advancement in sensors, wireless sensor network (WSN), WBAN, and the pervasive systems has taken healthcare to a new height. These technologies are aptly supported by Big Data technologies and different analytics that allows doctors for predictive diagnostics. The patients and medical care units are far more engaged as never before.

Considering the continuous enhancements in this area, WBAN will no longer be utilized in medical but will also move to military based applications, sports training and even for safeguarding human lives. WBAN has become a forefront area of research and development as it offers a huge potential for improvement in health care and monitoring.

The rest of the chapter is organized as follows. The basics of telemedicine and remote health monitoring and their differences are discussed in the next section. Section 3 discusses the components and architecture of WBANs. The difference between WBAN and WSN is also pointed out here. While mentioning the role of WBAN in remote health monitoring, a general architecture of remote health monitoring using WBAN is also illustrated in this section. Section 4 covers the communication architecture of WBAN. It talks about the network structure, topologies, network layers and standards of WBAN. Section 5 explores the WBAN MAC layer while mentioning the importance, properties, channel access and modulation techniques, and traffic adaptive MAC protocols. Several wireless technologies that are used in WBAN and remote health monitoring are reviewed in Section 6. The UWB-based MAC layer protocols and the standard medical radio services are also discussed. Finally, Section 7 concludes the chapter.

2 TELEMEDICINE AND REMOTE HEALTH MONITORING

2.1 Telemedicine

The rise of ICT-based healthcare gave birth to an innovative concept of practicing medicine and clinical services from a distance, using telecommunication, named as *telemedicine*. Telemedicine enables doctors to serve patients remotely by accessing health information through a telecommunication link.

Telemedicine has been a great relief for people in the rural and remote areas where there is a limited medical facility including doctors and infrastructure. Doctors diagnose and consult patients remotely with the help of the local medical staffs who administer the patients directly.

The key enabler of telemedicine is the telecommunication link used in/with different form/technologies such as mobile communication, video conferencing, fax, scanners, etc. for the purpose of communicating and exchanging medical documents (e.g., X-ray and sonography image, infection's photograph, previous prescriptions, pathological report, ECG report, etc.). Based on these documents, doctors assess patients' condition and make recommendations [5]. The widespread and omnipresence of the internet has broadened the scope of telemedicine.

The concept of telemedicine is implemented in the following three ways [6]:

i. Store and forward: This is an asynchronous process and does not need both parties to be in contact or online at the same time. Health information and medical records of the patient are sent to the doctors for assessment. The doctors examine the reports at their convenient time and give feedback/instructions to the local medical staffs.

ii. Remote monitoring: Physician monitors patient's vital statistics remotely.

iii. Real-time interaction: Physician and patient from remote place interact in real-time. As per the patient's convenience, either at home or a nearby medical facility, an interactive schedule with the doctor is pre-arranged [7].

2.2 Remote Health Monitoring

Remote health monitoring is a part of remote healthcare or e-healthcare. It is an approach for automated health monitoring from anywhere. This has been possible thanks to the advancement in sensors and WSN and other newer technologies such as WBAN [8] and IoT [9]. Different health sensors planted within and on the body sense different physiological data like body temperature, heart and pulse readings, blood pressure, blood sugar, brainwave, the oxygen level in blood, etc. [2]. These data are sent to the concerned health professionals who interpret them for assessing a patient's health status, diagnosis, and medication or treatment recommendation. On analyzing the data, it might be decided whether a preventative therapeutic intervention is required or the patient's prescription is needed to be changed. If the monitoring devices are connected to the internet directly, doctors can monitor patients in real-time. Furthermore, integrating the remote monitoring system with sophisticated analytical tools provides physicians with greater visibility and insights into a patient's health status.

2.2.1 Applications of remote health monitoring

With the growing usage of bio-sensors and wearable devices, remote health monitoring is getting very popular and has promoted several health care applications such as diabetes, heart diseases, cancer detection, Parkinson, asthma, Alzheimer, etc. [10,11]. Below some of the most prominent general applications of remote health monitoring are discussed:

Chronic patient care: Number of people suffering from chronic illness is on the rise. Treating chronic diseases in hospitals for longer durations increase the expenses considerably, which many of the times goes beyond the bearing capacity of the patient's family. Instead, taking care of these chronic conditions at home reduces the expense to a great extent. Doctors and nurses continuously keep in touch with the patient and guide, when needed. Remote health monitoring not only makes chronic disease management efficient and less expensive, but also improves the patient's quality of life. Patients enjoy a feeling of independence with the liberty of mobility [12].

Rehabilitating: Rehabilitation can be defined as the restoration of function [13]. In clinical terms, rehabilitation is restoration to a normal life after critical illness, serious injury, or major surgery. Basically, it is the recovering process of getting back to functioning at a level where the patient can live more or less regular lifestyle at home and hopefully at work [14]. Rehabilitation requires constant monitoring of physicians and the remedial professions. The effectiveness of the clinical treatments is strongly dependent on the patient's dedication and sincere adherence to the rehabilitating programs. For a successful rehabilitation, it is of utmost importance to maintain the correct therapy of the prescribed amount and intensity. Deviating from that can affect the rehabilitation process badly. The monitoring device tracks the patient's activity and checks for the adherence to the prescribed rehabilitation session. The relevant statistics about his performance are transmitted to the concerned clinician who can guide the patient if any irregularity is noticed.

Caring elderly people: As mentioned in Section 1, the number of aged people is growing continuously. Nearly half of all hospital treatments pertain to this elderly population. And in most cases, the elderly people are affected not just by one, but by multiple diseases. This results in an escalated frequency of hospitalization which inevitably leads to expensive healthcare. The remote health monitoring systems supporting aged people can significantly reduce the healthcare expenditure while boosting the quality of life and probably the lifespan of the aged people.

2.2.2 Benefits of remote health monitoring

Remote health monitoring exhibits several benefits as follows:

- Remote health monitoring ensures delivering constant and quality care to patients in remote locations.
- Healthcare becomes more available. Remote monitoring allows doctors to reach out to potential patients especially to those people who can't afford to visit a doctor or not been able to go to the hospitals, for certain reasons.
- Offers better quality of life, improved mobility and decrease the mortality rate for the unprivileged populating in terms of healthcare services.
- Reduction in healthcare expenditure by curtailing the long-stay at hospitals and reducing the frequency of re-hospitalization.
- Prevention of the deteriorating health condition through continuous monitoring through advanced and sophisticated devices.
- The pervasive access of patient data allows interdisciplinary collaboration and consultation that helps in the precise and overall better treatment process which leads to increased patient satisfaction and confidence.
- Faster access to relevant patient data enables quick treatment initiation and also shortens treatment duration.
- Doctors and patients both have better access to health information with up-to-date values of the vital parameters. It will be easier for doctors to assess the data collected from the monitoring devices in real-time with the help of visualization tools such as charts and diagrams.
- Monitored data are automatedly fed into the expert systems and advanced data analysis tools for better insight into the patient's health. It is possible to detect the deterioration of non-compliance and a patient's clinical condition early by analyzing the trend of change in physiological parameters [12]. This offers great help to the physicians to proactively manage a patient's treatment.
- Automated monitoring reduces the chance of erroneous diagnosis and incorrect treatment.

- Curtails the possibility of duplicate services (especially diagnostic).
- Automated monitoring can significantly simplify some of the complex clinical tasks such as haemodialysis and diabetes management [15].
- Notification for emergency situations can be set for the physician or family with the help of portable and intelligent devices such as smartphones.
- Real-time data provided by the monitoring devices improve the timeliness of care and boost treatment adherence [16].
- Time-saving for both doctors and patients. The doctors and medical staffs need not be completely preoccupied with a particular patient. They can balance better in attending other patients as well [15].
- Allows healthcare professionals to meet the demand for greater responsibility and accountability from patients and their families.
- Remote monitoring significantly improves the patient follow-up by enabling adjustment of treatment, diet or patient lifestyle.
- The cumbersome burden of patient transportation is avoided.
- Allows hospitals to downsize infrastructure and operational costs.

2.2.3 Challenges in remote health monitoring

To achieve the promised benefits of remote health monitoring, some obstacles, as mentioned below, are needed to be addressed [17].

- The philosophy of remote health monitoring is to provide health services to the people at remote places where there is not enough healthcare infrastructure. But even implementing and using remote monitoring, a minimum technical infrastructure is needed. For example, to process the sensed data, a minimum computing facility is required. Similarly, for communication, good broadband connectivity is needed. Sometimes, these minimum facilities are also not present at small healthcare institutions and in rural areas.
- People are not yet familiar with the wearables. Also, it takes time to be accustomed to implanted devices. The absence of the direct contact of the doctors makes patients skeptical and timid. Treatment based on technology only may upset the patient's confidence. They also might be concerned about the fact that some third parties can get hold of their private health data and use unethically.
- Likewise, healthcare professionals might have apprehension in relying on only technology-based healthcare systems especially for the higher-risk

patients. The sensing devices are not errored free. In a recent study on various physical activity tracking wearables, a large variation in accuracy across the devices is observed. The error margins are recorded as high as 25% [18]. To make remote monitoring effective and convincing, imprecision should be eliminated by improving the device accuracy.

- To get the maximum benefit from the remote health monitoring, it should be complemented by suitable efficient software. These software need to be customized for different use cases and should have compatibility with various third-party applications. The remote health monitoring systems should have access to electronic medical record systems (EMRs) from different clinics and hospitals to have an overall visualization of the patients' health history.

Despite these challenges, remote health monitoring has great potential in changing the traditional way of delivering healthcare services. It is still in its early stage, but it surely will have a promising role in future healthcare. Medical professionals need to pay more effort in effective utilization of remote monitoring and patients need to be motivated to be more engaged as reaping of the benefit of it is very much dependent on patients' ability to use the technologies involved [12]. The establishments providing remote health monitoring and the stakeholders need to adopt standard practices [15].

2.3 Difference Between Telemedicine and Remote Health Monitoring

There are no universal definitions for the terms "telemedicine" and "remote health monitoring". Consequently, there are no agreed and specific differentiating factors between the two. Both refer to the exchange of medical information electronically between two sites [19]. Both aim to help patients remotely. That is why people often use the terms interchangeably. Probably, the involvement of remote operation in both cases is the reason for the misperception.

But, as a matter of fact, they are quite different approaches to e-healthcare. In fact, remote health monitoring is the key differentiating factor between traditional telemedicine and today's e-healthcare which comprises both telemedicine and remote health monitoring. The major differentiating factors are discussed below. Table 7.1 summarizes the differences between telemedicine and remote health monitoring.

- The aim of telemedicine is to enable patients, especially from remote and rural areas, to interact directly with the distant doctors probably seating in a city hospital. Doctors asses the patient's condition by different means with the help of telecommunication and suggest accordingly. The healthcare

TABLE 7.1
Difference Between Telemedicine and Remote Health Monitoring.

	Telemedicine	Remote Health Monitoring
Approach	Reactive	Proactive
Role of doctors and staffs	Highly engaged	Less engaged, only for decision making.
Medical personnel's involvement period in treatment	Short-term and need basis	Continuous monitoring
Health care delivered by	Mainly physicians, with the help of local health technicians.	Physicians and general health professionals including nurses, clinicians, and others.
Scope of real-time patient observation	Less	High
Scope of health prediction	Less scope	Highly possible
Precision medicine	Not aimed for	Possible
Automated health care	No	Yes
Technical sophistication	Less	High
Suitable for high-risk patients	Not really	Yes
Reliability	Depends on the conversing and communing ability of the healthcare personnel at both ends; otherwise reliable.	Depends on the accuracy of the monitoring devices.
Security risk	Low	High
Patients' confidence	High	Low

Continued

TABLE 7.1
Difference Between Telemedicine and Remote Health Monitoring.—cont'd

	Telemedicine	Remote Health Monitoring
Doctor-patient communication	High	Less
Infrastructure needed at the receiver end	Not much	A minimum infrastructure is needed for processing the health data and to transmit it.
Patients responsibility	Not much. Follows the directions of the local medical professionals.	Patients need to be familiar with the devices if operated externally.
Patients' freedom	Less	High
Patients' dependency on healthcare professionals	High	Less
Need for healthcare professionals at the patient's end	Must	Not required
Healthcare service received at	Local clinics	Generally, at home
Pre-scheduling of patient-physician interaction	Most of the cases, must	Not required
Primary communication medium	Telecommunication, Internet	RFID, WBAN, WSN, Internet
Interdisciplinary treatment	Complicated and challenging	Effortless, due to smooth sharing of medical data both historical and real-time.

Continued

TABLE 7.1
Difference Between Telemedicine and Remote
Health Monitoring.—cont'd

	Telemedicine	Remote Health Monitoring
Advantageous to elderly people	Not particularly	Very much
Advantageous to rehabilitating patients	Not particularly	Ideally suitable
Tech-savvy healthcare professionals	Not necessarily	Must
Transparency	Low	High

professionals, local to the patient, guide him/her in treatment and medication. In contrast, remote health monitoring is an automated approach for observing and assessing patients' health status remotely. Evaluating the medical data, the physicians and staffs guide the patient directly for taking suitable action.

- In telemedicine, doctors are fully responsible for assessing the patients, making a decision and recommending treatment and medicine. In remote health monitoring, automated monitoring makes the doctor's job much effortless. Using modern analytical tools, the health assessment also somewhat can be automated.
- Monitoring devices capture health data in the continuum, the data recorded over the past could be analyzed for monitoring and predicting future health problem and progress in health fitness. In contrast to this, telemedicine is short time process where physicians in a short session assess a patient's current medical condition and recommend based on the present symptoms. As a result, remote monitoring has many capabilities in overall health care.
- Telemedicine is aftermath but using remote health monitoring doctors can pre-assess the patient's conditions. As monitoring devices constantly assess a person's health, in case of any abnormality, it is immediately reported without waiting for the doctor's check-up schedule. Remote monitoring paves the way for precision medicine.
- Remote monitoring is technological intensive development whereas telemedicine is a human/expert intensive process; doctors diagnose with their expertise.

3 WBAN AND REMOTE HEALTH MONITORING

3.1 What is WBAN?

According to IEEE 802.15 Task Group 6 [20], WBAN is defined as:

> Low power devices operating in or around the human body (but not limited to humans) to serve a variety of applications including medical, consumer electronics/personal entertainment and other.

A WBAN is a special type of WSN that is associated with the human body, where fewer sensor nodes are deployed on the patient's body as compared to traditional WSNs. The primary requirement of any WBAN is tiny, low-power, wearable or implantable sensor node. A WBAN comprises of a set of heterogeneous sensors and medical devices performing individual roles to perform monitoring of patient's health.

WBAN sensors are capable of monitoring, sampling, processing and communicating all vital signs of a patient's body to the medical server and provide real-time feedback to the user via doctor in a matter of minutes. Various sensors such as temperature sensor, ECG monitor, EMG monitor, heart rate sensor, pulse sensor, blood pressure sensor, etc. can be deployed to form a WBAN system [21]. A WBAN is based on radio frequency wireless technology that interconnects intelligent, low-power and tiny biosensor nodes which can be wearable or implanted on the human body to monitor the vital signs like stress, temperature, the oxygen level in various activities like normal day-to-day operations, sporting or any sort of training [22]. The biosensor nodes create a network with sensors and control device.

WBAN can be used in different types of applications like military, sports, etc., but primarily it is implemented in medical technologies. For medical applications, there is a need for varied sorts of information like heart rate, blood pressure, ECG, EEG and much more and system demands very low latency and high reliability.

A typical WBAN comprises tiny biosensors and gateway node connected to remote locations like hospitals etc. Gateway node can also be called as "Body Control Unit (BCU)" or "Central Control Unit (CCU)" or "Personal Control Unit (PCU)". The collected data can be aggregated and processed locally or they are sent for remote processing, usually to a cloud or a dedicated remote monitoring server [23].

In recent years, it is observed that interest in WBAN applications have increased manifold. According to ABIresearch (Allied Business Intelligence Inc.):

> Wireless market tends to double in 2020 and market for wireless devices to monitor patient's health is also going to explode and tend to touch $9 billion by 2020.

A number of the latest research efforts by researchers primarily focus on wearable healthcare systems for patient health monitoring. The latest innovation in WBAN is proposed by MIT Media Lab, where a WBAN based system termed as "MIThril" [24] is proposed, a complete wearable WBAN platform fully equipped with custom and off-the-shelf sensors. In addition to bio-sensors, MIThril is fully equipped with other body monitoring parameters like ECG, skin temperature, and galvanic skin sensors.

WBANs are expected to give a dramatic shift in the medical technology in terms of thinking, utilizing, monitoring and analyzing the patient in different diseases.

The quality of service (QoS) of WBAN depends on the type of applications being used, especially medical applications. For an efficient WBAN system, multiple sensors are required so that data transmission can be reliably transmitted over efficient medium access protocol.

3.2 WBAN Versus WSN

Though WSN has laid the foundation stone of developing WBANs they differ from each other in several aspects. WSNs are used typically in battlefield monitoring, traffic monitoring, environmental monitoring, industrial automation and control and many more. WSNs are regarded as self-organizing, distributed network comprising tons of sensor nodes sensing certain data from varied conditions and giving aggregated information to the base nodes. On the other hand, WBAN is highly short-range, less processing power consuming network and is limited to body parameters sensing only. Table 7.2 enlists the differences between WSN and WBAN based on technical and operational parameters:

3.3 WBAN System Attributes

The primary objective of any WBAN system is to acquire all types of crucial or non-crucial data from patient's different parts of the body and WBAN system should be designed intelligently and with accurate caution to deal with wide range of real-time challenges [25]. In real-time monitoring of the patient, the two most important parameters are "reliability" and "latency". All these parameters depend on the Physical and MAC layer of WBAN. Physical Layer and MAC layer are responsible for determining the energy consumption of WBAN device which is regarded as a foremost issue in the design of any type of WBAN system. The role of the MAC layer is to determine efficiency in network operations and efficient resource utilization. The Physical layer should be designed in such a manner to handle

TABLE 7.2
Difference Between WSN and WBAN.

	WSN	WBAN
Rate of data transmission	Variable	Variable
Real-time deployment	Random	Fixed
Density	High	Low
Mobility	Absolutely no mobility	Fixed mobility
Range of transmission	Few km	0.01 m to max 2 m
Operational accuracy	Average	High
Homogeneity of nodes	Homogenous	Heterogeneous
Latency	Variable	High
Replacement of nodes	Easy	Implanted sensor nodes are difficult to replace
Network topology	Fixed or static	Variable
Security	Less security	High security to prevent leakage of patient information
Wireless technology backbone	Bluetooth, ZigBee, RF, WLAN	UWB, BLE, ZigBee

all sorts of transmission issues to transmit the patient's data remotely without any hiccup.

The following points highlight the attributes of a typical WBAN system:

- **Reliability:** WBAN system should be highly reliable in packet transmission and should face minimal delays in packet transmission which is influenced by BER (bit error rate) of channel and MAC layer transmission process. WBAN physical layer should be highly efficient to reduce the BER rate via adaptive modulation and coding techniques to adapt to typical conditions of transmission channels. Reliability of WBAN system also depends on network interference situation.

- **Energy efficiency:** Energy efficient WBAN system are always reliable for precision health monitoring of the patient. MAC layer of WBAN system should be reliable enough by adopting various types of energy efficient techniques in terms of packet scheduling, channel access so that high rate of packet transmission can be possible for prolonged periods of time.

- **Scalability:** WBAN sensor nodes, deployed on the patient's body, collect vital statistics like physiological data. WBAN systems should be scalable enough to allow WBAN professionals to re-configure the entire WBAN operational system by adding or deleting sensor nodes without any impact on the reliability of WBAN system. It depends on the MAC protocol. Efficient MAC protocol will enable future ready and high scalable WBAN system.

In addition to reliability, energy efficiency, and scalability, WBAN systems should be designed considering the following other attributes [26]:

- WBAN systems should be equipped with fault-tolerant and self-healing capability to ensure secure and efficient transmission of data from wearable and implantable sensors to medical servers or online websites.
- Highly efficient to operate in power constraint environment and WBAN systems should be highly focused more on implantable rather than wearable sensors as implantable sensors are more critical to patient's health.
- Should support data transmission from Kbps to Mbps to transmit a wide range of vital signs data from a patient's body to online servers.
- WBAN systems should be competent enough to have compatibility with QoS parameters in terms of reliable delivery of sensor node data to online servers or mobile apps.
- WBAN systems should have strong coordination with other network nodes operating at different frequency spectrums and have different operational standards.
- The medical data should be secured by means of encryption etc. if the data goes beyond the local network.

3.4 Components of a WBAN System

The following components make up an efficient WBAN system:

- **Sensors:** Biosensors should be incorporated depending on the application of capturing real-world data and should consume very less power and should be equipped with efficient sleep and wake up modes to keep hardware efficient in operation.
- **Data communication:** Wireless communication is the main source of transmission but should be operational in those frequencies bands which are tolerant to interference, and no mix up of frequency can be there during real-time operation.

- **Security:** Protocols of WBAN should be such that, no data can be hacked in between by hackers or intruders and data should be transmitted via the WBAN system to a remote location with strong encryption.
- **Handover mechanism:** It is also desirable to have a handover mechanism in WBAN system using a gateway or router and should not make the design of a node in an overload state.
- **Antenna:** Highly small size antenna should be implemented in WBAN system to operate at high frequencies.
- **Gateway node:** Gateway devices should be designed to interact with wireless networks in medical systems with advanced algorithms having machine learning or deep learning technology so that efficient capturing and re-transmission of data can be made possible.
- **Backup mechanism:** Every WBAN system should be equipped with a backup support system to trigger an alarm on low power or node failure.

3.5 WBAN Sensor Node Categories

Every node in WBAN operate as autonomous device and is fully equipped with a communication system to relay the data back to the medical server. Nodes operating in WBAN are categorized on the basis of functionality, implementation, and roles [27].

On the basis of functionality, the nodes in WBAN are of following different types:

a. **Body control unit (BCU):** It collects all the information from sensors and actuators operating as part of a WBAN system.

b. **Sensors:** WBAN applications range from medical to non-medical applications. Non-medical applications comprise motion and various gestures for detecting one's fitness, social interactions and even medical assistance in varied situations like floods, earthquakes, fires, etc. Medical applications include healthcare-based applications. Depending on situations, varied sensors are available either wearable or implantable to transmit physiological signals.

Generally, the following body sensors are used for health monitoring (typically termed as biosensors):

- **Inertial motion sensors:** To measure body posture and movement patterns like accelerometers and gyroscopes.
- **Bioelectrical sensors:** To measure electrical variations in a patient with regard to the certain activity of organ-like ECG, EMG.
- **Electrochemical sensors:** Measure some chemical agents of the human body like glucose sensor.

- **Optical sensors**: To determine the degree of absorption of light passing through the patient's blood vessels like the oximetry sensor.
- **Temperature sensors**: Measure the normal body temperature of the patient.

Considering WBAN sensors technology, various wearable and implantable sensors are also available. The following enlists various wearable and implantable biosensors [2]:

- **Wearable sensors**: The primary objective of wearable sensors is to gather all types of physiological and patient's movement data to facilitate remote health monitoring of the patient. Wearable sensors are used both for diagnosing as well as monitoring applications which include biochemical sensing and motion sensing.
  - Electrocardiographic (ECG) sensor: ECG sensor monitors the electrical energy produced during a heartbeat. The changes in electrical energy are detected via two leads and references to a ground signal. All the energy changes are displayed via waveform. ECG Sensor is attached to the patient's body using disposable electrodes on the chest left side and right side. The ECG sensor gives an analogue signal as an output and is converted via ADC (analogue to digital converter) and send to a mobile phone using Bluetooth module.
  - Heart rate sensor: Heart rate sensor measures heart rate in beats per minute using an optical LED source and LED light sensor.
  - Electromyography sensor (EMG) sensor: EMG sensor is used to measure all patient's body muscle's electrical activity. It is used as a control signal for all types of prosthetic devices.
  - Electroencephalography (EEG) sensor: It is primarily used to sense and obtain all electrical activity of the brain. It measures all types of voltage fluctuations resulting via iconic current within brain neurons.
  - Body temperature sensor: It is used to measure the body temperature of the patient. This sensor is highly useful as different diseases create fluctuations in the patient's body temperature. It is used to monitor the patient's in critical conditions to have a precise observation.
  - Photoplethysmography: Often implemented using a pulse sensor or pulse oximeter and is used to measure non-invasive heart rate and also the blood oxygen levels.
- **Implantable sensors**: Implantable sensors are primarily used to determine force, torque, pressure, and temperature inside the human body. Very small in size and have compatibility with human tissue and can withstand all physical forces inside the human body. Implantable sensors have a self-powering mechanism and transmit data wirelessly. Examples of implantable sensors are:
  - Pacemaker: A pacemaker is a tiny device placed at the chest or abdomen to control all sorts of abnormal heart rhythms in a patient's body known as Arrhythmias. A Pacemaker makes use of electrical pulses to enable the heart to beat at normal rates. A pacemaker can control two types of arrhythmias - Tachycardia (fast heartbeat) and Bradycardia (slow heartbeat).
  - Cochlear implants: An electronic device to facilitate the working of inner ear damaged parts to supply sound signals to the brain.
  - Implantable cardioverter-defibrillator (ICD): Wearable device for detecting all sorts of rapid heartbeats and life-threatening situations. During abnormal heartbeats, ICD creates an electrical shock to the heart to restore a normal heartbeat.
  - Deep brain stimulator: Medical device implanted inside the human brain through neurosurgery to supply electrical impulses via implanted electrodes to specific areas of the brain for treating all types of movement and neuropsychiatric disorders.

Fig. 7.1 highlights the various wearable and implantable sensors for making up WBAN:

c. **Actuator**: This device primarily performs the task of interaction with the user on receipt of data via sensors. It provides feedback in the network based on sensor data.

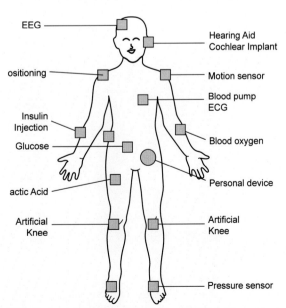

FIG. 7.1 A typical WBAN with attached bio-medical sensors.

On the basis of implementation, WBAN nodes can be classified as follows:

a. **Implant node:** Node implanted inside the human body either underneath the skin or body tissue.
b. **Body surface node:** Node primarily installed on the body surface of the patient.
c. **External node:** Node which doesn't have direct body contact but is placed to about 5−10 cm away from the body.

On the basis of role, WBAN nodes can be classified as follows:

a. **Coordinator node:** Acts as a gateway to the outside world, or another WBAN or Access coordinator.
b. **End node:** Performs certain functions prespecified in WBAN.
c. **Relay node:** Intermediate nodes between parent and child nodes.

3.6 WBAN Architecture

Fig. 7.2 demonstrates a typical WBAN architecture. In this architecture, a mobile phone or laptop are connected to communication access points to transfer the patient's data to secure medical server via wireless technology or the internet. In addition to this, all the access points are connected to patient's emergency family member number, emergency services via alarm notifications and health practitioners for emergency response in case of the critical condition of the patient. A WBAN consists of a three-tier architecture as discussed below:

3.6.1 Sensors/Intra-BAN communication/Tier 1

Sensor nodes are placed on the varied parts of patient's body depending on the patient's monitoring and depends on the disease vital statistics to be monitored. Patients are required to keep the sensors in place to the body to allow the doctors to monitor the patient's health continuously from anywhere and anytime. Sensors come under "Intra-BAN Communications" and are placed at about two meters near to the human body. The design of Intra-BAN is highly important as a patient, and body sensors have direct contact. In addition, MAC protocols should be designed in such a manner to keep the sensors working for long period of time.

3.6.2 Communications module/Inter-BAN communications/Tier 2

Like WSN's autonomous behavior, WBAN systems are required to work in a highly autonomous manner. In terms of WBAN architecture, Inter-BAN Communications are primarily concerned with communication modules. The communications link up the sensors to the Access Point. The access points form a typical part of WBAN infrastructure to connect WBAN with different networks to access in a routine manner like mobile communication networks or the internet. Inter-BAN Communication can operate in two main categories:

• **Infrastructure based:** High bandwidth with the centralized administration is provided, and WBAN system can be termed as highly robust and scalable.
• **Ad hoc based:** Fast deployment can be possible and dynamic architecture, and high mobility can be achieved.

3.6.3 Medical database servers/Beyond-BAN/Tier 3

To facilitate strong bridging of Tier 1 and Tier 2, i.e., Inter-BAN and Intra-BAN, a gateway device like a laptop

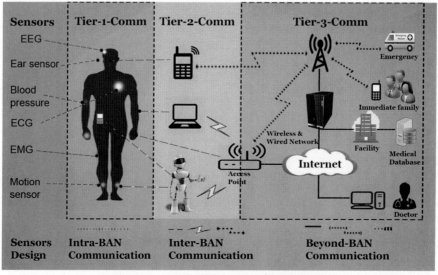

FIG. 7.2 A general architecture of WBAN [8].

or tablet is deployed to create a wireless link between the two. With Tier-2, WBAN system is provided with the power of scalability as the patient's data can be monitored by the doctor either via mobile app or a website or using secure login access to specific files of medical servers.

In beyond-BAN communication, the central point is a database which contains all record of the patient's data in terms of history, current medical status and other personal information. In addition to this, the database and other patient's data can also be accessed by other relatives of the user. The design of the Tier-3 component is based on applications. Mobile applications or IoT-based websites like thingspeak.com can be used to acquire the signal-based data and can be sent to the doctor via email or SMS to alert in case of emergency and also to nearby health services.

3.7 Role of WBAN in Remote Health Monitoring

As per the latest report by the World Health Organization (WHO), about 18 million people died in the year 2008 due to cardiovascular and heart diseases [28]. As per the estimation by WHO, it is predicted that around 25 million people will die due to cardiovascular disease by 2030. One of the instigating factors for cardiovascular diseases is diabetes. Both Type 1 and Type 2 diabetes are linked with cardiovascular diseases, which is also a major cause of deaths in people with diabetes. These statistics lay a strong foundation for the development of remote health monitoring using WBAN technology that can make a contribution toward saving the lives of people from several diseases and enhance the quality of living. Constant health monitoring via WBAN technology plays a crucial role to reduce risk factor of life threats against all types of diseases and early warnings, or critical signs can be reported to medical practitioners at an immediate interval of time.

WBAN is potentially the key enabler of remote health monitoring systems and extends the prospect to take e-healthcare beyond telemedicine [29]. Various health sensors in the WBAN acquire specific physiological parameters from the human body. These data are collected, aggregated, and sent through a central coordinator to the desired destination for effective processing and analysis to obtain reliable and accurate physiological estimations [11]. This helps doctors and medical staffs to diagnose and monitor patients remotely be it offline on in real-time.

WBAN systems are primarily designed for remote monitoring of vital signs of the patients to determine physiological signals and systems are required to have a high degree of reliability and low level of latency. All the sensors transmission rates are low, but with a high degree of energy efficiency will enable the WBAN system to work for a prolonged period of time. Patients can roam freely anywhere with the feeling that WBAN system will do the respective task side by side.

Present remote monitoring systems face several issues. For example, existing remote monitoring systems are location specific, i.e., they should be within reach of the network access points either wired or wireless. Another problem: in the case of short-range wireless communication, Bluetooth and Wi-Fi have been the popular standards for close communications in monitoring systems. But both of them eats up a substantial amount of energy and also cause interference. The WBAN gives remote monitoring systems a better edge than traditional e-monitoring systems in the following aspects:

- Thanks to portable health monitoring devices, patients enjoy the freedom of mobility and flexibility [26]. WBAN has enabled remote monitoring of patients not only when they are on the bed but also while they work, walk or play. Owing to WBAN, along with the wearables and smartphones, now health monitoring is not an additional activity; rather it has been assimilated and dissolved into people's daily life.
- WBAN offers location independent monitoring facility [26]. WBAN can make use of widely deployed mobile data networks, eliminating the dependence on the certain internet access point.
- The old remote monitoring systems, in spite of dedicated wireless link for medical data transmission, do not always guarantee persistent monitoring. A WBAN provides continuous gauging of different physiological parameters that allow a doctor, from a distant location, to make real-time assessment and diagnosis. The real-time monitoring provides faster detection in emergency cases and better revelation in the case of organ failures [11].
- Since WBAN are in general autonomous they are capable of finding a suitable communication network opportunistically. The advancement in the field has enabled WBAN employing cognitive networks [30] for optimal uses of the communication paths.
- WBAN is enriched by some latest power efficient wireless communication protocols such as UWB which makes the sensor batteries last longer [8].
- With the increased popularity of WBAN, the demand of health sensors is also in soaring. The mass

production of these sensors has reduced the cost of WBAN which in turn instigates a cost-effective and affordable healthcare.

In the recent past, WBAN has attracted considerable research interest from people from different domains focusing on issues related to energy aware and miniature sensor design, efficient and optimized sensor circuitry, effective signal processing, energy efficient communications, etc. [8]. As a result, today's remote monitoring systems are enriched with lightweight, tiny, ultra-low-power, and intelligent wearable sensors. The continuous advancement in WBAN and related fields will make remote health monitoring most suitable for delivering affordable, flexible, scalable, and robust health care [31]. The popularity of wearables and portable health sensors as components of WBAN will make the remote monitoring ubiquitous in the true sense by supporting a wide range of health applications.

3.8 A General Architecture of Remote Health Monitoring Using WBAN

A general architecture for the remote health monitoring system is presented in Fig. 7.3. Health attributes that require patients' self-input for collecting their values include weight, blood pressure, blood sugar level, etc. Value of health attributes like heart rate, blood pressure, etc. can be collected/measured using wearable devices like biosensors. All the collected data is sent to the remote patient monitoring server. A remote patient monitoring server is equipped with the necessary hardware and software required for complex analytics on the collected data and generate alerts according to the outcome of the analysis. Depending on the nature of outcomes/alerts, they are sent either to the patient or the attending medical staff (nurse, physician, etc.). In both cases, the output of the monitoring server is computerized feedback that assists the recipient in informed and effective decision making. Broadly, a remote health monitoring system consists of the following three components:

i. Patients' wearable and self-input measurement devices.

ii. A web/mobile interface for the physicians/authorized personnel to monitor patients' data.

iii. A processing and feedback unit called remote patient monitoring server to send alerts for both patients and clinicians.

All the wearable devices and measuring equipment at the patient's end aid accurate and effective data collection. Data collection serves as the basis for further triggering aggregation, processing, and visualization of data for the medical practitioner or the patient. The data collection module of the patient monitoring system further consists of the following components (Fig. 7.4):

- Sensors with connectors
- Application interface for manual data entry
- Pre-processor
- Communication manager

Sensors are the most important components of a remote health monitoring system as they help in collecting consistent data continuously from the patient's body. The network of sensors (WBAN) must be portable, lightweight, low power consuming, and autonomous. These body sensors must be optimized so that they operate on a suitable bandwidth and with latency requirements suitable for accurate collection of values of health attributes. For reducing the signal-to-noize ratio, the number of sensors must be the smallest

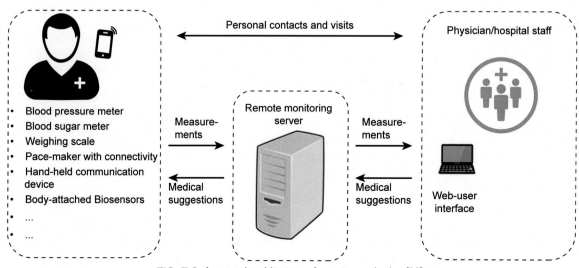

FIG. 7.3 A general architecture of remote monitoring [32].

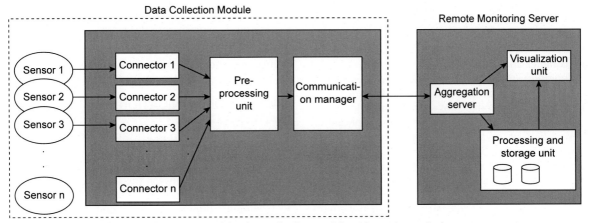

FIG. 7.4 Components of the data collection module and remote monitoring server.

possible. The location of sensors must be planned for accurate collection of health data.

Each sensor or a group of sensors has a connector that manages communication of collected data to the WBAN coordinator. This is often also referred as WBAN controller (WBANC). This controller acts as a sink of the network of sensors attached to the body of the patient. WBANC is responsible for collecting data which further sends collected data to the remote monitoring server over the internet. It is significant that the data collected by the sensors and sent to the WBANC is good in quality and is sent efficiently considering the limited battery power of the sensors nodes. These considerations are met using the appropriate routing techniques and accounting for the quality of service requirements of data transmission. The routing protocol used also determines the overall energy requirements of the network. Routing techniques of the sensor networks used for remote health monitoring are broadly classified into:

i. Flat routing protocols: Each sensor node plays exactly the same role. That is, no sensor node has designated access to more data than other(s).

ii. Hierarchical routing protocols: Each node has a different responsibility. Each node collects different data, and some nodes are dedicated for routing data packets to the controller. Hierarchical routing protocols are designed to achieve QoS-aware data transmission.

Some of the categories of extensively used routing techniques are thermal-aware routing protocols, cluster-based routing protocols, cross layer-based routing protocols, early-aware routing protocols, and delay-tolerant-aware routing protocols.

Though data collection through sensors seems to solve most of the problems, there are health attributes that cannot be captured through the use of sensors. Such attributes include weight, height, blood sugar, etc. To account for such health attributes, the data collection module must also include an application interface for manual data entry by the patient. Data collected from and entered manually by the patient reaches the preprocessing unit of the WBANC that performs integrity checks, groups the data and determines the usability of data through completeness checks. The preprocessor further sends its output to the communication manager that initiates a connection with the remote monitoring server.

The remote monitoring server has three major components (Fig. 7.4):
• Aggregation unit
• Visualization unit
• Processing unit

The data aggregation unit is responsible for collecting data from the patient's communication manager and getting it ready for processing or visualization. Data collected by the aggregation unit can be directly visualized through visualization unit by the doctor or the patient. An aggregation unit is responsible for the cancellation of noise in collected data and for verifying data integrity. Aggregation unit also performs some trivial computations and generate health alerts to both — doctors or the patient through the visualization unit. For example, on input, the weight and height of a patient, the aggregation unit can send the Body Mass Ratio (BMR) to the visualization unit for display. Aggregation unit can store extreme values of health parameters like blood pressure, blood sugar level, etc. Depending on the input value of these parameters, aggregation unit sends to the visualization unit, essential alerts intended for both doctors and patients.

Alternatively, the aggregation unit, after performing integrity and consistency checks, can send the output to the storage and processing unit that is a specialized hardware and software combination that helps carry out complex analyses on data.

Typically, a visualization unit is nothing more than a basic web service that can be accessed by the patient and the medical staff through a web-based application like a web browser or a dedicated mobile phone application. Presentation of results is very important for better understanding and informed decision making. It is required that the complex unrelated details about patient's data be hidden from him. This boosts the usability of the remote health monitoring system. Also, it is required that information that may be relevant for the patient but is trivially interpreted by the doctors were not presented to the medical practitioner. Therefore, it is sometimes the case that there are different visualization units for the patient and the doctors. Alternatively, efficient access management for accessing only relevant information by the respective users — patients and doctors, can also be present.

A processing unit is essentially a specialized hardware and software system that processes data collected by the aggregation unit. The complex pattern recognition and analysis algorithms form an important part of the processing unit. Typically, processing unit concerns detecting certain patterns in values of health parameters of a patient. Also, a processing unit records their current pattern and store them for future learning. A shift in values of health parameters of a patient, when matches a pre-recorded pattern in the database, triggers an alert for the patient and/or the medical staff. For this, the processing unit communicates the results of all the computations to the visualization unit after indicating the intended recipient of the alert. Real-time processing and communication require hardware that is expensive to both purchase and maintains. Therefore, services of a dedicated third-party may be utilized for carrying out complex computation and analysis on patient's preprocessed data. Issues to be taken care of are the data privacy and reliability of results.

One of the early works for architectural advancements of remote health monitoring system was performed in 2001–3 in the form of an integrated platform called MobiHealth [33] which was solely based on communication using GPRS/UMTS. Ever since the development of MobiHealth, numerous other architectures have been developed (see Ref. [34] for the survey). Recently, the focus has shifted from an integrated architecture to a general architecture with scope for efficient inclusion of more advanced analytics as and when needed (see Project SHERPAM [35]). An internet-like multi-layer protocol stack for better and more expressive access policies in each module of the remote health monitoring architecture [36] has also attracted the majority of attention.

4 WBAN COMMUNICATION ARCHITECTURE

4.1 Communication Architecture of WBAN

WBAN communication is divided into three major categories:

- Communication between the node on the body to an outside base station.
- Communication between two nodes on the body surface.
- Communication of node implanted inside the body to outside node.

These three communications are known as off-body communication, on-body communication, and in-body communication. In certain situations, WBAN needs to transmit the data across large distance and range of WBAN can be enhanced via multi-hop network via gateway or router to connect to external network resources. The range should not affect the design of a sensor node as large distance transmission is taken care of by a gateway node; so, a modern approach is required to design an efficient WBAN system.

4.1.1 WBAN network structure

A WBAN network structure consists of several interconnecting networks. There exists communication between sensor nodes and the base node under short-range. The short-range communication is less than 2 m, and sensors are characterized by the following features:

- All the sensor nodes are connected to a gateway node in a typical star topology.
- All sensor nodes have limited capabilities in terms of process and are highly energy efficient.
- Gateway nodes do all sorts of high processing and are equipped with high battery power.
- There is one-way communication between sensor nodes and gateway nodes.

A WBAN system can accommodate various wireless platforms and a multiple WBAN can have multiple wireless links.

The following points elaborate typical requirements of WBAN system:

- Limited range of transmission from less than 0.01 m to max 2 m.
- Ultra-low power consumption especially in sleep mode from 0.1 to 0.5 mW for retaining battery power.
- Data range from 1 Kbps to 10–15 Mbps.

- QoS support for efficiently managing all sorts of physiological signals.
- Low latency and lightweight.

4.1.2 WBAN topologies

In simple terms, topology is defined as a systematic arrangement of the body sensor network in an existing network, which fulfills all the basic requirements of the application. WBANs are generally asymmetric where at least one node is superior to others in the network. The superiority may be in terms of having a larger power supply and more processing power. This node and can take over energy-consuming tasks from the other low-power and inferior sensor nodes [4]. The size of the network size is an important factor for considering suitable topology for WBAN. Unlike environmental sensor networks, a WBAN is inherently small-scale. The maximal distance between two nodes is less than 10 m, and each network has less than 100 nodes. In such a small network, it is more power efficient to use single-hop communication instead of multi-hop. Since the network is small-scale, asymmetric and single-hop, a star topology is suitable, which is used by most WBANs. WBAN topologies are of following different types:

- **Point to point topology:** In a point to point topology, the nodes in a WBAN are connected in direct order and communication takes place via peer to peer.
- **Star topology:** In a star topology, the entire WBAN network is managed by a single node, i.e., Master node. Considering the WBAN network, the master node has larger power capacity as compared to other sensor nodes forming a WBAN network. The entire network is managed by Master node, and under WBAN-star topology, only one node can stay active and transmit data at a particular period of time to avoid any sort of data collisions among other sensor nodes [4]. In a star topology, all the sensor nodes attached to the patient's body communicate via the gateway for transmitting the sensed data. In WBAN systems, star topology is regarded as highly suitable topology for Inter WBAN communication.
- **Peer to peer topology:** A peer-to-peer WBAN network allows multiple hops to route messages from any device to any other device on the network. Such functions can be added at the higher layer, but they are not part of this standard. Each independent PAN selects a unique identifier. This PAN identifier allows communication between devices within a network using short addresses and enables transmissions between devices across independent networks [37].
- **Mesh topology:** In a mesh topology, every device in the radio range can communicate either in a direct fashion or via multi-hopping fashion. Mesh topology

can be implemented in those scenarios where there is a high requirement of the reliability of data transmission and communication should be facilitated in a flexible manner as compared to energy utilization and prolonged lifetime.

- **Tree-mesh hybrid:** A tree-mesh hybrid topology is regarded as a combination of tree topology and mesh topology in which nodes at the lower level have dedicated connection with nodes at a higher level and connected via full mesh connectivity. Tree-mesh topology is advantageous in several factors like best coverage, low latency, fault tolerance, and overall high throughput.

4.1.3 Network layers for WBAN

The actual implementation of different communication layers depends on the protocol/technology used for communication. But in general, the proper designing and implementation of the physical (PHY) layer and the media access control (MAC) layer are most important for any WBAN protocol. Because these two layers are combiningly responsible for improving WBAN operations in terms of power consumption, reliability, network efficiency, latency, resource utilization, operating cost, etc. in general, since sensors in WBAN are resource constraint, it is really challenging to design and implement efficient, reliable, and secure protocols for WBAN communication. Table 7.3 summarizes the challenges in each layer.

4.1.3.1 PHY layer. This is the layer that is responsible for providing WBAN with data transmission capabilities. It establishes a reliable physical link between communicating nodes for transmitting binary data. This layer deals with several important factors of communication such as specifying operating bands and channels, modulation and demodulation techniques, the structure of data units, signal encryption, etc. [38]. It also does characterizing the link quality/strength of a received signal known as Link Quality Indication (LQI). PHY layer specifies three different operational states for the devices: transmitting, receiving and sleeping. When the devices in WBAN are not active, they are put into sleep mode. This allows WBAN to save energy which is very crucial for battery operated sensors.

Challenges in WBAN physical layer: Following are some of the challenges related to the physical layer that need to be addressed while designing WBAN [39,40]:

- **Interoperability:** WBAN consists of various devices of different types having different properties and functionalities. Furthermore, they might be made by different vendors. This causes interoperability issues within WBAN devices.

TABLE 7.3
Designing and Implementation Challenges in Different Layers of WBAN Protocols.

Physical Layer	MAC Layer	Network Layer	Transport Layer	Application Layer
• Interoperability • Temperature control • Changing topology • Varying bandwidth needs • Interference • Fault acceptance • Constant signaling • Security • QoS • Varying data rates	• Dynamic channel assignment • Control packets overhead • Protocol overhead • Synchronization • Throughput • Consistency • Over-emitting • Packet scheduling • Error control • Overhearing • Calibration • Fault acceptance • Energy conservation • QoS • Multi-radio and multi-channel design • Data flow control • Idle listening • Security • Delay control	• Optimum routing • Network condition • Real-time streaming • Localization • Mobility • Temperature and heat control • Traffic control • Multi-path routing • Security • QoS • Fault tolerance	• Reliable transport • Congestion control • Self-configuration • Energy awareness • Biased implementation • Constrained addressing	• Efficient interface • Security • Congestion control • Flow control • Bandwidth allocation • Packet-loss recovery • Energy efficiency

- **Varying bandwidth needs:** For different medical applications, the bandwidth requirement depends on the type of data transmitted over a WBAN. So, it is desirable to choose the wireless radio band and the bandwidth wisely that can accommodate a range of application within a single WBAN.
- **Varying data rates:** Since different nodes in a WBAN may have different purposes and execute different tasks they may generate at different rates. WBAN physical layer should be capable of handling these varying data rates in the same network.
- **Temperature control:** The heat and radiation generated from the implanted devices may affect body tissues and other devices in close proximity. Hence, the designing goal of these devices should be taking care of this issue.
- **Changing topology:** The WBAN topology gets changed along with the movement of the body. This makes the connection between the devices unstable. WBAN should be able to accommodate the dynamic topology.
- **Interference:** The working of WBAN devices may be affected due to the interference caused by other surrounding devices, heat and radiation, body tissues, fluid, and chemicals, etc. Such types of

interferences should be minimized for the proper functioning of WBAN while co-existing with other devices and networks.
- **Susceptible to faults tolerance:** Different factors such as changing topology, environment influence, transmission power, etc. make WBAN prone to faults. To maintain faultless functioning and seamless connectivity the WBANs need to be fault tolerant.
- **Constant signaling:** Many remote health monitoring cases require continuous and real-time monitoring. The WBAN need to be able to carry out constant signaling task smoothly and seamlessly.
- **Security:** Security threats like eavesdropping, tampering, and jamming are usually targeted at the physical layer. Hence, it is desirable to detect these types of attacks and implement preventive measures at the physical level.
- **QoS:** Since WBAN is related to human health and life it is of utmost importance to maintain QoS in WBAN operations. In the physical layer, QoS is generally determined by communication throughput, reliability, and energy efficiency which can be achieved by adaptive channel coding, power scaling, and effective time slot division.

4.1.3.2 MAC layer. It is usually a part of the Data Link layer that provides the interface between the upper layers and the physical layer and is in charge of transferring data, at a low level, between the nodes within a WBAN. It is responsible for carrying out the jobs such as multiplexing of data streams, data frame detection, medium access and error control, etc. [41]. Both IEEE 802.15.4 and IEEE 802.15.6 standards suggest using *superframe* for MAC layer communication. The superframe is the time gap between one *beacon* to the next beacon. Beacon is a signal that is sent from the coordinator node in WSN to another node within the range. The purpose of using beacon is to synchronizing between the communicating devices, identify the WBAN, describe the superframe structure,[1] inform the WBANC about any pending data, etc. [42]. Basically, the time axis is divided into beacon periods; in other terms, the channel is divided into superframes. A superframe is divided into active and passive regions. If it is in the passive region, a node does not do anything. This inactive period helps in saving energy. If it is in the active region, the node connects to other nodes and transmit data. The active region of a superframe is allocated a number of equal duration slots that are used to transmit the patient's data [43]. The channel time can be bounded (optionally) by the coordinator in a WBAN using a superframe structure. A superframe is bounded by the transmission of equal length beacon frames. The communication can be either in beacon mode or non-beacon mode and with or without superframe boundaries. Section 5 discusses the MAC layer in details.

Challenges in WBAN MAC layer: Similar to the physical layer, there are some challenges that are specific to the MAC layer and need to be addressed. Some of them are as follows [39,40]:

- **Synchronization:** For real-time monitoring and data transmission, perfect synchronization between the two communicating ends are crucial.
- **Throughput:** The exchange of control packets and low duty cycles leads to low throughput in WBANs. Opting for self-adjustable duty cycles in the protocols may solve this issue.
- **Consistency:** The WBAN deals with crucial data. Hence, it is essential to minimize the data loss which directly affects the consistency or reliability of WBAN.
- **Calibration:** Due to several factors such as random noise, sensor node failure, node blockage, damage, random errors, aging, etc. calibration in WBAN is difficult. WBANs need self-calibration mechanisms to surmount this.
- **Packet scheduling:** In medical applications, different packets may have different priorities. For example, some packets may require the earliest delivery but to some packets reliable delivery if more importance. Therefore, MAC layer protocols of WBAN need to adopt efficient scheduling schemes.
- **Dynamic channel assignment:** The presence of noise and interference in the communication channel is the reason for low throughput, high delay, and high data loss. To deal with this, MAC layer protocols should support dynamic channel sharing and efficient bandwidth assignment.
- **Multi-radio and multi-channel design:** In addition to the dynamic channelling, to improve the network capacity as well as the overall WBAN performance, instead of a fixed channel, multiple channels should be used for data transmission, especially for emergency data. The MAC layer protocol should provide this facility.
- **Control packets overhead:** The size of control packets should be as low as possible because they do not carry any medical data but consume a significant amount of energy.
- **Protocol overhead:** Majority of the protocols include extra information, besides actual data, such as the addresses of the source, destination, and, packet type, sensor type, sequence number priority, checksum, message length, timestamp, etc. in each transmitted packet. This leads to transmission overhead and bandwidth consumption. The protocols need to minimize these overheads wherever possible.
- **Idle listening:** Idle listening happens either when a node is expected to receive some packets, but no packets are received or when a node listens to an idle channel. Handling idle listening is very crucial to WBAN because it consumes a lot of power unnecessarily.
- **Over-emitting:** Sometimes, due to lack of synchronization between the source and sink, the generated data gets overflown because the receiving end is not ready to receive data at the same rate. This should be avoided because it causes unnecessary power wastage and makes the channel needlessly busy which, in turn, degrades the overall performance of WBAN.
- **Overhearing:** When a node receives a packet that is destined not for it but for some other node. This involves unnecessary processing and wastage of power which should be avoided.

---

[1]The superframe structure in IEEE 802.15.4 and IEEE 802.15.6 standards defines the classification of data, frame format and schemes for multiple access of channels.

- **Energy conservation:** Power conservation is very crucial to WBAN. Hence it imperative to make optimum use of energy and take care of all the above-mentioned factors that cause unnecessary power consumption.
- **Data flow control:** As mentioned above, the lack of synchronization between transmitter and receiver causes data flooding at the receiver end. This causes propagation delay and transmission delay. The goal of the MAC layer protocol should be to minimize such delays by controlling the rate of data generation and data flow.
- **Delay control:** In a busy network, low duty cycles cause a delay in packet transmission. To reduce the transmission delay, data can be instantly forwarded instead of buffering.
- **Error control:** Due to several factors, the WBAN operations suffer from errors due to which health data get lost or corrupted. Hence error control schemes are to be adopted for ensuring reliability.
- **Fault tolerant:** In WBAN the fault tolerance in the MAC layer can be achieved by adopting priority management, adding new nodes, having redundant path and resources, etc.
- **Security:** Denial of sleep, spoofing, eavesdropping, sinkhole, sybil, collision, unfairness, exhaustion, etc. are the example of the security attacks specific to the MAC layer. The protocols should have proper security schemes to protect WBAN against these attacks.
- **QoS:** The QoS in the MAC layer can be achieved by the suitable application of adaptive channel coding, power scaling, and time slot division.

4.1.3.3 Network **layer.** The function of the network layer is to extend communication beyond the BAN. The network layer enables WBAN to be internetworked with external networks such as other WSNs, external servers and data centers, remote monitoring management systems, etc. [41]. Using suitable routing protocols, the WBAN data are sent efficiently to remote destinations fulfilling the purpose of remote health monitoring. The routing protocols and the functioning of the network layer are designed keeping in mind the factors such as low power consumption, data-centric approach, data aggregation and fusion, attribute-based addressing, etc. [44]. Often the routing in WBAN is based on data, not node address. This approach of routing is known as data-centric routing where instead of traditional IP-based global addressing, the attribute-based naming [45,46] is used. This naming scheme is used to query based on the attributes of a certain phenomenon, event or process.

Challenges in WBAN network layer: Following are the challenges that are to be addressed by the network layer of the WBAN protocols [40]:

- **Optimum routing:** Realizing efficient communication in WBAN is challenging mainly due to limited bandwidth that varies with the presence of fading, noise, and interference. Achieving optimum routing efficient and enhanced routing techniques are required.
- **Multi-path routing:** For reliable and robust communication, the routing protocols should adopt multi-path and multi-point routing.
- **Network condition:** For reliable transmission, the network conditions are needed to be stable. Hence it is important to check the issues such as channel, buffer information, priority measurement, etc. periodically.
- **Real-time streaming:** Many of the remote health monitoring scenarios require real-time data transmission. It is the network layers responsibility to make sure that in such cases the packets reach at receiving end in time. In case of delay, the worst transmission time and the probability of it are also should be estimated.
- **Localization:** In WSN the data transmission quality depends on the distance between the transmitting/receiving node and the nearest access point. Since WBAN may be mobile, it is required to adjust the localization of the nodes in operation whenever there is displacement of the nodes with respect to the access point.
- **Mobility:** The mobility in WBAN, especially in case of wearables and fitness devices, poses a challenge for identifying the best suitable route in such dynamic or mobile environment.
- **Traffic control:** To control congestion and transmission rate, designing and adopting efficient traffic control routing protocol is necessary.
- **Security:** The network is prone to several security attacks such as sybil, sink hole, neglect, homing, misdirection, black hole, selective forwarding attacks, etc. The Network layer protocols should be capable of mitigating these attacks.
- **Fault tolerance:** Like other layers in WBAN communication, the networks layer should also be tolerant to fault that is supposed to arise in this layer, e.g., faults in network topologies and routing.
- **QoS:** The QoS in network layer can be improved by addressing the factors such as improving path latency, controlling congestion, robust routing, taking proper action in case of lost and damaged packets contribute to improving QoS.

4.1.3.4 Transport layer. Like the network layer, the transport layer comes into play when there is remote accessing of a WBAN. As usual, the responsibility of the transport layer in the protocols for beyond-BAN are handling congestion control, flow control, fair allocation of bandwidth, reliability, packet-loss recovery, energy efficiency, and heterogeneous application support [41,47]. But contrasting to the traditional transport layer protocols such as TCP and UDP no global addressing is used for the end-to-end communication. Instead, attribute-based naming is considered for addressing the endpoints [44].

Challenges in WBAN transport layer: The designing of the transport layer protocols heavily depends on the specific WBAN application. In this respect, the design and implementation challenges faced by the transport layer are as follows [48]:

- **Reliable transport:** Typically transport layer protocols are responsible for ensuring reliable communication. In WBAN, the sensor data or the events should be reliably transmitted to the receiving end. Similarly, the management or administrative operations and queries on sensors should be reliably delivered to the target sensor node for the proper functioning of WBAN.
- **Congestion control:** Data flooding and channel congestion result in packet loss. Hence, to achieve reliable communication, intelligent congestion control is required for monitoring the channels and regulating the data transmission rate dynamically when congestion is detected or anticipated. It will increase network efficiency as well as conserve network resources.
- **Self-configuration:** The WBAN structure may be changed due to various reasons such as node mobility, node failure, random node deployment, temporary power down, the spatial variation of events, etc. The transport layer protocols must be able to be dynamically adaptive to these changes.
- **Energy awareness:** As usual, the operations (e.g., error and congestion control) in the transport layer also should be power efficient. The transmitting node should be allowed to conserve energy, wherever applicable, by controlling the data transmission rate or putting the sensor in sleep mode.
- **Biased implementation:** Most of the transport layer functionalities should be implemented at the sink (usually the WBANC or the WBAN server) while only the bare minimum portions run on the sensor end. This will be helpful for saving energy consumption at the low-powered sensors.

- **Constrained addressing:** As mentioned above, the transport layer protocols for WBAN use attribute-based naming and data-centric routing instead of end-to-end global addressing. This requires exercising different transport layer approaches.

4.1.3.5 Application layer. This layer provides the interface for managing the WBAN, managing and querying WBAN data, etc. In addition, this layer is also responsible for node localization and time synchronization. The implementation of the application layer in WBAN is application specific. But in the context of WBAN, the most crucial function of application layer is to ensure a secure environment for accessing sensitive medical data remotely. Subject to the purpose, the application layer of a typical WBAN can be divided as following [44]:

i. Sensor Management Protocol (SMP): This is the interface through which the system administrators interact with the WBAN. Using SMP a WBAN administrator should be able to perform the following tasks [48]:
   - Set up rules for attribute-based naming and data aggregation.
   - Assess and inspect sensor status, WBAN configuration.
   - Reconfigure WBAN, if required.
   - Synchronizing health monitoring devices.
   - Move those devices.
   - Turn the monitoring devices on and off as per requirement.
   - Enforce authentication, key distribution, and security in beyond-BAN communications.

ii. Task Assignment and Data Advertisement Protocol (TADAP): Using this protocol users generally express their interest, about a certain phenomenon or an event, to a particular or a group of the sensor device(s) in a WBAN or the whole WBAN. Alternatively, health monitoring devices advertise their corresponding health data, and the users query those data based on their interest.

iii. Sensor Query and Data Dissemination Protocol (SQDDP): This protocol enables users to issue location-based and attribute-based queries to the WBAN as a whole rather than a particular node. Users also can respond to other's queries and can collect incoming replies.

Challenges in WBAN application layer: The main challenge in designing application layer protocols for WBAN is to provide an efficient and swift interface that either can be used by the human or can be interfaced to other devices for automated diagnosis. The application layer is also the first level for enforcing security measures. In addition, most of the functionality of the transport layer protocols are overlapped that of with the application layer protocols. For example, functions like congestion control, flow control, bandwidth allocation, packet-loss recovery, energy efficiency, etc. are at times considered as part of the application layer as well [41]. Hence, designing application layer protocols also faces common challenges as discussed in Section 4.1.3.4.

4.1.4 Desirable properties of a WBAN communication architecture design

Below mentioned are some general properties that should be given importance while designing the communication architecture of any WBAN [26]:

Reliability: Since WBAN deals with very critical and sensitive data, reliability is the most important quality that every WBAN should maintain. The WBAN should ensure no packet loss and minimum noise induction. The WBAN data should be encapsulated from external interferences.

Power efficiency: Another factor of primary importance that every communication protocol in WBAN should take care of. Power usage should be minimized by optimizing the operations in every layer.

Scalability: For the successful application of WBAN in remote health monitoring, the WBAN system should have the flexibility to be scaled as per requirement. The devices in a WBAN may be added, removed or replaced at any point of time. The WBAN should be able to accommodate this change without affecting the usual monitoring process.

4.2 WBAN Standards

IEEE 802 standardization committee is regarded as an international organization whose main task is to propose international standards with regard to wireless communication. Several standards (in the IEEE 802.15 family) have been proposed for different wireless communication technologies. For example, the IEEE 802.15 (WG15) standard especially focus on wireless personal area networks (WPAN) [49]. WG15 has proposed several different wireless standards like IEEE 802.15.1 (Bluetooth); IEEE 802.15.4 (PHY) for low-rate WPAN and is applied to ZigBee; IEEE 802.15.4a (PHY) for ultra-wideband technology; IEEE 802.15.3c (High rate WPAN); IEEE 802.154d (Japanese WPAN); IEEE 802.15.4e (MAC improvement for IEEE 802.15.4), etc.

IEEE 802.15.4 standard was finalized by the 802.15 working group of the IEEE in the year 2003.

This standard was specifically designed for low range communications for wireless personal area networks (WPANs) [50,51]. A WPAN essentially deals with communication of locally situated sensors, typically in a small room, among themselves and over the internet. WBANs used WPAN standards in communicating the data collected by the body planted sensors to the central server for storage and processing. Later in 2012, a working group IEEE 802.15.6 was formed for standardizing WBAN communication protocols for sensors that strictly remain inside or on the surface of the human body and communicate over shorter ranges of the order of a few meters [52]. This standard addressed the problem of low range and battery limitations of sensor devices more specifically. Along with the communication efficiency, the standard also addressed security requirements for WBANs by introducing new physical layer (PHY) and MAC protocols for WBANs. A couple of standards that are most suitably applicable to WBAN are briefly discussed below.

4.2.1 IEEE 802.15.4a standard

This is an UWB-based standard on WPAN and an extension of the IEEE 802.15.4 (2006), IEEE 802.15.4a (2007) [53]. It was introduced with the specification of additional PHYs. These new PHYs enriched the newer version with enhanced scalability in data rates (from several Kbps to 10 Mbps) and operational range, lower power consumption, and precision ranging capabilities [54]. The MAC layer of this standard uses a guaranteed time slot (GTS) using which the QoS is achieved in the continuous transmission of vital data in health monitoring applications such as ECG and EEG [55].

4.2.2 IEEE 802.15.6 standard

IEEE 802.15.6 [56] is the first ever standard that exclusively defines the lower tiers of WBAN communication. IEEE 802 Task Group has published this standard that aims to govern the communications inside and around the human body. The objective IEEE 802.15.6 is to provide an international standard for ultra-lower power, short-range and reliable wireless communication within reach of the human body at a data rate up to 10 Mbps for a varied set of applications.

4.2.2.1 IEEE 802.15.6 standard requirements. The following are the main requirements of the IEEE 802.15.6 standard:

- WBAN link rates range in between 10 Kbps to 10 Mbps.
- Nodes removal and addition should be done within 3 s.

- Every WBAN system should be capable of supporting 256 nodes.
- Every sensor node should have reliable communication especially when the patient is moving from one place to another and doing various body movements like sitting, walking, twisting, running, arms and all sorts of leg movements.
- Packet error rate (PER) should be less than 10%. Latency rate should be less than 125 ms in medical and less than 250 ms in non-medical applications.
- Energy saving mechanisms should be integrated into WBAN to make them fully operational in the power-constrained environment.
- WBAN should be fully equipped with UWB technology to allow transmission in all sorts of environments.
- Best QoS features in terms of security and self-healing should be there in WBAN.

The number of communication layers depends on the protocol/technology used for communication. But the physical layer (PHY) and the media access control (MAC) layer are the two most important for any WBAN system. Hence, the IEEE task group obliges this importance and has come out with the specification for these two layers with the IEEE 802.15.6 standard.

4.2.2.2 PHY layer specifications of IEEE 802.15.6. The standard defines three PHY layers:

i. Narrowband (NB) PHY: Purpose of this layer is to handle communication between the sensors on and within the body [38]. It has mainly three responsibilities [56]:
  - Activation/deactivation of the radio transceiver
  - Performing Clear Channel Assessment (CCA)[2] before transmission
  - Transmitting and receiving data
  The working bands of NB PHY includes the 400, 800, 900 MHz and the 2.3 and 2.4 GHz.
ii. Ultra-wideband (UWB) PHY: Purpose of this layer is to provide efficient, power efficient, and robust WBAN communication. uses the 3.1–11.2 GHz [38]. According to the IEEE 802.15.6 specification, UWB PHY operates in two modes:
  - Default mode
  - High QoS mode
  The data rate of UWB PHY ranges from 0.5 Mbps up to 10 Mbps. The UWB WBAN is discussed in more details in Sections 6.5 and 6.6.

iii. Human body communication (HBC) PHY: HBC is the equivalent of Electrostatic Field Communication (EFC) specification of PHY which is the basis of physical realization. HBC PHY is responsible for dealing with the entire WBAN protocol that may include packet structuring, modulation, preamble/SFD, etc. [56]. The hardware specification has been simplified for optimal usage in WBAN, for example, the transmitters within the sensor nodes in a WBAN system that comply with this standard, do not need antenna; instead, they use a simplified circuit that consists of only one electrode [38]. Similarly, the receiver hardware does not require RF modules which makes the devices power efficient and easier to implement. The communication frequency of the specification conforms within the range of 1050 MHz.

4.2.2.3 MAC layer specifications of IEEE 802.15.6. The patient data in IEEE 802.15.6 specification for WBAN has been categorized as follows:

- Type-I: represents life-critical data
- Type-II: represents normal and regular health monitoring data.

The superframe structure of IEEE 802.15.6 is divided into four access phases [43]:

i. Exclusive access phases (EAP - I & II): EAPs are used to transfer high-priority or emergency traffic, i.e., Type-I data. They offer guaranteed data transfer service.
ii. Random access phases (RAP - I & II): The RAPs are used for nonrecurring traffic and are reserved for transmitting Type-II data.
iii. Managed access phase (MAP): This phase is also used to transfer Type-I data. It guarantees high priority data transfer.
iv. Contention access phase (CAP): This phase is also used for nonrecurring traffic and is reserved for Type-II data.

The IEEE 802.15.6 network operates in one of the following modes [56]:

i. Beacon mode with superframe boundaries: The coordinator in a WBAN transmits beacons in the active region of superframes. The beacon is used for synchronizing the communicating nodes in a WBAN. In between two active superframes, there might be multiple inactive superframes whenever there is no scheduled transmission [43]. CSMA/CA or Slotted Aloha is used for channel allocation in this mode.
ii. Non-beacon mode with superframe boundaries: Here, communicating nodes need not be synchronized with the coordinator. Instead of setting a

---

[2]CCA is used to determine whether the communication medium is free or not.

superframe boundary by using beacons, the WBANC takes help of polling [26] for scheduling the data transmission of each individual sensor node. This ensures that the communication between the sensors falls within the superframe structure [57]. The non-requirement of synchronization saves energy. But the problem is that the coordinator cannot transmit data directly to the sensor nodes within the WBAN. If a coordinator wants to communicate to a node, it must send an activation alert signal to the recipient to initiate the communication. Another serious issue is there in this mode that at a time only one type of health data is allowed per slot. And since the WBANC operates only during the MAP period in this mode, this exclusion policy might not be ideal for life-critical situations of a patient when different vital status need to be accessed simultaneously by the doctors [58].

iii. Non-beacon mode without superframe boundaries: In this mode, a pre-defined superframe structure is not required. That means the superframe structure does not have any predefined slot allocated to any sensor node for data transmission. The slot allocation is done by contention or post-contention methods. The coordinator opts for polling or posted allocation for data communication where a certain number of timeslots are openly allocated which can be accessed by any sensor node waiting for data transmission. This mode of operation is designed for transmitting Type-II data. The advantage of this mode is that it avoids the unnecessary interruption by the health sensors that are sensing not so critical health status in contention of acquiring communication slots that are required by the sensors that are employed for critical health monitoring [58].

5 WBAN MAC LAYER

5.1 Importance of MAC Layer Protocols for WBANs

In any computer networking system, the protocols adhering to the MAC layer contributes significantly to improving the overall performance of the network as well as extending the network lifetime [59]. The same applies to the MAC layer protocols in WBAN. Different sensors and actuators, implanted on or within the human body, which collects critical and non-critical physiological information throws a major challenge in designing MAC layer protocols in terms of adaptability, dynamicity, flexibility, reliability, and power efficiency [59]. It is essential to adopt an efficient MAC layer protocol for effective implementation and working of WBAN.

5.2 Properties of WBAN MAC Layer Protocols

The inherent challenging environment of WBAN implementation demands the MAC layer protocols to be able to handle these challenges effectually. The working of the MAC layer becomes more challenging in WBAN compared to a regular WSN mainly because of the following facts:

- To save energy consumption, the sensors' radios are usually low-powered, and hence the line of sight is also limited.
- Depending on the position of the sensor (within or on the human body), the line of sight varies significantly. For example, the sensors planted on the chest and the back of a patient find it difficult to communicate. Similarly, in a sitting position (on chair/sofa), communication between sensors planted on the chest and the ankle is not possible.
- If the line of sight is to be improved, the propagation of the radio waves should be broadened. This can be achieved by making the radios more powerful. But that will consume more energy which is not acceptable in WBAN.

The sensor nodes' proximity to the human body, in WBAN, makes the things more complicated. Below, some of the adversities that WBAN MAC layer protocols are expected to address are mentioned.

- **Energy consumption:** The WBAN have small batteries with limited power capabilities, and in most of the cases they are neither rechargeable nor replaceable. That's why power consumption is a major design issue for every protocols and application in WBAN. In the networking stack, MAC is the key layer which can save a great deal of energy consumption by maintaining an optimum ratio of packet delivery and energy efficiency. The major reasons for large power consumption in WBAN communication are [60,61]:
  - Collision between packets: After detecting a collision the transmitting node retransmits the packet that was involved in a collision. This retransmission of data packets causes extra energy consumption.
  - Overhearing of nodes: Sometimes, the sensor nodes receive packets that are not destined for them but some other nodes. On receiving such packets, they are dropped which unnecessarily consumes power.
  - Idle listening to receive possible data packets: Sensor nodes optimistically continue to listen to idle channel in anticipate to receive packets possibly transmitted by other nodes. This also results in needless energy consumption.

- Communication control packet overhead: Usage of control packets in communication results in added energy consumption mainly due to energy wasted in transmitting and receiving those control packets.
- Transceiver state switching: In practice, the transceiver of the sensor nodes is kept in sleep mode when no data to transfer or receive. When required, it is switched on to active mode. This is done to avoid idle listening and overhearing. This may actually decrease the power consumption, but if the switching between the sleep mode and the active mode is frequent, the energy consumption rises sharply. An optimal balance of the number of switching can improve overall energy consumption.

  A good MAC layer protocol can either eliminate or minimize the effects of these sources of power dissipation.

- **Limited bandwidth:** WBANs have limited bandwidth capacity due to the inherent physical constraints. So, the bandwidth requirement of WBAN applications should be low. That is why it is crucial to concentrate on the designing of the applications and services low data generating and transferring rate.
- **Limited device capability:** Due to size constraint, the WBAN nodes have very limited computing and storage capacity. This prevents implementing heavy and complex protocols. That's why the protocols in WBAN need to be lightweight.
- **QoS:** Since WBAN is related to human health, it is crucial to maintaining QoS at each service criteria of WBAN. There are various parameters that determine the QoS such as minimum latency, high reliability, energy efficiency, negligible electromagnetic interface with the human body, and effective communication are to be taken into consideration, etc. [59]. Among these, latency and reliability are the most crucial for patient health monitoring systems. The reliability of a WBAN depends on two factors: transmission delay and packet loss. The MAC protocols play an important role in minimizing both. The proper selection of the medium access technique at the MAC layer significantly influences the reliability. For instance, in the case of TDMA and polling, the packet loss and packet delay are deterministic hence easy to set and adhere to the QoS level. But the accurate transmission time cannot be determined in the case of contention-based media access protocols like CSMA because transmission takes place only when the shared channel is free. Similarly, in the case of other protocols which follow

random media access techniques, the packet loss and delay varies from time to time and not predictable [61]. In both cases, QoS is compromised severely.

QoS is also improved by adopting the MAC layer protocols which come up with optimal packet size, suitable packet re-transmission schemes, and enhanced scheduling schemes.

Also, many often the different wireless services like Bluetooth, Wi-Fi, and ZigBee coexist in a single band. This results in significant interference. A QoS-aware MAC protocol, essentially, should be able to tackle these conditions.

5.3 Channel Access Techniques for MAC Protocols in WBAN

In general, MAC protocols based on Frequency Division Multiple Access (FDMA) and Code Division Multiple Access (CDMA) are not suitable in WSNs mainly because of the requirement of high computing capacity and complex hardware [62]. That's why the MAC protocols for WSNs use either Time Division Multiple Access (TDMA) or CSMA/CA for implementing shared medium access policies. Compared to TDMA the performance of CSMA/CA is much better in dynamic networks. But, in most of the cases, communications in WBANs are not dynamic. Hence, the TDMA approach is preferred in WBANs. However, one drawback of using TDMA-based MAC protocols in WBANs is that TDMA consumes extra energy for synchronization [61].

5.4 Modulation Techniques for MAC Protocols in WBAN

For the same reason as mentioned above, the simple and efficient modulation techniques such as Gaussian FSK (GFSK), Gaussian minimum shift keying (GMSK), differential phase-shift keying (DPSK), offset quadrature phase-shift keying (OQPSK), and phase silence shift keying (PSSK) are preferred in WBAN [63]. These modulation techniques are either bandwidth or power efficient or both.

5.5 Traffic Adaptive MAC Protocols

In WBANs, different sensors are used for various purposes. Some of them require periodical monitoring that results in low but consistent traffic load. Whereas, some sensors produce a huge amount of data suddenly and irregularly. That is why it is important for the MAC protocols for WBANs to be able to adjust the data traffic dynamically according to the varied traffic loads. And because of this quality, these MAC protocols are called traffic adaptive protocols. Surely, designing traffic adaptive MAC protocols are not straightforward.

Below, some of the properties are listed that are required to be considered for designing traffic adaptive MAC protocols [64]:

- **Traffic classification:** In WBAN, for better management, the data traffic is generally be classified into four categories viz. emergency traffic, critical traffic, reliability traffic, delay traffic, and normal traffic. The traffic classification is generally done based on delay and packet delivery ratio. Though, in practice, generally, this classification depends on the context of the traffic.

- **Traffic prioritization:** WBAN includes various implanted nodes with different purposes and priorities which generally access the shared communication path. Hence, it is essential for the MAC protocols to be flexible and dynamic to adjust traffic according to priority. The time slots for data communication are allocated dynamically as per the traffic priority.
  Prioritization of data traffic in WBAN is determined either statically or dynamically. In the first case, priority is set beforehand for each type of data. For example, the emergency traffic is assigned as the highest priority, the critical traffic may be assigned as a second highest priority whereas the normal traffic is assigned as the lowest priority [65]. In the second case, the traffic class value and the data generation rate of the particular WBAN application determines the prioritization of traffic [66].

- **Traffic load estimation:** It is important to carefully estimate the traffic load in a WBAN environment especially where the data traffic is variable. And this estimation process should be light, for a dynamic environment, in terms of the computation required for calculating the traffic load in. Based on the traffic load estimation the variable and heterogeneous traffic loads are adjusted dynamically. There are different approaches for estimating the traffic load in WBAN:
  a. The channel usage of the nodes is measured and communicated to the Body Coordinator (BC) periodically and based on that statistics the BC estimates the traffic load.
  b. The data generation rate is taken into account for each WBAN application, and accordingly, the traffic load is estimated.

Considering the above properties, the traffic adaptive protocols are classified into three categories [64]:

i. Traffic load estimation (TLE) based MAC protocols: The energy efficiency is achieved through attaining low delay by estimating the traffic load.

ii. Adaptive wake-up interval (AWI) based MAC protocols: The sensors remain sleeping by default. When there is a requirement of data transfer, they are awakened either autonomously or by some external entity, generally the coordinator node. This reduces the energy consumption.

iii. Adaptive time slot allocation (ATSA) based MAC protocols: These protocols manage the traffic load diversity dynamically. The traffic allocation is dynamically adjusted based on the traffic load.

6 WIRELESS TECHNOLOGIES FOR REMOTE HEALTH MONITORING

Several wireless technologies are present which are used for communication in remote health monitoring applications. Some of the popular wireless technologies are discussed in this section.

6.1 Bluetooth

Bluetooth [67,68] has been a very popular wireless communication technology for short-range. The Bluetooth protocol enables devices to communicate between them even if they are not in the line-of-sight. Multiple devices (the official claim is 7) can have physical connections simultaneously with a satisfactory data rate (up to 25 Mbps, with the help of WLAN, in Bluetooth 3.0 [69]). Bluetooth has been widely used in wireless health monitoring applications.

Despite the success and popularity, Bluetooth has few serious issues which make it incompetent to be the preferred communication technology for WBAN:

- The major problem is the power consumption. Bluetooth drains the battery very quickly.
- Suffers from frequent connection loss for which frequent pairing and re-pairing is required.
- Does not support automatic network formation. If the master moves away or goes down, the network will collapse.
- The number of simultaneous active connection is very small.
- Connection establishment process is slow.
- Suffers from several security flaws.

6.2 Bluetooth Low Energy (BLE)

As the name suggests, BLE [70,71] is the more energy efficient version of Bluetooth. BLE comes with the package Bluetooth 4.0 that also includes classic Bluetooth and high-speed Bluetooth. BLE addresses some of the drawbacks of classic Bluetooth. In addition to low power consumption, the other advantages of BLE in using WBAN are [72]:

- **Cost efficient:** To minimize cost, BLE technology is optimized for small battery-operated devices. For instance, in BLE, the number of channels is reduced to 40 2-MHz from Bluetooth's 79 1-MHz wide channels.

- **Robust transmission with minimized interference:** In medical applications, it is crucial to have a transmission technology that can provide a transmitting environment that is not affected by the "noisy" RF environments. Alike, Bluetooth, the use of adaptive frequency hopping (AFH) technology enables BLE to achieve this. Due to AFH, BLE also achieves low interference with other wireless transmissions.
- **Extended connection range:** Having a slightly different modulation, in comparison to classic Bluetooth, BLE (with a 10 dBm radio chipset) can cover a range up to 300 m.
- **Ease of use and integration:** In a master-slave arrangement, in Bluetooth, the slave communicates only after initiated by the master. In BLE, a slave can announce itself if it has something to transmit to the master. This is really important in case of an event that occurred at the slave end or if the health sensor (slave) has to send any physiological data to the controller (master).

6.3 ZigBee

ZigBee [73] is a wireless networking protocol suite that defines communication functionality within a personal area network (10−100 m). Zigbee communication relies on small and low-power digital radios. It is based on the IEEE 802.15.4 specification [73]. The ZigBee 3.0 [74] allows interoperability among different applications. Features and advantages of ZigBee are as following [75,76]:

- Short-range operation
- Significantly less power consuming
- Robust and reliable data transfer (its advanced routing feature provides self-healing routes between the communicating devices)
- Reliable data transfer (because it uses mesh topology that has no single point of failure)
- Secure
- Scalable
- Low-cost
- Easy to implement

6.4 WLAN

A wireless local area network (WLAN) [77,78] is a LAN that does not involve physical wired connections. It is standardized by the IEEE 802.11x. Combining WLAN with WBAN can give an effective local communication system for remote monitoring. The WLAN can be used to transmit the health data collected by WBAN to a local centralized server or the nearest internet access point. Since both are defined by different standards, they need to be bridged for working in conjunction [79−81]. Security is the major concern in WLAN because the wireless signal is broadcasted to every reachable and compatible devices.

6.5 Ultra-Wideband (UWB)

UWB [82,83], standardized by IEEE 802.15.3, is a data transmission method in a wireless network that offers, as the name suggests, a very large bandwidth. This high-rate throughput capability is really a great support to WBAN considering the number of wearables and sensors are used for remote monitoring these days. But the winning property of UWB that is applicable in WBAN is the significant power efficiency that it provides. The lower energy consumption increases the sensor operational period and life even transmitting more data.

There are two approaches to implement UWB: i) multi-band (MB) OFDM UWB, and ii) impulse radio (IR) UWB [84]. Characteristically, the second one is more suitable for WBAN because IR-UWB radios are generally of low-complexity by design and consume less power. The reason behind that is, to transmit and receive data, IR-UWB systems use pulses of very short duration (typically 2−3 ns) [85]. That is why they are preferable for the WBAN applications which are energy constrained and require short-range communication. In addition to that, the UWB physical layer is designed to provide WBAN robustness and the capability to implement high-performance operations [38].

UWB suffers from an important drawback. To achieve the short pulse width and power efficient signal transmission, the designing of the front-end circuitry of a UWB receiver is somewhat complex and also consume more power [57].

6.6 UWB-Based MAC Layer Protocols for WBAN

This section highlights some interesting UWB-based MAC layer protocols that can be used in WBAN. For extensive details, interested readers may refer to the book [57] that studies UWB WBAN in particular.

6.6.1 PSMA-based MAC

The PSMA-based MAC protocol is based on Preamble Sense Multiple Access (PSMA) [86] medium access mechanism where a preamble sequence is appended at the starting of a data packet sent by a sensor node to indicate a busy channel condition [57]. For WBAN with a large number of sensors, as the study [87] suggests, PSMA-based MAC protocol has an edge in terms of throughput and energy efficiency in comparison to slotted ALOHA based IEEE 802.15.4a standard.

But the PSMA-based protocols face a problem with the possibility of collision when multiple sensor nodes perform preamble sense simultaneously [57].

6.6.2 UWB²

The Uncoordinated Wireless Baseborn Access for UWB Networks ((UWB)²) [88,89] is a multi-channel MAC protocol. A multi-channel protocol divides the overall available resource such as timeslot (TDMA), a frequency band (FDMA) or a code (CDMA) to form separate communication channels. (UWB)² is based on CDMA. In comparison to usual multi-channel MAC protocols, it does not require Clear Channel Assessment (CCA) [90,91]. Instead, it uses orthogonal time-hopping (TH) CDMA [92,93] for achieving multiple access in a shared medium. It reduces energy consumption that makes THCDMA a suitable contender for WBAN applications. It is claimed that (UWB)² does not require complex mechanisms for interference control and supports a wide range of data rates (hundreds to thousands of bps) and a number of users (tens to hundreds) [88].

(UWB)² has a serious reliability loophole. In case of the loss of a Link Control (LC)³ frame, it does not have any provision for reinitializing the transmission [57]. Also, there are possibilities of collisions in TH CDMA. (UWB)² does not address this issue.

6.6.3 Multi-band UWB MAC

In a multi-band approach, the available UWB bandwidth is divided into multiple simultaneously usable bands. Multi-band systems provide more efficient use of the allocated spectrum. In a WBAN system, using multiband UWB MAC [94], the coordinator node can simultaneously communicate to multiple sensors through unique frequency bands allocated to each communication channel. This allows concurrent data transmissions without the possibility of collision. The capability of communicating concurrently results in low latencies and increased throughput, which are important for the WBAN applications demanding high data rate [57].

The considerable drawback of multi-band UWB MAC is that in creating the environment for concurrent communication makes the hardware design complex which may negate its application value in WBAN.

6.6.4 U-MAC

Rather than expending a fixed power level for data transmission, the U-MAC protocol [95] opts for an adaptable approach by adjusting the need for power for the transmission. The sensor node, on receiving

data packets from its neighbor, assesses the ranging information of the neighboring device [57]. Based on this information, it determines the power required to transmit data to that particular neighbor and adjust the transmit power levels accordingly. This dynamic adjustment of power expense saves unnecessary energy wastage.

The problem with this approach is that determining the transmitting power requirement of a sensor node and dynamically adjusting to that level involves heavy processing which is an additional overhead as well as more power consuming. In fact, this may overdo the gain of power saving achieved by the U-MAC protocol, especially in WBAN.

6.6.5 DCC-MAC

DCC-MAC [96] allows getting away with the restriction of mutual exclusion in wireless communication where no simultaneous communication is allowed within the same collision domain. It provides a communication environment among sensors where they can simultaneously communicate. This will obviously increase the chance of interference, but that can be mitigated [97,98]. But this mitigation comes at a cost as the physical layer operations become more complex and power consuming.

6.6.6 Transmit-only MAC

Though the above mentioned UWB-based MAC protocols have been successful in minimizing the power consumption, there are still more rooms to do that. In most of the cases, the functioning of the WBAN sensors is limited to data transmission only. The receiver part is redundant. But it consumes power in trying to detect incoming signals continuously. The transmit-only MAC protocol [99,100] supports the sensors which have transmit-only hardware in their design. This saves the power that was otherwise eaten up by the receiver part of the hardware.

The drawback of this protocol is that it is of asynchronous nature which throws the challenge to the receiving devices in terms of synchronization and collision avoidance. To overcome these issues, several remedies have been suggested [57,101].

6.7 Medical Radio Services

All the above mentioned wireless technologies are optimized for short-range communication and designed for general purpose sensor network applications. None of them is specifically designed for medical applications. This section mentions some radio services that are regulated specifically for medical monitoring and applications.

---

³LC message is sent by the coordinator in response to a Link Establishment (LE) request received from a sensor node.

6.7.1 Industrial, scientific and medical (ISM) radio bands

ISM [102] is the first special radio band that is aimed only for industrial, scientific, and medical purposes [103]. Internationally, a subset of radio spectrum has been set aside that is not meant to be used for telecommunications. 2.54 GHz is the commonly accepted band for ISM. The operations in ISM band is defined by the ITU Radio Regulations [104] for global uses, but the actual implementation of this band in different countries depends on the national radio regulations of the individual countries and regions [105]. Though initially, ISM was not meant for telecommunication, recently it is allowed for the same and widely used for short-range wireless communications, e.g., Wi-Fi, Bluetooth, RFID, etc. It, of course, causes interference but the communication technologies and the devices which use ISM band are responsible for mitigating it.

6.7.2 Medical implant communications service (MICS)

MICS [106] is an ultra-low power, short-range (2 m), high-data-rate, unlicensed, mobile radio service recommended by different telecommunication regulatory bodies for transmitting data in implanted medical devices [107]. The 402–405 MHz frequency band is allocated for MICS to operate. The same band is shared by the other primary users of this spectrum - Meteorological Aids Service (Medaids), the Meteorological Satellite Services, and the Earth Satellite Service [108].

Though MICS is designed for supporting diagnostic or therapeutic functions of the implanted medical devices, it is not absolutely free from interferences.

6.7.3 Wireless medical telemetry service (WMTS)

Wireless Medical Telemetry Service (WMTS) [109,110] is a set of specific spectrum bands that are used for remote health monitoring. WMTS was established by the Federal Communications Commission (FCC) [111] in 1999 to safeguard the use of licensed medical telemetry [112]. FCC has allocated dedicated bands in the range of 608–614, 1395–1400, and 1427–1432 MHz which can be used by the licensed physicians, healthcare facilities, and certain trained and supervised technicians. The allocation of the fixed bands allows interference-free operation of medical telemetry systems. The American Society for Healthcare Engineering (ASHE) [113] is in charge as the *Frequency Coordinator* for the WMTS bands and to ensure interference-free operation, it is mandatory for all transmitters operating in the WMTS bands to register with ASHE. In comparison to ISM band, WMTS has a much smaller broadcast arena.

7 CONCLUSION

The traditional healthcare systems are becoming seriously insufficient in the wake of increasing demand for health care. Innovations in ICT is assisting the healthcare sector to efficiently deliver advanced healthcare services. Telemedicine is being one of them has been successfully used to deliver healthcare services to remote locations. But it has its own limitations. Remote health monitoring has emerged as a suitable alternative for many traditional healthcare services. This enhances the quality of patient's life and reduces the number of hospital visits and rehospitalization and, in turn, the medical expenses. Patients can be monitored continuously by the medical staffs from anywhere. Patients and relatives can contact emergency healthcare services in an instant manner. WBAN has high potential to improve the performance of remote health monitoring system and can bring revolution in the healthcare sector as it integrates smart body sensors, latest communication devices, protocols and secured medical server. It significantly reduces the effort and error in diagnosis and prescribing. WBAN technology has significantly improved in recent times with more technological advancements and intelligent sensor integration. The high-end health sensors read different physiological parameters which are sent to medical staffs who take a decision based on those readings. WBAN data (in, on, or outside the human body) is collected, sent to an external medical server where it is processed and analyzed, and the outcome is used for treatment decision making. Since the physical layer and the MAC layer are combiningly responsible for improving WBAN operations in terms of power consumption, reliability, network efficiency, latency, resource utilization, operating cost, etc., they are the two most important part of any WBAN protocol. Because these two layers are Standards like IEEE 802.15.6 has been proposed to develop and standardize physical layer and MAC layer protocols for short-range, low power, highly efficient, and secured wireless communication schemes to operate on, inside, and off the human body. UWB provides high data rates in WSNs without consuming much power. Therefore, this is the preferred communication technology for WBAN. Considering the increasing usage of newer technologies such as the Internet of Nano Things (IoNT) in healthcare, it is expected to witness further advancement in efficient and power efficient WBAN communication technologies.

REFERENCES

[1] P. Campbell, Current Population Reports (Population Projections: States, 1995–2025, Census Bureau, United States, 2005.

[2] P.K.D. Pramanik, B.K. Upadhyaya, S. Pal, T. Pal, Internet of things, smart sensors, and pervasive systems: enabling the connected and pervasive health care, in: N. Dey, A. Ashour, S.J. Fong, C. Bhatt (Eds.), Healthcare Data Analytics and Management, Elsevier, 2018, pp. 1–58.

[3] P.K.D. Pramanik, S. Pal, M. Mukhopadhyay, Healthcare Big data: a comprehensive overview, in: N. Bouchemal (Ed.), Intelligent Systems for Healthcare Management and Delivery, IGI Global, 2018, pp. 72–100.

[4] M. Lont, D. Milosevic, A. v. Roermund, Wake-up Receiver Based Ultra-low-power WBAN, Springer, 2014, pp. 7–28.

[5] P.K.D. Pramanik, G. Pareek, A. Nayyar, Security and privacy in remote health care: issues, solutions and standards, in: D.J. Hemanth, V.E. Balas (Eds.), Telemedicine Technologies: Big Data, Deep Learning, Robotics, Mobile and Remote Applications for Global Healthcare, Elsevier, 2019.

[6] Wikipedia, Telemedicine, Wikipedia, August 12, 2017 [Online]. Available: https://en.wikipedia.org/wiki/Telemedicine.

[7] M. Rouse, "telemedicine," TechTarget, [Online]. Available: http://searchhealthit.techtarget.com/definition/telemedicine.

[8] M. Chen, S. Gonzalez, A. Vasilakos, H. Cao, V.C. Leung, Body area networks: a survey, Mobile Netw. Appl. 16 (2) (2011) 171–193.

[9] P.K.D. Pramanik, P. Choudhury, IoT data processing: the different archetypes and their security & privacy assessments, in: S.K. Shandilya, S.A. Chun, S. Shandilya, E. Weippl (Eds.), Internet of Things (IoT) Security: Fundamentals, Techniques and Applications, River Publishers, 2018, pp. 37–54.

[10] M. Maksimović, V. Vujović, B. Perišić, Do it yourself solution of internet of things healthcare system: measuring body parameters and environmental parameters affecting health, J. Inf. Syst. Eng. Manag. (2016) 25–39.

[11] R. Negraa, I. Jemilia, A. Belghitha, Wireless body area networks: applications and technologies, in: The Second International Workshop on Recent Advances on Machine-to-Machine Communications, 2016.

[12] K. Rojahn, S. Laplante, J. Sloand, C. Main, A. Ibrahim, J. Wild, N. Sturt, T. Areteou, K.I. Johnson, Remote monitoring of chronic diseases: a landscape assessment of policies in four European countries, PLoS One 11 (5) (2016).

[13] M.D. Warren, The need for rehabilitation, in: Rehabilitation Today, Springer, 1977, pp. 6–10.

[14] R.C. Senelick, Why Rehabilitation Is Important, March 6, 2012 [Online]. Available: https://blogs.webmd.com/chronic-conditions/2012/03/why-rehabilitation-is-important.html.

[15] S. Kumar, Advantages of Remote Patient Monitoring in Healthcare Systems, September 8, 2016 [Online]. Available: https://healthable.org/advantages-of-remote-patient-monitoring-in-healthcare-systems/.

[16] B. Noah, M.S. Keller, S. Mosadeghi, L. Stein, S. Johl, S. Delshad, V.C. Tashjian, D. Lew, J.T. Kwan, A. Jusufagic, B.M.R. Spiegel, Impact of remote patient monitoring on clinical outcomes: an updated meta-analysis of randomized controlled trials, npj Digit. Med. 1 (2018).

[17] H. Bogdanova, Pros and Cons of Remote Patient Monitoring, February 7, 2018 [Online]. Available: https://www.healthitoutcomes.com/doc/pros-and-cons-of-remote-patient-monitoring-0001.

[18] L. Piwek, D.A. Ellis, S. Andrews, A. Joinson, The rise of consumer health wearables: promises and barriers, PLoS Med. 13 (2) (2016).

[19] C.R. Doarn, S. Pruitt, J. Jacobs, Y. Harris, D.M. Bott, W. Riley, C. Lamer, A.L. Oliver, Federal efforts to define and advance telehealth—a work in progress, Telemed. J. e Health 20 (5) (2014) 409–418.

[20] IEEE 802.15 WPAN™ Task Group 6 (TG6), Body Area Network, IEEE, June 9, 2011 [Online]. Available: http://www.ieee802.org/15/pub/TG6.html.

[21] T. Penzel, B. Kemp, G. Klosch, A. Schlögl, J. Hasan, A. Varri, I. Korhonen, Acquisition of biomedical signals databases, IEEE Eng. Med. Biol. Mag. 20 (3) (2001) 25–32.

[22] H.-B. Li, K.Y. Yazdandoost, B. Zhen, Wireless Body Area Network, River Publishers, 2010.

[23] P.K.D. Pramanik, S. Pal, A. Brahmachari, P. Choudhury, Processing IoT data: from cloud to fog. It's time to be down-to-Earth, in: Applications of Security, Mobile, Analytic and Cloud (SMAC) Technologies for Effective Information Processing and Management, IGI Global, 2018, pp. 124–148.

[24] R. DeVaul, M. Sung, J. Gips, A. Pentland, MIThril 2003: applications and architecture, in: *7th IEEE International Symposium on Wearable Computers*, NY, USA, 2003.

[25] P. Maiti, S. Addya, B. Sahoo, A. Turuk, Energy efficient wireless body area network (WBAN), in: Handbook of Research on Advanced Wireless Sensor Network Applications, Protocols, and Architectures, IGI Global, 2017, pp. 413–432.

[26] J.Y. Khan, M.R. Yuce, Wireless body area network (WBAN) for medical applications, in: D. Campolo (Ed.), New Developments in Biomedical Engineering, IntechOpen, 2010.

[27] S. Movassaghi, M. Abolhasan, J. Lipman, D. Smith, A. Jamalipour, Wireless body area networks: a survey, IEEE Commun. Surv. Tutorials 16 (3) (2014) 1658–1686.

[28] WHO, 10 Facts on Noncommunicable Diseases, [Online]. Available: http://www.who.int/features/factfiles/noncommunicable_diseases/en/.

[29] S. Warren, J. Lebak, J. Yao, J. Creekmore, A. Milenkovic, E. Jovanov, Interoperability and security in wireless body area network infrastructures infrastructures, in: Annual International Conference of the IEEE Engineering in Medicine and Biology Society, 2005.

[30] P.K.D. Pramanik, S. Pal, P. Choudhury, Beyond automation: the cognitive IoT. Artificial intelligence brings sense to the internet of things, in: A.K. Sangaiah, A. Thangavelu, V.M. Sundaram (Eds.), Cognitive Computing for Big Data Systems Over IoT: Frameworks, Tools and Application, Springer, 2018, pp. 1–37.

[31] C.A. Otto, E. Jovanov, A. Milenkovic, A WBAN-based system for health monitoring at home, in: 3rd IEEE-EMBS International Summer School and Symposium on Medical Devices and Biosensors, MIT, Boston, USA, 2006.

[32] A.-L. Vuorinen, J. Leppanen, H. Kaijanranta, M. Kulju, T. Heliö, M. v. Gils, J. Lähteenmäki, Use of home telemonitoring to support multidisciplinary care of heart failure patients in Finland: randomized controlled trial, J. Med. Internet Res. 16 (12) (2014).

[33] K. Wac, D. Konstantas, A.V. Halteren, R. Bults, I. Widya, N. Dokovsky, G. Koprinkov, V. Jones, Herzog, Rainer, Mobile patient monitoring: the mobihealth system, J. Inf. Technol. Healthcare 2 (5) (2004) 365–373.

[34] H. Alemdar, C. Ersoy, Wireless sensor networks for healthcare: a survey, Comput. Network. 54 (15) (2010) 2688–2710.

[35] M. Bagot, P. Launay, F. Guidec, A flexible architecture for mobile health monitoring, in: IEEE 18th International Conference on e-Health Networking, Applications and Services, Healthcom, 2016.

[36] R.K. Pathinarupothi, M.V. Ramesh, E. Rangan, Multi-layer architectures for remote health monitoring, in: IEEE 18th International Conference on e-Health Networking, Applications and Services, Healthcom, 2016.

[37] IEEE Standard Association, Part 15.4: Low-Rate Wireless Personal Area Networks (LR-WPANs), IEEE Computer Society, New York, USA, 2011.

[38] M. Li, M. Zhuang, An overview of physical layers on wireless body area network, in: International Conference on Anti-Counterfeiting, Security and Identification, ASID, Taiwan, 2012.

[39] A. Singla, S. Malik, Review on "A priority adaptive routing to optimize WBAN", Int. J. Enhanced Res. Sci. Technol. Eng. 4 (6) (2015) 616–620.

[40] A. Sangwan, P.P. Bhattcharya, A study on various issues in different layers of WBAN, Int. J. Comput. Appl. 129 (11) (2015).

[41] L. Filipe, F. Fdez-Riverola, N. Costa, A. Pereira, Wireless body area networks for healthcare applications: protocol stack review, Int. J. Distributed Sens. Netw. 11 (10) (2015).

[42] Rudiyanto, R.F. Sari, Analysis of the effect of beacon order and Superframe Order value to the performance of multi-hop wireless networks on IEEE 802.15.4 protocol, in: *International Conference on Advanced Computer Science and Information Systems (ICACSIS)*, Depok, Indonesia, 2012.

[43] S. Ullah, M. Mohaisen, M.A. Alnuem, A review of IEEE 802.15.6 MAC, PHY, and security specifications, Int. J. Distributed Sens. Netw. 9 (4) (2013).

[44] I.F. Akyildiz, W. Su, Y. Sankarasubramaniam, E. Cayirci, Wireless sensor networks: a survey, Comput. Network. 38 (4) (2002) 393–422.

[45] J. Mirkovic, G.P. Venkataramani, S. Lu, L. Zhang, A self-organizing approach to data forwarding in large-scale sensor networks, in: *IEEE International Conference on Communications (ICC'01)*, Helsinki, Finland, 2001.

[46] C. Shen, C. Srisathapomphat, C. Jaikaeo, Sensor information networking architecture and applications, IEEE Pers. Commun. (August 2001) 52–59.

[47] A.J.D. Rathnayaka, V.M. Potdar, Wireless sensor network transport protocol: a critical review, J. Netw. Comput. Appl. 36 (1) (2013) 134–146.

[48] W. Su, Ö.B. Akan, Communication protocols for sensor networks, in: Wireless Sensor Networks, Kluwer Academic Publishers Norwell, MA, USA, 2004, pp. 21–50.

[49] A.W. Astrin, H.B. Li, R. Kohno, Standardization for body area networks, IEICE Trans. Commun. 92 (2) (2009) 366–372.

[50] S. Saleem, S. Ullah, K.S. Kwak, A study of IEEE 802.15.4 security framework for wireless body area networks, Sensors 11 (2) (2011) 1383–1395.

[51] S.A. Salehi, M.A. Razzaque, I. Tomeo-Reyes, N. Hussain, IEEE 802.15.6 standard in wireless body area networks from a healthcare point of view, in: *22nd Asia-Pacific Conference on Communications* (APCC), 2016.

[52] K.S. Kwak, S. Ullah, N. Ullah, An overview of IEEE 802.15.6 standard, in: *3rd International Symposium on Applied Sciences in Biomedical and Communication Technologies* (ISABEL), 2010.

[53] IEEE Standards, in: Part 15.4: Wireless Medium Access Control (MAC) and Physical Layer (PHY) Specifications for Low-Rate Wireless Personal Area Networks (LR-WPANS): Amendment to Add Alternate Phy (Amendment of IEEE Std 802.15.4), IEEE, 2014.

[54] E. Karapistoli, F.-N. Pavlidou, I. Gragopoulos, I. Tsetsinas, An overview of the IEEE 802.15. 4a standard, IEEE Commun. Mag. 48 (1) (2010) 47–53.

[55] K. Takizawa, H.-B. Li, K. Hamaguchi, R. Kohno, Wireless patient monitoring using IEEE802.15.4a WPAN, in: *International Conference on Ultra-Wideband*, Singapore, 2007.

[56] K.S. Kwak, S. Ullah, N. Ullah, An overview of IEEE 802.15.6 standard, in: *3rd International Symposium on in Applied Sciences in Biomedical and Communication Technologies* (ISABEL), 2010.

[57] T.K.M. Silva, R. Jean-Michel, Y.M. Rasit, Ultra Wideband Wireless Body Area Networks, Springer, 2014.

[58] F. Ullah, A.H. Abdullah, O. Kaiwartya, S. Kumar, M.M. Arshad, Medium access control (MAC) for wireless body area network (WBAN): superframe structure, multiple access technique, taxonomy, and challenges, Human-centric Comput. Inf. Sci. 7 (34) (2017).

[59] S. Bhandari, S. Moh, A priority-based adaptive MAC protocol for wireless body area networks, Sensors 16 (401) (2016).

[60] H. Li, J. Tan, Body sensor network based context aware QRS detection, in: Pervasive Health Conference and Workshops, 2006.

[61] A. Rahim, N. Javaid, M. Aslam, Z. Rahman, U. Qasim, Z.A. Khan, A comprehensive survey of MAC protocols for wireless body area networks, in: *7th International Conference on Broadband, Wireless Computing, Communication and Applications (BWCCA)*, Victoria, Canada, 2012.

[62] S. Marinković, E. Popovici, C. Spagnol, S. Faul, W. Marnane, Energy-efficient low duty cycle MAC protocol for wireless body area networks, IEEE Trans. Inf. Technol. Biomed. 13 (6) (2009) 915–925.

[63] C. Lee, J. Kim, H.S. Lee, J. Kim, Physical layer designs for WBAN systems in IEEE 802.15.6 proposals, in: *9th International Symposium on Communications and Information Technology (ISCIT)*, Icheon, South Korea, 2009.

[64] F. Masud, A.H. Abdullah, G. Abdul-Salaam, F. Ullah, Traffic adaptive MAC protocols in wireless body area networks, Wireless Commun. Mobile Comput. 2017 (2017).

[65] M.M. Monowar, M.M. Hassan, F. Bajaber, M. Al-Hussein, A. Alamri, McMAC: towards a MAC protocol with multi-constrained constrained QoS provisioning for diverse traffic in Wireless Body Area Networks, Sensors 12 (11) (2012) 15599–15627.

[66] I. Anjum, N. Alam, M.A. Razzaque, M.M. Hassan, A. Alamri, Traffic priority and load adaptive MAC protocol for QoS provisioning in body sensor networks, Int. J. Distributed Sens. Netw. 9 (3) (2013) 1–9.

[67] J. Haartsen, The Bluetooth radio system, IEEE Pers. Commun. 7 (1) (2000) 28–36.

[68] B.A. Miller, C. Bisdikian, Bluetooth Revealed: The Insider's Guide to an Open Specification for Global Wireless Communication, Prentice Hall, 2001.

[69] A. Bit, M. Orehek, W. Zia, Comparative analysis of Bluetooth 3.0 with UWB and certified wireless-USB protocols, in: IEEE International Conference on Ultra-Wideband (ICUWB2010), Nanjing, China, 2010.

[70] E. Mackensen, M. Lai, T.M. Wendt, Bluetooth Low Energy (BLE) based wireless sensors, in: IEEE SENSORS, Taipei, Taiwan, 2012.

[71] R. Heydon, Bluetooth Low Energy: The Developer's Handbook, Prentice Hall, 2012.

[72] A. Gupta, I. Mohammed, The Basics of Bluetooth Low Energy (BLE), October 14, 2016 [Online]. Available: https://www.edn.com/5G/4442859/The-basics-of-Bluetooth-Low-Energy–BLE–.

[73] P. Baronti, P. Pillai, V.W.C. Chook, S. Chessa, A. Gotta, Y. Fun Hu, Wireless sensor networks: a survey on the state of the art and the 802.15.4 and ZigBee standards, Comput. Commun. 30 (7) (2007) 1655–1695.

[74] What's New in ZigBee 3.0, Texas Instruments, February 22, 2018 [Online]. Available: http://processors.wiki.ti.com/index.php/What%27s_New_in_ZigBee_3.0.

[75] P. Kinney, ZigBee technology: wireless control that simply works, Home Toys (October 2003).

[76] ZigBee 3.0 – Facilitating the Internet of Things, NXP, 2016.

[77] S. Banerji, R.S. Chowdhury, On IEEE 802.11: wireless LAN technology, Int. J. Mobile Netw. Commun. Telemat. 3 (4) (2013).

[78] R.A. Dayem, Mobile Data & Wireless Lan Technologies, Prentice Hall, 1997.

[79] R. d. Francisco, L. Huang, G. Dolmans, Coexistence of WBAN and WLAN in medical environments, in: IEEE 70th Vehicular Technology Conference, 2009.

[80] S. Rashwand, J. Mišić, Two-tier WBAN/WLAN healthcare networks; priority considerations, in: *IEEE Global Communications Conference* (GLOBECOM), Anaheim, CA, USA, 2012.

[81] N. Bradai, L.C. Fourati, L. Kamoun, WBAN data scheduling and aggregation under WBAN/WLAN healthcare network, Ad Hoc Netw. 25 (2015) 251–262.

[82] Ultra-Wideband (UWB) Technology: Enabling High-Speed Wireless Personal Area Networks, Intel, 2004.

[83] J.H. Reed, Introduction to ultra wideband communication systems, in: An Introduction to Ultra Wideband Communication Systems, Prentice Hall, 2005.

[84] J.R. Fernandes, D. Wentzloff, Recent advances in IR-UWB transceivers: an overview, in: *IEEE International Symposium on Circuits and Systems*, Paris, France, 2010.

[85] M. Singh, A. Ranganathan, S. Capkun, IR-UWB Based Indoor Localization System, Microsoft, 2016.

[86] J. Haapola, L. Goratti, I. Suliman, A. Rabbachin, Preamble sense multiple access (PSMA) for impulse radio ultra wideband sensor networks, in: *6th International Conference on Embedded Computer Systems: Architectures, Modeling, and Simulation*, Samos, Greece, 2006.

[87] L. Kynsijärvi, L. Goratti, R. Tesi, J. Iinatti, M. Hämäläinen, Design and performance of contention based MAC protocols in WBAN for medical ICT using IR-UWB, in: *21st International Symposium on Personal, Indoor and Mobile Radio Communications Workshops*, Instanbul, Turkey, 2010.

[88] M.D. Benedetto, L.D. Nardis, M. Junk, G. Giancola, (UWB)2: uncoordinated, wireless, Baseborn medium access for UWB communication networks, Mobile Netw. Appl. 10 (5) (2005) 663–674.

[89] M.-G. Benedetto, L. Nardis, G. Giancola, D. Domenicali, The aloha access (UWB)2 protocol revisited for IEEE 802.15.4a, ST J. Res. 4 (1) (2007) 131–141.

[90] P. Vrancken, IEEE 802.11 Medium Access Control (MAC), Wireless Communication, 2004 [Online]. Available: www.wirelesscommunication.nl/reference/chaptr01/wrlslans/80211_page2.htm.

[91] I. Ramachandran, S. Roy, Clear Channel assessment in energy-constrained wideband wireless networks, IEEE Wirel. Commun. 14 (3) (2007) 70–78.

[92] L.-L. Yang, Time-hopping multicarrier code-division multiple access, IEEE Trans. Veh. Technol. 56 (2) (2007) 731–741.

[93] C.Y. Jung, J.W. Chong, Y.J. Hong, B.C. Jung, D.K. Sung, Orthogonal time hopping multiple access for UWB impulse radio communications, in: *Asia-Pacific Conference on Communications*, Perth, Australia, 2005.

[94] I. Broustis, S. Krishnamurthy, M. Faloutsos, M. Molle, J. Foerster, Multiband media access control in impulse-based UWB Ad Hoc networks, IEEE Trans. Mob. Comput. 6 (4) (2007) 351–366.

[95] R. Jurdak, P. Baldi, C.V. Lopes, U-MAC: a proactive and adaptive UWB medium access control protocol, Wireless Commun. Mobile Comput. 5 (5) (2005) 551–566.

[96] J.-Y.L. Boudec, R. Merz, B. Radunovic, J. Widmer, DCC-MAC: a decentralized MAC protocol for 802.15.4a-like

UWB mobile ad-hoc networks based on dynamic channel coding, in: *First International Conference on Broadband Networks (BROADNETS'04)*, San Jose, CA, USA, 2004.

[97] F. Cuomo, C. Martello, A. Baiocchi, C. Fabrizio, Radio resource sharing for ad hoc networking with UWB, IEEE J. Sel. Area. Commun. 20 (9) (2002) 1722−1732.

[98] S. Kolenchery, J. Townsend, J. Freebersyser, A novel impulse radio network for tactical military wireless communications, in: *IEEE Military Communications Conference* (MILCOM'98), 1998.

[99] H. Keong, M. Yuce, Analysis of a multi-access scheme and asynchronous transmit-only UWB for wireless body area networks, in: *31st Annual International Conference of the IEEE Engineering in Medicine and Biology Society* (EMBC'09), 2009.

[100] H. Keong, K. Thotahewa, M. Yuce, Transmit-only ultra wide band body sensors and collision analysis, IEEE Sens. 13 (5) (2013) 1949−1958.

[101] K.M.S. Thotahewa, J.Y. Khan, M.R. Yuce, Power efficient ultra wide band based wireless body area networks with narrowband feedback path, IEEE Trans. Mob. Comput. 13 (8) (2014) 1829−1842.

[102] R. Singh, Significance and utility of industrial, scientific and medical band, AKGEC Int. J. Technol. 8 (2) (2017) 53−60.

[103] S. Ungureanu, V. Sontea, V. Vidiborschii, C. Lepadatu, N. Sipitco, D. Fosa, Using of ISM radio bands for wireless charging of medical implants, in: *9th International Conference on Microelectronics and Computer Science*, Chisinau, Republic of Moldova, 2017.

[104] Radio Regulations Articles, ITU, 2012.

[105] ERC Report 83, Interference from Industrial, Scientific and Medical (ISM) Machines, European Radiocommunications Committee (ERC), Luxembourg, 2000.

[106] H.S. Savci, A. Sula, Z. Wang, N.S. Dogan, E. Arvas, MICS transceivers: regulatory standards and applications, in: IEEE SoutheastCon, FL, USA, 2005.

[107] M. NoorIslama, M.R. Yuce, Review of medical implant communication system (MICS) band and network, ICT Express 2 (4) (2016) 188−194.

[108] T. Gee, Medical Implant Communications Service Tutorial, March 3, 2006 [Online]. Available: https://medicalconnectivity.com/2006/03/03/medical-implant-communications-service-tutorial/.

[109] Wireless Medical Telemetry Systems, U.S. Food and Drug Administration, March 8, 2018 [Online]. Available: https://www.fda.gov/MedicalDevices/DigitalHealth/WirelessMedicalDevices/ucm364308.htm.

[110] Wireless Medical Telemetry Service (WMTS), Federal Communications Commission, March 8, 2017 [Online]. Available: https://www.fcc.gov/wireless/bureau-divisions/broadband-division/wireless-medical-telemetry-service-wmts.

[111] Federal Communications Commission, 2018 [Online]. Available: https://www.fcc.gov/.

[112] J. Temple, P. Sherman, Wireless medical telemetry service is changing, not going away, Biomed. Instrum. Technol. 50 (s6) (2016) 52−53.

[113] ASHE: Optimizing Health Care Facilities, The American Society for Healthcare Engineering of the American Hospital Association, [Online]. Available: http://www.ashe.org/.

CHAPTER 8

Remote Monitoring of Children With Chronic Illness Using Wearable Vest

R. JANSI • R. AMUTHA • S. RADHA

Department of Electronics and Communication Engineering, SSN College of Engineering, Chennai, India

1 INTRODUCTION

In recent years, the number of children suffering from chronic illness is increasing at a steady rate. The common conditions involving such illness include obesity, epilepsy asthma, mental illness and neurodevelopmental disorders [1]. It was shown in Ref. [2] that children with epilepsy have a high probability of dying compared to children without epilepsy. Obesity and overweight conditions were estimated to cause around 3.4 million deaths in 2010 [3]. In general, chronic illness is a permanent illness that cannot be completely cured. However continuous monitoring is very important for such children to ensure immediate treatment in case of emergency conditions. Immediate identification of abnormal conditions can facilitate doctors to give efficient treatment on time. Action monitoring plays an important role because certain activities may lead to serious abnormalities for such children. It is also not possible for parents to be with the child all the time. Hence, a smart system for such a monitoring is required. A number of smart action monitoring systems are being developed recently. Some of them include smartphone-based, wearable sensor-based etc. Smart phone-based action recognition systems [4,5] utilize the sensors that are embedded in smartphones like accelerometers, gyroscopes, orientation sensor, magnetometer, GPS etc. These sensors can be used to identify the actions performed by the subjects. A system was designed to identify the change in state for bipolar disorder patients [6]. However, the main drawback of such systems is that the smartphone must be carried by the subjects all the time. Although the level of the feasibility of such a system is high in the case of adults, it is less for children. Hence such a system cannot be used. Wearable sensor-based systems [7–11] involves sensors that are worn by individuals. These systems are more popular due to the large-scale manufacture of miniaturized

sensors and development of wireless technologies. Wearable sensor-based systems are being used popularly in real-world applications like rehabilitation, physical activity monitoring, sports activity monitoring, assisted living etc.

In this paper, we present the design of a smart vest for remote monitoring of children with chronic illness. Using the sensor data from the sensors that are embedded in the smart vest, children are being continuously monitored. The identification of activities performed by the children is performed inside the Raspberry Pi module. This module processes the raw sensor data which is gathered using a Zigbee transceiver. This data is first segmented and then the features are extracted and the classification is performed. The values of the pulse oximeter sensor, temperature sensor and the activity being classified are then transmitted to the cloud. These are then displayed to the parents/caretakers using a smart mobile application.

The remainder of the article is organized as follows. We present a few related works present in the literature in Section 2. In Section 3, we describe our system, that includes data acquisition, transmission, feature extraction and classification. Section 4 presents the experiments conducted. Section 5 shows the simulation results in the developed mobile app. Finally, the conclusion is shown in Section 6.

2 RELATED WORKS

Recently, many health care systems are being developed continuously to guarantee the wellbeing and also to improve the standard of living of individuals. Efforts are being taken to design systems that produce high-quality health services with reduced cost. A system for recognizing movements of children suffering from Autism spectrum disorders was proposed [12]. In this

Telemedicine Technologies. https://doi.org/10.1016/B978-0-12-816948-3.00008-8

work, accelerometers were utilized to collect the movement-based data from the wrist. The collected data was transferred to the computer where the processing was done. Various features were extracted from the collected data that included root mean square, standard deviation, peak and valleys. Using these features, the classification was done to distinguish a stereotyped gesture from a normal gesture. A survey on context recognition systems for healthcare applications was presented [13]. Continuous health monitoring involves continuous sampling and monitoring of sensor data. This results in heavy energy requirement in implementing these systems. To address this issue, this paper did a detailed survey on various energy-efficient approaches that are existing in this domain. In addition, the authors also proposed a new classification for such mechanisms that are related to the health-based applications.

A health monitoring vest for monitoring the elderly people was developed [14]. The designed vest was embedded with various biomedical sensors for monitoring parameters like Electrocardiogram, Galvanic skin response, temperature and position of the body. These signals were then transferred to the microcontroller using UART radio frequency protocol. This data was then processed by the microcontroller and was sent to the server using a 3G/4G connection. Finally, the data was displayed using GUI on a website. A detailed survey on various smart home-based health monitoring systems has also been published [15]. This paper highlighted the current technology, along with the challenges and requirements involved in the development of systems for monitoring the health. This paper gave an overview of three types of sensor systems namely personal sensor networks, body sensor networks and multimedia devices. The complete overview of context-aware health environment was also presented in this paper. An adaptive monitoring system was presented [16]. This system was capable of learning the behavior of the monitored person by continuously monitoring their daily activities. A novel approach to forecasting risky behavior was built using a Gray model. The Markovian model was used to evaluate its performance. This system was capable of accurately evaluating the dependency of a person, estimate health condition and was also able to detect abnormal scenarios.

The main concern in healthcare applications is its implementation. It is tedious to design systems that can be used in real-time. To tackle this problem, smart wearable systems are rapidly emerging in recent years. A smart wearable system for monitoring the chronic illness of children was presented [17]. This system included a smart algorithm that could automatically detect if any parameter exceeded a predefined threshold so that immediate actions could be taken. Parameters like heart rate and body temperature were measured. In addition, the authors also developed a secure mechanism to ensure secure blue-tooth connectivity. It was also capable of performing data fusion to increase the performance of the protocol. A smart vest system was designed and presented [18] for the safety and well-being of children. This system was built using Lilypad Arduino and Adafruit Flora platforms. The various sensors used were GPS, temperature and accelerometer sensor. This vest was used to collect information regarding the location using the GPS module. If the child moved across the restricted area, alert signals would be automatically sent to the teachers. The collected data was transferred to the sensor bus and then to the sensor web enablement services that included a sensor observation service, sensor planning service and sensor alert service.

A wearable vest was developed for monitoring postures in real-time [19]. In this model, the data acquisition was done from the accelerometer sensor and wireless data transfer was performed using a Blue-tooth module. The identification of the tilting angle was performed using low complexity CORDIC-based algorithm. This system also involved the detection of abnormal postures and generation of warning signals if they exist continuously for an extended period of time. Wearable sensor-based gait cycle recognition system was proposed [20]. In this paper, abnormality in the walking pattern was predicted using analysis of the center of force of a human. A mapping system based on pressure was used to collect the force and pressure values during the gait cycle. A new segmentation approach was also proposed in this paper that involved regression model-based modeling of segments. The transition between various segments was modeled using logistic functions. Since logistic functions were used to model the transitions, this system could detect both the abrupt and smooth transitions. A detailed review on wearable devices and multimodal interface for human action recognition were given [21]. This paper discussed various forms of wearable sensors that can be used. A detailed description of the wearable smart shirt was also presented. The basic requirements of developing a wearable system were also listed and described. It includes aesthetics, size, water tolerance, power consumption, wireless communication and the operating system. The complete architecture of the wearable system involving the power supply, wireless transmission, inertial sensors, output display and processor were also discussed in detail.

Wearable sensors were used for assessing the tremor activity in the patients with Parkinson's disease [22]. Here, accelerometers were placed on different parts of the patient's body. Hidden Markov model was used to estimate the type of tremor, its severity level and the type of posture. Five basic postures were considered for classification namely, rest in bed, rest in a chair, standing up, standing up with extended hands and moving hands. A smart vest was designed and presented for rehabilitation application [23]. This system was used to monitor various physiological postures while performing rehabilitation exercises. An inductive sensor was embedded in the designed vest. The designed vest was capable of varying its impedance value based on its physical deformation. The proposed wearable system was designed such that the patient's posture could be continuously monitored and any incorrect posture could be indicated in the form of feedback. An Internet of Things (IoT) based fall detection system was proposed using energy efficient wearable sensor node [24]. A tiny, flexible, energy efficient device for fall detection was proposed. Different factors that impact the energy consumption of wearable sensor devices such as sampling rate, transmission rate, transmission protocol etc., were investigated. A descriptive analysis of the energy consumption of these devices in different operating conditions and various configurations was also presented.

A remote physiological multi-parameter monitoring system was presented [25]. A wearable physiological monitoring system in the form of a washable shirt was developed. This system included sensors for monitoring the physiological parameters, data acquisition using a wearable unit, hardware for processing the data and a remote monitoring system. These signals were acquired at a sampling rate of 250 samples/s. The sampled signal was digitized using a resolution of 12 bit. This digitized signal was then transmitted to the monitoring station. Along with the digitized signal, the geological location of the user was also transmitted. This location was obtained using GPS modules that were interfaced with the wearable system. A wearable shirt-based system for activity and health monitoring was presented [26]. ECG data and physical activity data were acquired and transmitted using an Ad-hoc network in IEEE 802.15.4 communication standard. The artifact noise from the fabric electrodes was removed using adaptive filtering. The shirt was designed such that the wearable sensor node was at the chest of the subject who wears the vest. The main steps involved in data processing were the acquisition of data from electrodes, differential AC coupling, amplification using AC-coupled instrumentation amplifier, filtering using low/high pass filter, final gain stage amplification and analog to digital conversion using Microcontroller A/D converter. The current status and future challenges of the smart wearable systems were discussed in detail [27]. This paper gave a detailed analysis of the assessable parameters, the location of the sensing system, sensor network and smart biomedical wearable system. The assessable parameters discussed in this paper included when, where and how this system could be used. It further included a descriptive analysis of the various diseases and disabilities that can be monitored using this system. Location analysis involved the various ways in which this system could be worn by the subject. This included accessory type, implantable, portable, furniture etc. The sensor network analysis included both a wireless body area network and personal area network. The current status and ongoing research in the field of smart biomedical wearable systems were also included in this paper.

In the past few decades, in addition to communication, smartphones are widely being used in various healthcare monitoring applications due to its accessibility and ubiquitous nature. A human action recognition system using deep learning was developed by utilizing data from smartphone sensors [28]. Using the accelerometer and gyroscope sensor of smartphones, tri-axial angular velocity and linear acceleration signals were first collected using a sampling frequency of 50 Hz. These signals were then pre-processed using median and low-pass Butter-worth filter to reduce noise levels. Then features such as mean, median, correlation, signal magnitude area, AR coefficients etc were extracted from the filtered signal. The dimension of the extracted features was reduced using PCA. Finally, the action recognition was performed using Deep Belief Network. A system for monitoring health parameters using smartphones has been proposed [29]. An extensive comparison of various health-based applications using smartphones and using traditional medical devices was performed in this paper. Three categories of smartphone sensors namely, kinematics, camera and microphone sensors and their corresponding applications were discussed in this paper. Azimuth, Pitch and Roll angles of the kinematic sensors were utilized to compute actimetric data. The applications of Kinematic sensors included actimetric monitoring, fall detection and posture monitoring. Cameras were utilized in applications like blood pressure evaluation, heart rate extraction and eye examination. Microphones were used for Spirometric diagnosis, nose-related symptoms recognition and cough analysis.

Various classification techniques are continuously being proposed for classifying different activities performed by individuals. A technique for action recognition in body sensor networks using string matching technique was developed [30]. Here motion transcripts are constructed from biomedical signals using motion primitives. These primitives are then labeled using unique symbols to form a symbol sequence called as motion template. Thus, in this paper action recognition was done using a distributed algorithm based on edit distance. The main advantage of this method was the reduction in the number of active notes after every classification decision. A novel approach for recognizing walking and gait patterns using adaptive Bayesian inference system was presented [31]. This system uses a sequential analysis method based on probabilistic formulation. This system makes use of the decisions made by the inference system over a period of time. Three types of walking activities were considered namely, level walking, ramp ascent and ramp descent. Data for this recognition was collected from inertial measurement units (IMU) that were placed at three locations namely, the thigh, the shank and the foot. This system utilized angular velocity signals generated by the IMU. These signals were collected at a sampling rate of 100 Hz. The collected signals were filtered using a second-order Butterworth filter and were then used for classification. A phonological approach for construction of spatial decision trees was proposed for physical movement monitoring [32]. To reduce the number of nodes for classification, a decision tree (DT) model was used. The quality of action recognition in each node was used as a criterion while constructing the tree. Similar activities were grouped using clustering. The important set of nodes that contribute to action recognition was then identified by solving the optimization problem.

Sparse representation is popularly being exploited nowadays as a powerful tool in solving various classification-based applications. Its applications include human action recognition, face recognition, person recognition etc. A sparse representation-based system for human action recognition was proposed using the concept of group sparse coding [33]. The geometry of action components was explored using structured sparse coefficients. Sparse coding was used to represent the video using the histogram of coefficients. Hence, discriminative rich feature representations were performed to reduce the quantization noise of the system. Further, the computational complexity of this system was also low. An action recognition system based on pose using sparse representation has

also been proposed [34]. In this work, human actions were treated to be composed of numerous body poses over time. These poses were represented using human skeleton coordinates. Here, the skeletal data acquired using depth cameras were utilized for pose representations. The dissimilarity space in the sparse-based representation is utilized to recognize the activities. A technique for human action recognition using a sparsity-inducing shared dictionary for video sequences was presented [35]. A new classification metric called Class Selection Index was introduced for action classification using the ratio of correlation coefficient to Euclidean distance.

Sparse representation has also been used for face recognition [36]. This system involved learning class-specific dictionaries for each class and also a projection matrix. This system can be used for predicting faces from corrupted test data. Here, a projection matrix is constructed between original training data and the class-specific dictionaries. Then, the Eigenface method is used to extract features of the class-specific dictionaries and the corrected test data. Finally, the classification is performed using sparse representation-based classification. Sparse dictionary learning based systems were utilized in face recognition [37]. Here a discriminative dictionary was developed from the training data. Here a novel dictionary learning technique was proposed that used sparse embedded learning process to improve the classification capability of the learned dictionaries. This dictionary learning technique was designed such that the orthogonality of the projection matrix was preserved. Person recognition using discriminative sparse projections (DSP) has also been proposed [38]. Here, binary body mask was first extracted using background subtraction. The extracted mask was then projected to a low dimensional space using DSP and simultaneous clustering was done. Finally, orthogonal ensemble metric learning technique was utilized to obtain the distance metric that was used for person identification.

The main motivation of our work was to develop an algorithm that could accurately recognize activities performed by children. Since it is not possible to monitor the activities performed by children all the time manually, there is a necessity to accurately classify their activities using machine learning algorithms. Furthermore, energy consuming actions like running, if performed continuously for a long duration may result in fatigue conditions that may cause severe consequences especially for children with chronic illness. Though there are a number of smart wearable vest based-systems in the literature, the main contribution of this work lies in the proposal and implementation of a new sparse representation-based algorithm in the designed smart vest, that produces a very high accuracy of about 97.49%. We have exploited a new technique wherein, shared dictionary created from different action classes is coupled with a k-NN classifier for classification and monitoring of children activities. In addition to this, other physiological parameters like oxygen saturation level, heart rate and body temperature are being continuously acquired and transferred to the cloud for continuous monitoring of the children by the parents/caretakers.

The main contributions of this paper are twofold:

- A new design for a textile-based wearable smart vest to monitor children suffering from chronic illness has been presented.
- We have proposed a novel sparse representation-based classification algorithm for classifying daily activities performed by children.

3 PROPOSED SYSTEM

The complete block diagram of the proposed system is shown in Fig. 8.1. Data acquisition is done using the sensors embedded in the designed vest that is worn by the child. The embedded sensors include an accelerometer, pulse oximeter and a temperature sensor. The acquired data is continuously transmitted using XBee transmitter module embedded in the garment. The transmitted data is received by the XBee receiver that is connected to the Raspberry Pi module. Raspberry Pi is a credit card sized computer that can be programmed using the Raspbian software. This module was utilized for classifying activities using the received data. Three main steps are performed in the Raspberry Pi module, namely, feature extraction, dictionary generation and classification. The classified activity label and the values of oxygen saturation level, heartbeat and body temperature are then transferred and stored in the cloud. These data are then retrieved from the cloud and displayed

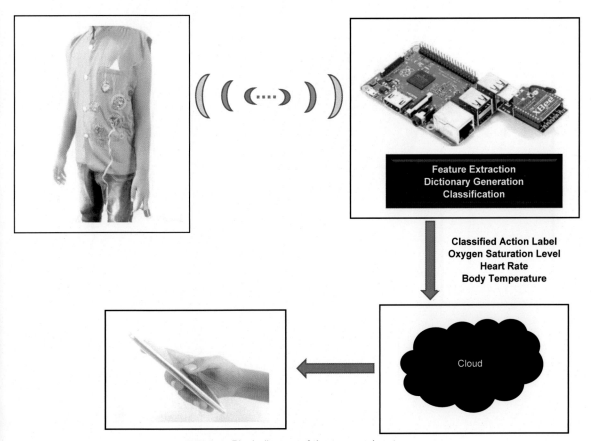

FIG. 8.1 Block diagram of the proposed system.

to the parents/caretakers using a customized mobile application for real-time monitoring.

3.1 System Design

The architecture of the designed vest is shown in Fig. 8.2. In this design, three sensors namely, the pulse oximeter, the accelerometer and the temperature sensor were interfaced to the Arduino Lilypad main board. Arduino Lilypad is based on ATmega 168V. Lilypad accelerometer is a three-axis accelerometer based on ADXL335 MEMS device. This device is designed especially for textile-based applications. It has three output leads namely x, y and z. Each of these is connected to the *a0*, *a1* and *a2* pin of the Lilypad Arduino board respectively. Lilypad temperature sensor has MCP9700 thermistor that can be used to sense the body temperature. The output lead of this module was connected to the *a3* pin of Lilypad Arduino board. The pulse

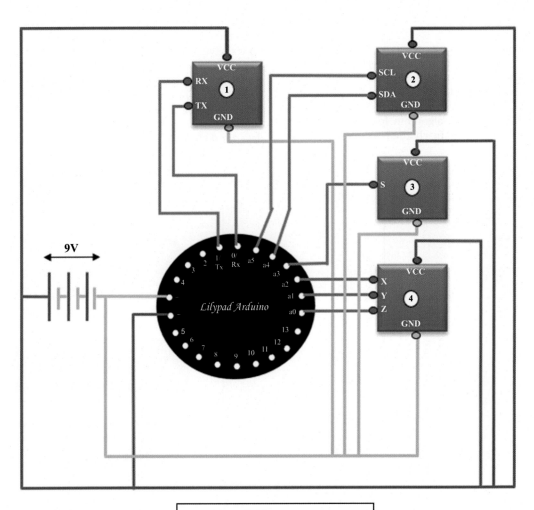

1 - Lilypad X-bee

2 - Pulse Oximeter

3 - Lilypad Temperature Sensor

4 - Lilypad Accelerometer sensor

FIG. 8.2 Architecture of the designed vest.

oximeter sensor used was MAX30100. It has 2 LEDs, optimized optics, analog signal processing unit and a photodetector module. This module was used to measure the oxygen saturation level of blood and heartbeat. It has two output leads namely SDA and SCL which were connected to the *a4* and *a5* of the Lilypad Arduino board. Data transmission was accomplished using XBee transceiver. Lilypad XBee board was used to interface the XBee module into the fabric. The transmitter and receiver pins of the Lilypad XBee were connected to the receiver and transmitter pins of the Lilypad main board respectively. These sensors were stitched to the fabric of the vest using conductive thread. Conductive thread is a thin, strong and smooth material that is made from stainless steel fiber. The power supply for the entire setup was provided using three Lithium coin batteries each of 3 V that were connected in series to provide a cumulative voltage of 9 V. These coin batters were sewed to the fabric using Lilypad coin battery holder module. The snapshot of the designed vest is shown in Fig. 8.3.

3.2 Data Acquisition and Transmission

The designed vest is worn by the child such that, the accelerometer is around the waist. The pulse oximeter is worn around the finger of the child using an adhesive strap. Pulse oximetry is an indirect method which is used to measure the blood oxygen level. In this sensor, the oxygen level is calculated by measuring the skin's absorbance of wavelength of the red and IR light. This sensor can also be used for measuring the heart rate

since the quantity of blood flowing through blood vessels varies in accordance with the rate of blood pumped from the heart. Thus, pulse oximeter can also be used to measure the heart rate without the need for connecting any ECG electrodes. The adhesive strap is a Velcro tape which is needed to adhere the pulse oximeter to the finger and also to prevent external radiation. It does not provide any interference to the child. Also, the temperature sensor is stitched in the vest such that it is under the ankle to sense the body temperature. These sensors enable real-time monitoring of the children. The data acquisition and transmission unit is depicted in Fig. 8.4. Lilypad accelerometer is a three-axis accelerometer which was used to monitor various activities performed by the child. The accelerometer provides the values of acceleration along three axis directions. Data from the accelerometer was collected for a duration of 20 s at a sampling rate of 20 Hz. Thus the size of the acquired data was 400. The dataset for this system was collected from 15 children with age group ranging from 8 to 12. These children were asked to perform 6 different kinds of daily activities. These activities included walk, run, jump, hop sit and stand. Pulse oximeter provided the values of oxygen saturation level (in %) and heart rate (in bpm). The normal range of pulse oximeter is 95–100%. The values of heart rate for normal condition ranges from 70 to 100 bpm. Any deviation from this normal range indicates an abnormality. Lilypad temperature sensor was used to monitor the body temperature level of the child. The acquired data is transmitted using the XBee transmitter.

FIG. 8.3 Snapshot of the designed vest.

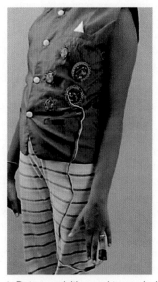

FIG. 8.4 Data acquisition and transmission unit.

FIG. 8.5 Data reception and processing unit.

XBee modules are used for long-range wireless communication and are based on IEEE 802.15.4 standard. It is a suite of high-level communication protocols used in creating personal area networks with small, low-power digital radios and is designed for applications which need a wireless connection. They can approximately cover a range of about 1600 m in line of sight and about 90 m in indoor or urban regions. The data reception and processing unit are shown in Fig. 8.5. The transmitted data is received using the XBee receiver module that is attached to Raspberry Pi. The Raspberry Pi module processes the received data and identifies the activity performed by the child using the proposed classification algorithm. The classified activity along with the values of oxygen saturation level, heart rate and body temperature are then transferred to the cloud.

3.3 Feature Extraction

The received dataset was segmented with a window size of 40 samples (2s) with 50% overlapping. Thus, the total number of segmented data collected from each subject was 19. Since 15 subjects were used for data collection, the total number of training samples were 285 per class. The segmented data was then used for feature extraction that involved extraction of features belonging to time and frequency domain [39]. The time domain features extracted were mean along each axis, standard deviation along each axis, Euclidean norm of mean vector, Euclidean norm of standard deviation, correlation among each pair of axis and signal magnitude area. To extract the frequency domain features the fast Fourier transform was applied to the raw accelerometer data. From the obtained coefficients, the energy and entropy of each band were computed. Thus, 14 features were extracted from each segmented data. Thereby, the size of the feature matrix extracted for each class was 14×285.

3.3.1 Time domain features

- The mean vector for the three axes is calculated using

$$\overline{a_{cc_i}} = \frac{1}{N} \sum_{t=1}^{N} a_{cc_i}(t) \tag{8.1}$$

where i represents the axes ($i = x,y,z$) and N is the size of the window.
- The standard deviation of each axial component is calculated as follows

$$\sigma_i = \sqrt{\frac{1}{N-1} \sum_{t=1}^{N} \left[a_{cc_i}(t) - \overline{a_{cc_i}} \right]^2} \tag{8.2}$$

- Euclidean norm of mean vector is computed for the three axes using

$$\left\| \overline{a_{cc}} \right\| = \sqrt{\left(a_{cc_x} \right)^2 + \left(a_{cc_y} \right)^2 + \left(a_{cc_z} \right)^2} \tag{8.3}$$

- Euclidean norm of standard deviation is computed using

$$\| \sigma \| = \sqrt{(\sigma_x)^2 + (\sigma_y)^2 + (\sigma_z)^2} \tag{8.4}$$

- The correlation helps in identifying the direction that produces maximum variation. The correlation between axis x and axis y is calculated as follows

$$\rho_{xy} = \frac{1}{N-1} \sum_{t=1}^{N} \frac{\left[a_{cc_x}(t) - \overline{a_{cc_x}} \right]}{\sigma_x} \frac{\left[a_{acc_y}(t) - \overline{a_{cc_y}} \right]}{\sigma_y} \tag{8.5}$$

Similarly, correlation is computed between the other two pair of axes.
- The signal magnitude area can be calculated as follow

$$SMA = \sum_{t=1}^{N} \left[a_{cc_x}(t) + a_{cc_y}(t) + a_{cc_z}(t) \right] \tag{8.6}$$

3.3.2 Frequency domain features

A fast Fourier transform is used to compute the discrete Fourier transform and its inverse. It converts time domain signal to frequency domain and vice versa. Let x_n represent time domain signal, then its coefficients are calculated using

$$X_k = \sum_{n=0}^{N-1} x_n e^{-j\frac{2\pi}{N}kn}, \quad k = 0, 1, ..., N-1. \tag{8.7}$$

- The spectral energy of the signal is calculated as

$$E = \sum_{k=0}^{N-1} |X_k|^2 \qquad (8.8)$$

- The spectral entropy of the signal is given by

$$S = - \sum_{f=-\frac{f_n}{2}}^{f=+\frac{f_n}{2}} \rho_n(f) \log_2[\rho_n(f)] \qquad (8.9)$$

Here, ρ_n refers to normalized power spectral density. It is computed as

$$\rho_n(f) = \frac{\rho(f)}{\sum_{f=-\frac{f_n}{2}}^{f=+\frac{f_n}{2}} \rho(f)} \qquad (8.10)$$

where ρ is the power spectral density which is calculated as

$$\rho(f) = \sum_{k \in f} |X_k|^2 \qquad (8.11)$$

3.4 Online Dictionary Learning (ODL) Algorithm

ODL is a computationally efficient algorithm for learning dictionaries [40]. It is an online version of the k-SVD algorithm [41]. Let us consider a finite set of training features to be represented by $Y = \{y_i\}_{i=1}^N$. This algorithm alternates between two steps namely sparse coding and dictionary update steps. For sparse coding, the sparse vector is computed using LARS-Lasso algorithm [42,43].

$$\alpha_t \triangleq \underset{\alpha}{\arg\min} \frac{1}{2}\|y_t - D_{t-1}\alpha\|_2^2 + \lambda\|\alpha\|_1 \qquad (8.12)$$

where λ is the regularization parameter.
The dictionary update is done using

$$\underset{D \in C}{\arg\min} \frac{1}{t} \sum_{i=1}^{t} \frac{1}{2}\|y_i - D_{t-1}, \alpha_i\|_2^2 + \lambda\|\alpha_i\|_1 \qquad (8.13)$$

where C denotes the convex set of matrices. The constraint of C is given as,

$$C \triangleq \{D \in R^{m \times K} s.t. \forall_i = 1,...,k \quad d_i^T d_i \leq 1\} \qquad (8.14)$$

3.5 k-nearest Neighbor (k-NN) Algorithm

k-NN [44] is an instance-based classification technique that is computationally efficient and also produces good classification results. This algorithm determines the k nearest neighbors of the test data and utilizes a simple majority voting principle to identify the class label of the test data. This algorithm uses Euclidean distance as the distance metric for identifying its k nearest neighbors. Considering two vectors $X = (x_1, x_2, ... x_m)$ and $Y = (y_1, y_2, ...y_m)$, their Euclidean distance is computed as

$$d = \sqrt{\sum_{i=1}^{m} (x_i - y_i)^2} \qquad (8.15)$$

3.6 Proposed Classification Algorithm

In this proposed scheme, a single shared dictionary is first created from the features of all the classes using the ODL algorithm. Then the features belonging to each class along with the shared dictionary is used to form a sparse coefficient matrix pertaining to each class. The dispersion of non-zero coefficients in the sparse co-efficient matrix completely depends on the distribution of the training features in the shared dictionary. Thus, these sparse coefficient matrices act as discriminative secondary features in classifying the test activity. The test sparse coefficient vector for the test feature vector is also computed using the shared dictionary. This vector is then classified with k-NN algorithm using the sparse coefficient matrices generated using training features. The creation of a single shared dictionary using features of all action classes instead of individual dictionaries for each class aids in decreasing the storage requirements.

The explanation of the proposed algorithm is as follows. Let $Y_i \in R^{m \times n}$ represent the feature matrix comprising of features of class i. Here, where m is the dimension of the feature vector and n is the number of training samples in each class. In the first step, the features extracted from all the action classes are used to create a single concatenated feature matrix (Y). In the second step, ODL dictionary learning algorithm is utilized to create a single shared dictionary (D_s) from the concatenated feature matrix (Y) created from Step 1. In the third step, sparse coefficient matrices (α_i) corresponding to each class is computed using the individual feature matrix (Y_i) of each class and the shared dictionary (D_s) created in Step 2. In the fourth step, the test feature vector (y_{test}) to be classified is used along with the shared dictionary (D_s) to compute the test sparse coefficient vector (α_{test}). In the final step, the test class label (i) is identified by utilizing the k-NN algorithm. Here, the test sparse coefficient vector (α_{test}) created in Step 4 is used as a testing sample and class-specific sparse coefficient matrices (α_i) computed in Step 3 are used as training samples.

The steps for the proposed classification algorithm are as follows.

Step 1: Form a concatenated feature matrix by concatenating the features from all the action classes.

$$Y = [Y_1|Y_2|...|Y_C] \in R^{m \times N} \quad \text{or} \quad Y = \{\gamma_i\}_{i=1}^{N} \qquad (8.16)$$

where $N = nC$ and C refers to the total number of classes.

Step 2: Generate a single shared dictionary $D_s \in R^{m \times K}$ with size $K < N$ using the features from all the classes (Y) using ODL algorithm.

Step 3: For each feature matrix Y_i belonging to class i, find its corresponding sparse coefficient matrix $\alpha_i \in R^{K \times n}$ over shared dictionary D_s using LARS-Lasso algorithm.

Step 4: Using the test feature vector $\gamma_{test} \in R^m$ and shared dictionary D_s, find the corresponding test sparse coefficient vector $\alpha_{test} \in R^{K \times 1}$ using LARS-Lasso.

Step 5: By using the sparse coefficient matrices $\alpha_i, i \in 1, 2, ..., C$ as secondary training features, the test sparse coefficient vector α_{test} is classified based on majority voting technique using k-NN algorithm.

4 EXPERIMENTS

4.1 Parameter Settings

The value of regularization parameter λ was set as 0.1. The number of iterations in the ODL algorithm for training the dictionaries was set to 40. The neighborhood size k in the k-NN algorithm was fixed to be 1. Prior to training the size of the concatenated feature matrix created from features of all the classes was 1710. To identify the appropriate dictionary size, the dictionaries were trained with various values of K ranging from 10 to 100 with a step size of 10. The overall accuracy was determined using each of the dictionaries. The variation of overall accuracy with respect to different dictionary size is shown in Fig. 8.6. From Fig. 8.6, it is clear that the accuracy of the system increases as the size of the dictionary increases from 10 to 50 and achieves the highest accuracy of 97.49% at $K = 50$. This is due to the fact that as the size of dictionary increases, the dictionary accommodates more atoms to efficiently represent all the action classes. Beyond a particular size, accuracy starts to decrease. This is because of the over-clustering issue. Hence in our experiments we have set $K = 50$.

4.2 Results and Discussions

To validate and compare the performance of the proposed classifier we have also used k-NN, DT and sparse representation based classifier employed in Ref. [45]. In this paper, classification is done based on l_1 − norm of the sparse coefficient vector obtained using test feature vector and a concatenated dictionary formed by appending the features from all the action classes. Let tp, tn, fp and fn indicate the number of true positives, true negatives, false positives and false negatives produced by the classifier. These values are utilized in the calculation of the below-mentioned performance evaluation metrics.

Overall accuracy (O) is a metric that portrays the overall effectiveness of a classifier.

$$O = \frac{tp + tn}{tp + tn + fp + fn} \qquad (8.17)$$

Recall (R) or sensitivity indicates the ability of a classifier to predict the positive instances.

$$R = \frac{tp}{tp + fn} \qquad (8.18)$$

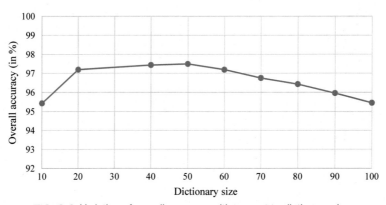

FIG. 8.6 Variation of overall accuracy with respect to dictionary size.

The precision (*P*) indicates the agreement of actual class labels with the positive labels predicted by the classifier.

$$P = \frac{tp}{tp + fp} \qquad (8.19)$$

Specificity (*S*) indicates the ability of a classifier to predict the negative instances.

$$S = \frac{tn}{tn + fp} \qquad (8.20)$$

F-score (*F*) is the harmonic average of precision and recall.

$$F = 2 \times \frac{P \times R}{P + R} \qquad (8.21)$$

The overall accuracy of the various classifiers is shown in Table 8.1. From Table 8.1, it is evident that the proposed classifier produces excellent result since it is 97.49% which is high compared to 87.02%, 94.33% and 95.73% produced by DT, k-NN, sparse [45] respectively.

The confusion matrices obtained using various classifiers namely DT, k-NN, sparse [45] and proposed classifier are shown in Tables 8.2–8.5 respectively.

Table 8.2 indicates the confusion matrix obtained for DT algorithm. From Table 8.2, we see that major misclassification occurs for critical activities like run and jump. Hence the DT algorithm cannot be utilized for our application. Table 8.3 indicates the confusion matrix obtained using the k-NN algorithm. From Table 8.3, we find that stationary activities like sit and stand gets misclassified among themselves. Thus, the overall performance of this classifier is low. Table 8.4 shows the confusion matrix obtained using sparse representation algorithm used in Ref. [45]. From Table 8.4, we again find that the major misclassification is for the action jump. Since actions like jump must be monitored accurately for children with chronic illness, this classifier also cannot be reliably used. Table 8.5 shows the confusion matrix obtained using the proposed classifier. From Table 8.5, it is evident that most of these misclassifications are drastically reduced in the proposed classifier. Hence, we infer that the proposed classifier outperforms other techniques.

The performance of various classifiers in terms of sensitivity, specificity, precision and F-score for different classifiers are shown in Tables 8.6–8.9. From Table 8.6 we observe that the sensitivity of DT classifier is very low for activities like jump and run. In particular, it produces a sensitivity of about 77.89% and 73.33% for run and jump respectively. Children performing these type of higher energy consuming activities for a prolonged duration must be monitored accurately to avoid exhaustive situations. Hence this makes DT classifier unfit for our application.

From Table 8.7, we find that the sensitivity and F-score of the sit activity is less than 90%. In addition, the precision of stand is also less than 90%. That is, it produces precision of about 88.77%. Thus, the overall performance of the k-NN classifier is not too good to be reliably used for our application.

TABLE 8.1
Comparison of Overall Accuracy.

Classifier	Overall Accuracy (in %)
DT	87.02
k-NN	94.33
Sparse [45]	95.73
Proposed	97.49

TABLE 8.2
Confusion Matrix Obtained Using DT.

Activities	Walk	Run	Jump	Hop	Sit	Stand
Walk	278	0	0	4	0	3
Run	44	222	0	19	0	0
Jump	4	20	209	16	2	34
Hop	16	0	0	269	0	0
Sit	0	0	0	4	239	42
Stand	3	1	0	4	6	271

TABLE 8.3
Confusion Matrix Obtained Using k-NN.

Activities	Walk	Run	Jump	Hop	Sit	Stand
Walk	279	0	0	2	0	4
Run	3	272	0	10	0	0
Jump	1	6	270	4	2	2
Hop	7	0	0	278	0	0
Sit	3	0	2	0	253	27
Stand	0	0	0	0	24	261

TABLE 8.4
Confusion Matrix Obtained Using Sparse [45].

Activities	Walk	Run	Jump	Hop	Sit	Stand
Walk	284	0	0	0	0	1
Run	1	265	0	19	0	0
Jump	3	3	257	6	0	16
Hop	13	0	0	270	0	2
Sit	0	0	0	0	276	9
Stand	0	0	0	0	0	285

TABLE 8.5
Confusion Matrix Obtained Using Proposed Classifier.

Activities	Walk	Run	Jump	Hop	Sit	Stand
Walk	284	1	0	0	0	0
Run	4	270	0	8	2	1
Jump	0	0	283	2	0	0
Hop	13	4	0	268	0	0
Sit	0	0	0	0	280	5
Stand	0	0	0	1	2	282

TABLE 8.6
Performance Metrics for DT Classifier.

Activities	Recall/Sensitivity	Specificity	Precision	F-score
Walk	97.54	95.29	80.57	88.25
Run	77.89	98.52	91.35	84.09
Jump	73.33	100.00	100.00	84.61
Hop	94.38	96.70	85.12	89.51
Sit	83.85	99.43	96.76	89.84
Stand	95.08	94.45	77.42	85.35

TABLE 8.7
Performance Metrics for k-NN Classifier.

Activities	Recall/Sensitivity	Specificity	Precision	F-score
Walk	97.89	99.01	95.22	96.53
Run	95.43	99.57	97.84	96.62
Jump	94.73	99.85	99.26	96.94
Hop	97.54	98.87	94.55	96.02
Sit	88.77	98.17	90.68	89.71
Stand	91.57	97.68	88.77	90.15

TABLE 8.8
Performance Metrics for Sparse [45].

Activities	Recall/Sensitivity	Specificity	Precision	F-score
Walk	98.24	98.94	94.91	96.55
Run	94.38	99.29	96.41	95.39
Jump	90.87	99.71	98.47	94.52
Hop	94.03	98.52	92.73	93.37
Sit	97.54	98.59	93.28	95.36
Stand	98.24	99.57	97.90	98.07

TABLE 8.9
Performance Metrics for Proposed Classifier.

Activities	Recall/Sensitivity	Specificity	Precision	F-score
Walk	99.64	98.80	94.35	96.92
Run	94.73	99.64	98.18	96.42
Jump	99.29	100.00	100.00	99.64
Hop	94.03	99.22	96.05	95.03
Sit	98.24	99.71	98.59	98.41
Stand	98.94	99.57	97.91	98.42

From Table 8.8, we infer that classifier used in Ref. [45] produces good results for almost all the activities. However, when comparing with Table 8.9, we find that the proposed system produces more accurate results. For instance, let us consider two crucial activities namely run and jump. The sensitivity of classifying activity jump is raised from 90.87% to 99.29% using the proposed classification algorithm. Similarly, F-score is raised from 94.52% to 99.64%. Also, the proposed

system produces excellent specificity and precision of 100% for this activity. For activity run, precision is raised from 96.41% to 98.18%. Also, there is an increase in F-score of around 1.03%. These aspects make the proposed classifier as our choice for classification.

Figs. 8.7–8.10 shows the graphical comparison of different classifiers in terms of recall, specificity, precision and F-score respectively. From Fig. 8.7, it can be clearly observed that for the critical activity jump, the

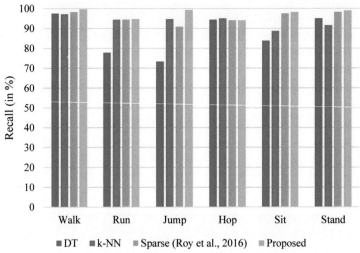

FIG. 8.7 Graphical comparison in terms of recall.

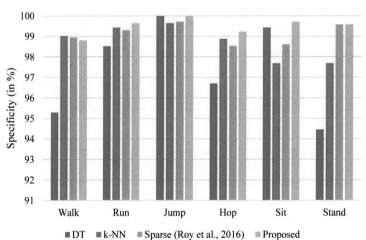

FIG. 8.8 Graphical comparison in terms of specificity.

performance of the proposed classifier is very high compared to other classifiers. In particular, it is 21.96%, 4.56% and 8.41% greater than that of DT, k-NN and sparse [45] respectively.

Fig. 8.8 shows the comparison of various classifiers in terms of specificity. Here, we find that for almost all the activities the performance of the proposed classifier is higher than that of other classifiers. For instance, when considering the action hop, the proposed classifier produces an accuracy of about 99.22% which is 2.52%, 0.35% and 0.7% higher than that of DT, k-NN and sparse [45] respectively.

Fig. 8.9 pictures the graphical comparison of various classifiers in terms of precision. For the action jump, the proposed system produces excellent precision of 100%.

Although the DT classifier also produces the same precision value for the action jump, its performance is low for other activities.

Fig. 8.10 depicts the graphical comparison of various classifiers in terms of F-score. From the graph, it is obviously seen that the proposed system produces exceedingly high performance in terms of F-score for all the activities.

5 CHILD MONITORING USING MOBILE APP DISPLAY

The data transferred to the cloud by the Raspberry Pi unit are retrieved and displayed in a customized MQTT app. The MQTT protocol is recently widely being

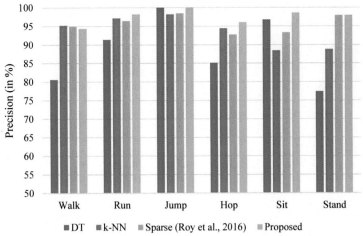

FIG. 8.9 Graphical comparison in terms of Precision.

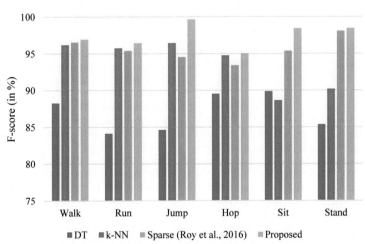

FIG. 8.10 Graphical comparison in terms of F-score.

used in IoT. It is a message-based Machine to Machine protocol. This protocol is user-friendly and can be easily customized. This protocol is supported by many cloud platforms for transmitting and receiving data from smart devices. This app can be accessed by the parents/caretakers. Thus, real-time monitoring of the children is feasible. The dashboard of the mobile app is shown in Fig. 8.11. It can be observed that the app displays temperature in degree Celsius, heart rate in beats/minute, oxygen saturation level in percentage and also the activity performed by the child. In this way, the parents/caretakers can continuously monitor the health conditions of their children, so that immediate actions can be taken in case of an emergency.

6 CONCLUSION

Monitoring of children with chronic illness is a crucial task. In this paper, we have presented a design model of wearable vest that can be used to monitor such children in real-time. This model ensures timely medical treatment of such children in case of emergency. Parameters like oxygen saturation level, heartbeat, body temperature and the activity performed by the child are constantly indicated to the parents/caretakers in the form of a mobile app. Thus, the children can be monitored at any time. The designed vest is suitable for both indoor and outdoor usage. High accuracy action recognition was achieved by exploiting a new shared dictionary with k-NN classifier for classification

FIG. 8.11 Dashboard of the customized mobile app.

of activities. Various performance evaluation metrics like sensitivity, specificity, precision and F-score were used to evaluate the performance of the proposed classifier.

REFERENCES

[1] J.M. Perrin, L.E. Anderson, J. Van Cleave, The rise in chronic conditions among infants, children, and youth can be met with continued health system innovations, Health Aff. 33 (12) (2014) 2099–2105.

[2] C.S. Camfield, P.R. Camfield, P.J. Veugelers, Death in children with epilepsy: a population-based study, Lancet 359 (9321) (2002) 1891–1895.

[3] M. Ng, et al., Global, regional, and national prevalence of overweight and obesity in children and adults during 1980–2013: a systematic analysis for the Global Burden of Disease Study 2013, Lancet 384 (9945) (2014) 766–781.

[4] Y.-S. Lee, S.-B. Cho, Activity recognition using hierarchical hidden Markov models on a smartphone with 3D accelerometer, Hybrid Artif. Intell. Syst. (2011) 460–467.

[5] S. Dernbach, B. Das, N.C. Krishnan, B.L. Thomas, D.J. Cook, Simple and complex activity recognition through smart phones, in: 2012 Eighth International Conference on Intelligent Environments, 2012.

[6] A. Grünerbl, et al., Smart-phone based recognition of states and state changes in bipolar disorder patients, IEEE J. Biomed. Health Inform. 1 (19) (2015) 140–148.

[7] L. Chen, J. Hoey, C.D. Nugent, D.J. Cook, Z. Yu, Sensor-based activity recognition, IEEE Trans. Syst. Man Cybern. C Appl. Rev. 42 (6) (2012) 790–808.

[8] R. Jansi, R. Amutha, A novel chaotic map based compressive classification scheme for human activity recognition using a tri-axial accelerometer, Multimed. Tool. Appl. 77 (23) (2018) 31261–31280.

[9] R. Jansi, R. Amutha, Sparse representation based classification scheme for human activity recognition using smartphones, Multimed. Tool. Appl. (2018) 1–19.

[10] Y. Liu, L. Nie, L. Liu, D.S. Rosenblum, From action to activity: sensor-based activity recognition, Neurocomputing 181 (2016) 108–115.

[11] A.Y. Yang, R. Jafari, S.S. Sastry, R. Bajcsy, Distributed recognition of human actions using wearable motion sensor networks, J. Ambient Intell. Smart Environ. 1 (2) (2009) 103–115.

[12] J.L. Rodrigues, N. Gonçalves, S. Costa, F. Soares, Stereotyped movement recognition in children with ASD, Sensors Actuat. A Phys. 202 (2013) 162–169.

[13] T. Rault, A. Bouabdallah, Y. Challal, F. Marin, A survey of energy-efficient context recognition systems using wearable sensors for healthcare applications, Pervasive Mob. Comput. 37 (2017) 23–44.

[14] H.M. Adam, P.J. Soh, Development of a health monitoring vest for the elderly, in: 2016 IEEE MTT-S International Conference on Numerical Electromagnetic and Multiphysics Modeling and Optimization, NEMO, 2016.

[15] H. Mshali, T. Lemlouma, M. Moloney, D. Magoni, A survey on health monitoring systems for health smart homes, Int. J. Ind. Ergon. 66 (2018) 26–56.

[16] H. Mshali, T. Lemlouma, D. Magoni, Adaptive monitoring system for e-health smart homes, Pervasive Mob. Comput. 43 (2018) 1–19.

[17] S. Sendra, L. Parra, J. Lloret, J. Tomás, Smart system for children's chronic illness monitoring, Inf. Fusion 40 (2018) 76–86.

[18] M. Jutila, H. Rivas, P. Karhula, S. Pantsar-Syväniemi, Implementation of a wearable sensor vest for the safety and well-being of children, Procedia Comput. Sci. 32 (2014) 888–893.

[19] W.Y. Lin, M.Y. Lee, W.C. Chou, The design and development of a wearable posture monitoring vest, in: Digest of Technical Papers – IEEE International Conference on Consumer Electronics, 2014.

[20] S. Mohammed, A. Samé, L. Oukhellou, K. Kong, W. Huo, Y. Amirat, Recognition of gait cycle phases using wearable sensors, Robot. Autonom. Syst. 75 (2016) 50–59.

[21] P. Kumari, L. Mathew, P. Syal, Increasing trend of wearables and multimodal interface for human activity monitoring: a review, Biosens. Bioelectron. 90 (2017) 298–307.

[22] G. Rigas, et al., Assessment of tremor activity in the Parkinsons disease using a set of wearable sensors, IEEE Trans. Inf. Technol. Biomed. 16 (3) (2012) 478–487.

[23] E. Sardini, M. Serpelloni, M. Ometto, Smart vest for posture monitoring in rehabilitation exercises, in: 2012 IEEE Sensors Applications Symposium, SAS 2012 - Proceedings, 2012.

[24] T. Nguyen Gia, et al., Energy efficient wearable sensor node for IoT-based fall detection systems, Microprocess. Microsyst. 56 (2018) 34–46.

[25] P.S. Pandian, et al., Smart Vest: wearable multi-parameter remote physiological monitoring system, Med. Eng. Phys. 30 (4) (2008) 466−477.

[26] Y.D. Lee, W.Y. Chung, Wireless sensor network based wearable smart shirt for ubiquitous health and activity monitoring, Sensor Actuat. B Chem. 140 (2) (2009) 390−395.

[27] M. Chan, D. Estève, J.-Y. Fourniols, C. Escriba, E. Campo, Smart wearable systems: current status and future challenges, Artif. Intell. Med. 56 (3) (2012) 137−156.

[28] M.M. Hassan, M.Z. Uddin, A. Mohamed, A. Almogren, A robust human activity recognition system using smartphone sensors and deep learning, Future Gener. Comput. Syst. 81 (2018) 307−313.

[29] F. Lamonaca, G. Polimeni, K. Barbé, D. Grimaldi, Health parameters monitoring by smartphone for quality of life improvement, Measurement 73 (2015) 82−94.

[30] H. Ghasemzadeh, V. Loseu, R. Jafari, Structural action recognition in body sensor networks: distributed classification based on string matching, IEEE Trans. Inf. Technol. Biomed. 14 (2) (2010) 425−435.

[31] U. Martinez-Hernandez, A.A. Dehghani-Sanij, Adaptive Bayesian inference system for recognition of walking activities and prediction of gait events using wearable sensors, Neural Network. 102 (2018) 107−119.

[32] H. Ghasemzadeh, R. Jafari, Physical movement monitoring using body sensor networks: a phonological approach to construct spatial decision trees, IEEE Trans. Ind. Informa 7 (1) (2011) 66−77.

[33] F. Moayedi, Z. Azimifar, R. Boostani, Structured sparse representation for human action recognition, Neurocomputing 161 (2015) 38−46.

[34] I. Theodorakopoulos, D. Kastaniotis, G. Economou, S. Fotopoulos, Pose-based human action recognition via sparse representation in dissimilarity space, J. Vis. Commun. Image Represent. 25 (1) (2014) 12−23.

[35] R. Jansi, R. Amutha, A. Gokulakrishnan, A novel framework for action recognition based on histogram of oriented gradients and sparsity-inducing shared dictionary, in: 2017 International Conference on Wireless Communications, Signal Processing and Networking (WiSPNET), 2017, pp. 493−497.

[36] F. Cao, X. Feng, J. Zhao, Sparse representation for robust face recognition by dictionary decomposition, J. Vis. Commun. Image Represent. 46 (2017) 260−268.

[37] Y. Chen, J. Su, Sparse embedded dictionary learning on face recognition, Pattern Recogn. 64 (2017) 51−59.

[38] H. Yan, Discriminative sparse projections for activity-based person recognition, Neurocomputing 208 (2016) 183−192.

[39] F. Duarte, A. Lourenço, A. Abrantes, Classification of physical activities using a smartphone: evaluation study using multiple users, Procedia Technol. 17 (2014) 239−247.

[40] J. Mairal, F. Bach, J. Ponce, G. Sapiro, Online dictionary learning for sparse coding, Proc. 26th Int. Conf. Mach. Learn (2009) 1−8.

[41] M. Aharon, M. Elad, A. Bruckstein, K-SVD: an algorithm for designing overcomplete dictionaries for sparse representation, IEEE Trans. Signal Process. 54 (11) (2006) 4311−4322.

[42] M.R. Osborne, B. Presnell, B.A. Turlach, A new approach to variable selection in least squares problems, IMA J. Numer. Anal. 20 (3) (2000) 389−403.

[43] B. Efron, et al., Least angle regression, Ann. Stat. 32 (2) (2004) 407−499.

[44] T. Cover, P. Hart, Nearest neighbor pattern classification, IEEE Trans. Inf. Theory 13 (1) (1967) 21−27.

[45] D. Roy, M. Srinivas, C. Krishna Mohan, Sparsity-inducing dictionaries for effective action classification, Pattern Recogn. 59 (2016) 55−62.

CHAPTER 9

A Predictive Model for Hypertension Diagnosis Using Machine Learning Techniques

MUHAMMAD ALI JAVED TENGNAH • RAGINEE SOOKLALL •
SOULAKSHMEE DEVI NAGOWAH

Department of Software and Information Systems, Faculty of Information, Communication and Digital Technologies,
University of Mauritius, Réduit, Mauritius

1 INTRODUCTION

Being one of the major causes of cardiovascular disease, hypertension has caused much more ravage worldwide, taking several lives and increasing death toll [1]. Hypertension is actually the persistent rise in systemic arterial pressure above a certain minimum value [2]have defined hypertension as *"a progressive cardiovascular syndrome arising from complex and interrelated etiologies"*. Another definition proposed by Geoffrey Rose is *"The level of blood pressure for which investigation and management do more good than harm"* [3]. Based on international health guidelines, hypertension has an average threshold value of 140 mm Hg for systolic blood pressure or 90 mm Hg for diastolic blood pressure, or both and it can be categorized into two types namely Primary and Secondary hypertension [3].

Hypertension can be classified into three stages. The first stage, also known as pre-hypertensive stage, is about having occasional blood pressure elevation with some early cardiovascular disease risks which can arise due to circulatory, vascular or renal adaptations to environmental or genetic stimuli [2]. The blood pressure level is above 115/75 mm Hg, which might vary due to environmental stress [2]. In the second stage, the blood pressure level often reaches a value of 140/90 mm Hg, considered as the individual's resting blood pressure level with some elevations and is induced by psychological or physiological factors [2]. Here, the disease has developed further and target organ damages can be detected. The third stage is an advanced stage where the blood pressure level has a sustained threshold value of 140/90 mm Hg, considered as the

individual's resting blood pressure level with an elevation of up to 160/100 mm Hg with medical evidence of clear target organ damage [2]. This stage is considered to be very critical as damages to the organs are quite severe and there is an increased risk of death.

Hypertension can be caused by several factors. One of the factors is genetic, that is, heritable from parents, which is one of the causes of Primary Hypertension [3]. Other causes are sedentary lifestyles, that is consuming unhealthy foodstuffs, increased intake of salt, high alcohol consumption, anxiety, psychosocial stress and weight gain in addition to doing nearly no physical activities [3–6]. Smoking as well is a risk factor for hypertension [4]. Moreover, maternal undernutrition can also lead to a raise in the systolic blood pressure and a protein under nutrition can raise the diastolic blood pressure [3]. Added to this, infants born with a heavier weight usually experience high blood pressure sooner or later in their childhood [3].

In addition to being one the greatest silent killer, hypertension also contributes in cardiovascular diseases and events such as strokes, heart failure and cardiovascular deaths [7]. According to Ibrahim and Damasceno [5], two-third of all strokes and half of all coronary diseases are attributed to non-optimum blood pressure. Episodic hypertension with a low systolic blood pressure contributes to cardiovascular events and poor hypertension monitoring and control gives rise to cardiovascular diseases [3,5].

Hypertension is also a common disorder during pregnancy whereby the lives of both the mother and the baby are at stake [8,9]. Based on certain research

Telemedicine Technologies. https://doi.org/10.1016/B978-0-12-816948-3.00009-X

works, black women, obese woman, older woman and diabetic women are more prone to suffer from this condition [8,9]. According to Health Line (2013), normal blood pressure is usually around 120/80 mm Hg [66]. A systolic blood pressure of 140 mg Hg or greater, and/or a diastolic reading of 90 mm Hg or greater on at least two separate occasions with a gap of minimum 4 hours while resting can be concluded as hypertensive [8,9]. However, a systolic blood pressure of 160 mm Hg and above and/or, diastolic blood pressure of 110 mm Hg or above taken of two separate occasions is considered as severe hypertension [9].

There is therefore the need to come up with predictive models that can help to diagnose hypertension. Embedding these models in mobile applications will surely help to improve the health of people who are likely to suffer from hypertension. Additionally, the combination of intelligent techniques to diagnose hypertension can be very beneficial. This chapter therefore proposes a predictive model that takes into account nearly every possible factor that might influence the blood pressure in a human being. Considering most of the factors might give a more accurate prediction of the blood pressure status of the individual [10]. A number of models have been implemented [11−14]. However, none of them have combined the factors proposed by the predictive model in this chapter to aid in the diagnosis of hypertension.

The rest of the chapter is structured as follows: Section 2 describes the concept of knowledge discovery and intelligent techniques that can be used to predict hypertension. Section 3 presents a predictive model for hypertension. The model is evaluated in Section 4. Details of user acceptance of the model are presented in Section 5. Finally, Section 6 concludes the chapter.

2 KNOWLEDGE DISCOVERY AND INTELLIGENT TECHNIQUES FOR HYPERTENSION

To be able to come up with a predictive model for hypertension and perform accurate diagnosis, it is of utmost importance to understand the knowledge discovery process and the different techniques and algorithms that can be used to extract the different patterns and relationship between the stored data. Details of the algorithms are given in this section.

2.1 Knowledge Discovery

Knowledge Discovery describes the overall process of transforming the low-level data into high level data. Moreover it refers to retrieving and finding useful

knowledge from large amount of data [15]. Knowledge Discovery is mostly used with databases which stores enormous amount of data thereby making it difficult for human to extract useful information and pattern from it [16]. Knowledge Discovery in Databases (KDD) comprises of different steps such as selecting the appropriate data, pre-processing, transforming the data and applying data mining in order to find patterns [19]. Data mining is an integral part of Knowledge Discovery and is also known as Knowledge Discovery and Data mining (KDDM). Data integration may be required if data from different sources need to be used to uncover new relationship and patterns between the data.

To carry out KDDM, the following steps have to be followed [17]:

- *Domain understanding and KDD goals*
  In this phase, the developers must understand the need for KDD in the respective domain and also what is expected by the end users. Experts are consulted to verify and check the knowledge stored in the database. The next step is the data selection.

- *Data selection, pre-processing and data transformation*
  After effectively understanding the knowledge area and having formulated the goals of the end user, the data to be used for knowledge discovery has to be checked and analyzed. The data can actually come from various sources such as large databases or datasets as most of the tools such as Weka, Rapid miner and R programming offer database connectivity service. Attribute selection is a very important aspect in data mining. It directly guarantees the success or failure of the model as unnecessary attributes can reduce the accuracy of the model [15]. Pre-processing such as filtering and data cleaning have to be applied on the dataset for the removal of inconsistencies. Data transformation is carried out through different methods available such as feature selection, extraction and discretization. Discretization process is a process whereby the numerical values are converted into nominal class [1].

- *Data mining*
  Based on the pre-processed dataset, an appropriate data mining method can be selected keeping in mind the defined goal of KDD and also the nature of the attributes that constitute the dataset. Data mining is the use of mathematical analysis in order to find patterns and trends based on the current data. Several techniques such as classification, regression, association rules can be used to build a predictive model [15]. The algorithm to be used for discovering

unknown pattern in the data is selected, for example, selection of decision tree algorithm in the classification technique. All the different algorithms will provide different options to be used for constructing and testing the model. For testing the model, there are different options such as "10 fold cross validation", the training set or simply uploading a dataset to be used solely for testing.

- *Evaluation and Interpretation*
  After the pattern has been produced by the data mining algorithm, evaluation is required based on the different goals defined. The impact of the pre-processing step is assessed for the output produced by the data mining step. Modification may be made to the pre-processing step of the dataset to find ways to yield output with greater accuracy [18]. To increase the confidence about the discovered knowledge, triangulation process can be carried out by having two or more different models and checking whether they converge toward comparable conclusions [19].

2.2 Intelligent Techniques
There are numerous intelligent techniques that can be used for diagnosis and knowledge discovery namely fuzzy logic expert system, decision tree, artificial neural network, k-nearest neighbor and support vector machine.

2.2.1 Fuzzy logic
Fuzzy logic is an improvement to the traditional knowledge based system in the sense that it can respond to unexpected conditions. It does not lay emphasis only on "yes and no" but also considers the degree to which something is correct or not [20]. An example of its use for hypertension is an expert system making use of fuzzy logic to predict risk of hypertension based on the user's blood pressure reading, the body and mass index and age [21]. There are three main processes required in the fuzzy logic expert system namely fuzzification, rulebase and defuzzification. Fuzzification is the stage whereby the degree to which a value belongs to the fuzzy set is estimated. The rule base comprises of a set of fuzzy representations derived from the knowledge base. Defuzzification is the process whereby the risk of a person being hypertensive is determined. These three processes describe how an expert system makes use of fuzzy logic functions.

2.2.2 Artificial neural network
Artificial neural network (ANN) was developed based on neurobiology in order to allow a computer program to imitate the decision-making style of a human brain, which involves massive processing of information in parallel by hundreds of billions of interconnected neurons [22]. In other words, it is *"the generalization of mathematical models of biological nervous system"*, involving multifactorial analysis [23]. The neural network basically comprises of three types of neuron layers: the input layer, the hidden layer which can consists of one or more layers and the output layer, which follows a strict feed-forward direction [24]. These neural nodes are interconnected using weighted connection lines [22]. The neural network architecture can be defined using the following mathematical formula [3]:

$$h_i = \sigma \left(\sum_{j=1}^{N} V_{ij} x_j + T_i^{hid} \right)$$

The output, h_i, of neuron i in the hidden layer is, where $\sigma()$, is known as the activation function or transfer, N is the number of neurons in the input layer, V_{ij} the weights of the connection lines, x_j inputs to the input neurons, and T^{hid}_i the starting terms of the hidden neurons [3].

ANN can play a vital role in medical decision support due to the effectiveness of the multifactorial analysis [22]. It is suitable for diagnosis purposes. It has been used for hypertension diagnosis with a large dataset with around 185,000 patients and has yielded an estimated precision of 82% [4].

2.2.3 Multilayer perceptron
Multilayer perceptron is a feed-forward artificial neural network which is defined as a finite directed acyclic graph implying that there is no direct cycle [25]. It comprises of three different types of neurons mainly input, hidden and output neurons. The neurons are organized into different layers. The input neurons that will represent the different attributes will be in the first layer. This layer will be followed by the hidden neuron layers. Finally, the layer which consists of the output neurons, represents the different class values that will be predicted by the network [62]. All the neurons are interconnected and each connection is assigned a weight. All the hidden and the output neurons will have a bias weight which is denoted by W_{jo} for neuron j. There are two variables that all the hidden and output neuron must have namely the network input variable and the activation variable which represents the output of each neuron [67]. Depending on the application there are different activation that are used such as logistic, linear and tanh activation [25]. For hypertension prediction, tanh activation can be used because classification is

required and there are more than two output classes. The following pseudocode summarizes the steps involved in multilayer perceptron:

```
for all input neurons i do
        set aᵢ ← xᵢ
end for
for all hidden and output neurons i in topological order
do
        set netᵢ ← wᵢₒ + Pⱼ ∈ ₚᵣₑ𝒹(i) wᵢⱼaⱼ
        set aᵢ ← tanh (netᵢ)
end for
for all output neurons i do
        assemble aᵢ in output vector Y
end for
return vector Y
```

After the above steps, the neural network has to be trained for all the instances given to it and the output will be in terms of the class label that is 0, 1, 2 and 3. An important aspect of the training process is to minimize the error by using the formula [25] illustrated below:

$$E(\vec{w};D) = \frac{1}{2}\sum_{i=1}^{p}\left\| \gamma\left(\vec{x}^{(i)};\vec{w}\right) - \vec{d}^{(i)}\right\|^2$$

This is how the multilayer perceptron algorithm works and how it makes use of weight of the different neurons to classify the particular instance.

2.2.4 Decision tree

Decision tree is one of the most used classification methods and it uses a divide and conquer approach for searching [2]. The decision tree consists of different nodes and each node is in the form of a decision. The mostly used decision tree classification algorithms are ID3, C4.5 and CART [26]. ID3 algorithm involves two main steps namely construction of the decision tree and pruning. The main purpose of the pruning process is to make the decision tree smaller. This process involves removing redundant branches which increase the overall size of the tree but do not provide higher precision [27]. For the tree construction process information gain is computed for the different nodes and the highest gain is used as the root node. One particular advantage of a decision tree is that the classification can be easily understood.

2.2.5 C4.5 algorithm

C4.5 algorithm is an improvement of ID3 since it makes use of gain ratio for the selection of a node which will be used as root and it eliminates useless branches in order to make the classification more accurate [28].

2.2.6 C5 algorithm

C5 algorithm is an improvement of the C4.5 in terms of accuracy and efficiency as it reduces the size of the decision tree by eliminating useless features [29]. C5 also has the ability to distinguish between relevant and non-relevant features for classifying the records. In comparison with ID3 and C4.5, CART makes use of GINI index in order to reduce size of decision tree by removing unnecessary branches [29]. CART additionally makes use of cost pruning and does not rely on probability assumption [29].

2.2.7 Support vector machine

Support Vector machine (SVM) is mainly used for analyzing data and for finding patterns which are used for classification and regression analysis [30]. In SVM, the data is classified into two classes and the hyper plane lies between those two classes. The advantage of SVM is that it also considers data being close to the opposite class and thus gives a reliable classification. It makes use of a Kernel version and can be used as substitute to multilayer perceptron classifiers. SVM is both accurate in classification, rapid in learning and is also useful for big datasets and when there are lot of features. SVM is applicable for diagnosis purposes especially in cases where classification needs to be done in exactly two classes, for example, for classifying whether the user has hypertension or not. One example of SVM used for classification in health care is its use to identify cardiac patients with risky condition [24].

3 PREDICTIVEMODEL FOR HYPERTENSION

In this section, a predictive model for hypertension is described. The predictive model is composed of the *Hypertension Diagnosis* model (HD), *Psychological* and *Stress Level* model (PSL) and the *Pregnancy Hypertension* (PH) model. To be able to perform proper diagnosis, user profile details are stored such as the age, gender, body and mass index, family history among others. Additionally, BP readings are recorded. In order to reduce the risk of wrong prediction, the diagnosis is not be based on a single reading but on an average of readings taken on a specific period of time, for example, an average of the reading for three consecutive days.

A mobile application has been developed based on the predictive model. The parameters entered by the user via the mobile application are then processed using the Weka software in order to provide a prediction by finding patterns in the raw data. Knowledge Discovery process is carried out on the available dataset to uncover relationship between the different attributes and hypertension causal. To perform data mining, the multilayer perceptron model, which is a feed-forward artificial neural network, is used in order to find the relationships between the different attributes found in the dataset.

3.1 Factors Considered for Prediction Purposes

This section describes the different factors that affect the blood pressure of a person and the different factors have been categorized in term of main diagnosis, psychological and pregnancy.

3.1.1 Main diagnosis model factors

There are various factors that affect a person's blood pressure such as age, bmi, stress level, alcohol level and the salt intake. These are the most common factors. However, increased blood pressure does not result only from those factors. Stress is also an important factor that affects hypertension. Stressful situations affect a person's mind and cause a peak in a person's blood pressure readings. In order to cope with the challenge of providing an accurate diagnosis, an extensive research and analysis has been carried out on the different factors that impact blood pressure reading.

3.1.1.1 Age and gender. Age and gender play an important role when diagnosing whether a person is hypertensive or not. A number of studies have been carried out that demonstrate that as a person's age increases, the risk of hypertension also increases [31,32]. In a survey which includes the reading of different regions in the world, 22.7% of men in the age group of 30−39 were hypertensive while the rate was 48.1% for men and 50.1% for women in the age group of 50−59 in Sub-Saharan Africa [31].

3.1.1.2 Weight. Weight is a factor that is directly linked with hypertension. Weight reduction is one of the treatments of hypertension. A weight loss of 1% of the body weight decreases the blood pressure of a person by an average of 1 mm Hg [33]. Therefore, people who are overweight and experiencing high blood pressure can follow a diet whereby losing 10 kg can have a consequent reduction of 6−10 mm Hg in their blood pressure. According to the National Heart, Lung and blood association, the body mass index (BMI) is used to determine whether the person is underweight, healthy, overweight or obese [34].

3.1.1.3 Salt intake. Salt intake is the major factor that increases blood pressure compared to the other life style factors such as the stress level, smoking and exercises [35]. According to American Heart Association, an ideal

limit of no more than 2,300 mg of sodium per day is recommended for non-hypertensive whereas for hypertensive people, the ideal limit is set to 1,500 mg per day [65]. Following a DASH diet plan can reduce a person's systolic blood pressure reading by 8−14 mm Hg [36]. The major types of foods that significantly rise the sodium level are bread, bakery, cereals, processed meat, cheese and dairy product, canned vegetables, sauces and pizzas [37].

3.1.1.4 Alcohol. Alcohol is a significant factor that affects blood pressure of a person. As a result, it has been observed that it constitutes 16% of risk to make a person hypertensive [38]. The blood pressure rises proportionally with the alcohol intake being > 20 g for women and >30 g for men. The blood pressure becomes normal once all the alcohol in the body has been drained out. However, if the person consumes excess alcohol on a regular basis, it will not only affect his blood pressure but will also lead to other health issues such as cerebral thrombosis and coronary artery disease [38].

According to a research conducted on 491 men who consumed alcohol, only their systolic blood pressure was affected [39]. This research also concluded that the effect of alcohol on the systolic reading was not based on factors such as age, obesity and physical exercises. Another study carried out with treated hypertensive patients consuming alcohol shows a reduction of systolic reading of 2 mm Hg and diastolic 0.5 mm Hg when reducing their alcohol intake though they were under hypertension medications [40]. It has been determined that the blood pressure increases by 1 mm Hg for each 10 g of alcohol intake [38].

3.1.1.5 Medication. There are various types of medication that can increase the blood pressure of a person. According to Mayo clinic the different types of medication are pain killers, antidepressants and hormonal birth control pill [41]. The most common antidepressants that increase blood pressure are venlafaxine, bupropion and desipramine [42]. There are also medications such a cough and migraine medication which have the side effect of raising blood pressure (WebMD). NSAID is a substance that is present in some medications for pain relief. Normally, NSAID can raise the diastolic reading by 5−6 mm Hg and increase the overall risk of stroke [43].

3.1.1.6 Family history. There are genetic factors as well that are considered to predict whether the person has more risk of being hypertensive such as the family history, that is, whether the parent of a person suffers from the disease [44].

3.1.1.7 Smoking level. When a person smokes a cigarette, the nicotine substance in it causes an immediate rise in the blood pressure. However, the effects of smoking on the blood pressure of the person is effective only within the 30 min he last smoked. There is an average increase of 20 mm Hg in systolic reading and an average of 7.4 mm Hg in the diastolic reading after smoking the first cigarette [45]. However, if the person has smoked too many cigarettes, the increase in blood pressure reading can be around 5.2 mm Hg during the whole day [45]. The risk of hypertension becomes higher if a person smokes greater than 14 cigarettes every day since this will induce the increased blood pressure over a prolonged period [46]. There is also an interactive effect between smoking cigarette and coffee, as it increases a person's blood pressure by an average of 7 mm Hg during the day [46].

3.1.1.8 Coffee level. Coffee is another factor that can be considered when it comes to hypertension. For the consumption of coffee, there is a significant rise in the blood pressure reading within the 30–120 min after consumed [47]. Its effect is not long lasting but it can be coupled with other factors such as cigarette which contains another pressor namely nicotine. The increase in blood pressure readings vary depending on the amount consumed and also the body sensitivity to caffeine. For various people who consume coffee on a daily basis, the increase in blood pressure can raise in a range of 5–15 mm Hg for systolic and 5–10 mm Hg for the diastolic reading [48]. Even if the impact on the blood pressure is not that significant, a mixture of coffee and high stress level will significantly increase the risk of having hypertension [49]. Therefore, large doses will definitely have a negative impact for people who already have hypertension and also those who are at risks.

3.1.2 Psychological and stress factors
Most studies showed a longitudinal association between blood pressure levels and several psychological factors such as stress, anger, depression and job strain [50,51]. Though, it is still subjective to determine the impact of certain psychological factors on blood pressure level, it can still be concluded that they are quite effective in determining whether the person is tending toward hypertension [51].

3.1.2.1 Anger. Anger is considered to be another great factor in sustaining a persistent rise in the blood

pressure level and contributing in the development of hypertension [15,51]. In terms of medical science, anger and mental stress activate the sympathetic nervous system and hypothalamic-pituitary-adrenal axis, which increases heart rate and blood pressure [15]. Several studies have shown that the expression of anger (Anger-out) is much more related to the development of sustained high blood pressure compared to suppression of anger (Anger-in) [15]. This is because expressing anger can lead to repeated episodes of conflicts which in turn sustain or increase the feeling of anger and also accommodate grudges, which in turn continues on yielding a higher blood pressure level [15]. This condition might become disastrous as anger can also lead to cardiovascular diseases and even death [15,17].

3.1.2.2 Stress. It is quite subjective to estimate the level of stress of someone based on a particular cause(s) as it is difficult to determine how stressful a given event is to the latter [52]. According to Dimsdale [52], stress can be defined as *"organism's responses to challenge"*. Another definition of stress given by Michie [28] is *"the psychological and physical state that results when the resources of the individual are not sufficient to cope with the demands and pressures of the situation"*. A constant exposure to stress can lead to hypertension development [53]. Psychological stress might also impair the vagal tone of the vagus nerve in the autonomic nervous system which contributes to the development of cardiovascular diseases [54,55].

Work stress is one major psychological stress dominating nowadays. Often people go through pressures and tensions at their workplaces which can seriously impact their health [28,56]. Several factors leading to stress at work were examined in some studies. Firstly, work overload and time pressure are quite stressful as time is very limited and work load is high in this competitive world [28,57]. Secondly, there is usually the problem of high effort and low reward [28,56]. Relationships between superiors and colleagues are also important as bad relationship(s) can create regular conflicts at work [28,57]. Other factors that cause stress at work are job insecurities, physical danger at work and lack of training [28].

Students, on the other hand, undergo a lot of stress due to academic work load [54]. They are mostly affected by study stressors compared to social ones [54]. According to some studies, it has been concluded that the main stressor during a student's life is the examination period and the phase that comes afterward, which is the results of their examinations [54,58].

Followed by this, come deadlines for assignment and dissertation [54]. It was also deduced in a study that final year students experience more stress and depression compared to non-final ones [58]. Added to this, other studies show that girls experience more stress and depression compared to boys concerning their academic performance [58,68].

Other stress factors are life changing event or stressful events that persist over an extended period of time such as death of someone close, caring for a sick person in the family, an accident and so on [55]. Unemployment and poverty can also lead to severe stress as it means that there is a lack of financial support [50,55]. Marital issues can also be very stressful for an individual [10].

3.1.2.3 Depression.
Depression is a mental disorder that is affecting many people by causing severe mental and even physical disabilities [25,59]. Sharma and Wavare [58] defined depression as a disorder of mood involves changing levels of sadness and despair in association with stress. Depression is usually triggered by a stressful event which then impacts blood pressure [50]. Depression is related to hypertension as, firstly, they have a common risk factor which is stress, and secondly, depression itself is a risk factor of hypertension [25,50]. It is important to take depression into consideration during prevention and treatment of hypertension [25].

3.1.2.4 Exercise.
Exercise can be used as a counter measure or intervention to reduce psychological effects such as anxiety, depression and stress [60]. Lack of physical exercises might lead to increased risk of cardiovascular diseases and hypertension. Therefore, it is important to exercise in order to reduce psychological effects on the body and blood pressure [61].

3.1.3 Pregnancy model factors
After a first diagnosis based on factors in Sections 3.1.1 and 3.1.2, a female patient will have to carry out another diagnosis to confirm whether she is hypertensive and what type of hypertension she has [8,9,62]. The criteria for diagnosis are explained as follows:

3.1.3.1 Blood pressure reading.
The blood pressure reading is one of the primary factors in determining whether the person is hypertensive or not. As mentioned in Section 1, a woman's systolic blood pressure must be equal to or above 140 mm Hg and/or diastolic blood pressure must be equal to or above

90 mm Hg in two separate readings to fall in the hypertensive category [8,9].

3.1.3.2 Number of weeks of pregnancy.
Number of weeks of pregnancy is crucial for determining the type of hypertension the woman is suffering from, whether it is chronic hypertension, gestational hypertension, chronic with superimposed preeclampsia or preeclampsia itself [62].

3.1.3.3 History of hypertension before pregnancy.
If the person had hypertension before pregnancy, this factor can help in diagnosing whether she has chronic hypertension or that the chronic hypertension might evolved with superimposed preeclampsia [62].

3.1.3.4 Proteinuria.
Proteinuria is another primary criteria of determining whether a woman has preeclampsia or eclampsia. The presence of proteinuria of equal to or above 300 mg per 24-h urine collection helps in diagnosing preeclampsia/eclampsia [9,62]. However, recently it has been considered that it is not primordial to have proteinuria to be diagnosed of preeclampsia [9,62]. The rest of the factors listed below can also be considered in the diagnosis of the type of hypertension [62].

3.1.3.5 History of preeclampsia.
Family history is considered as a risk factor of preeclampsia and/or eclampsia. If the individual has a first-degree family history of preeclampsia, the risk of suffering from preeclampsia is increased by twofold to fourfold, and it is increased to sevenfold if she had a previous preeclamptic pregnancy, based on the likelihood ratio used in clinical diagnosis of preeclampsia [62–64].

3.1.3.6 Dysfunctioning organs and thrombocytopenia.
In the absence of proteinuria, dysfunctioning organs can be considered as a criteria. It englobes malfunctioning of certain organs such as impaired liver functions, renal insufficiency and pulmonary edema while suffering from preeclampsia/eclampsia [62]. Thrombocytopenia is another condition that occurs due to this disorder which means that platelets count has fallen to less than 100,000 per microliter of blood [62].

3.1.3.7 Diabetes.
Diabetes is another risk factor that affects blood pressure in general. It is also considered

during the prediction and diagnosis of hypertension in pregnant women [62,63,69].

3.1.3.8 Visual and cerebral disturbance. Lastly, visual and cerebral disturbance is another criteria considered and used to diagnose whether the person is suffering from preeclampsia or not [62].

3.2 Description of the Model Workflow

Based on the different factors described in Section 3.1, a predictive model for diagnosing hypertension has been designed. The predictive model is composed of the *Hypertension Diagnosis* model (HD), *Psychological* and *Stress Level* model (PSL) and the Pregnancy Hypertension (PH) model. The PSL takes different factors such as stressful event, anger, financial support, marital status, work overload and time pressure, high effort low gain, examination, exercise, project deadline and depression to estimate the stress level. After the estimation of the stress level, the different factors such as salt intake, alcohol, smoke, caffeine are estimated. Once all the different factors have been estimated, the data is now ready to pass through the HD model. The HD model takes the parameters such as systolic, diastolic, age, gender, family history, alcohol intake, salt intake, smoke level and caffeine intake and medication. Based on these attributes, the model predicts whether user has low blood pressure, is normal, is hypertensive or is facing hypertensive crisis. After the prediction of the main model, the system checks whether the patient is female and asks if she is pregnant. If the user is pregnant and the prediction of the HD model is hypertensive, the user is then asked details specific to pregnancy. The PH takes the different attributes such as history of pre-eclampsia, proteinuria, pre-dated pregnancy, number of weeks, visual disturbance, organ dysfunction and diabetes. The overall process is illustrated in Fig. 9.1.

3.3 Artificial Neural Network Main Model

For the HD model, a dataset that consists of the blood pressure level, age, gender, BMI and other attributes such as smoking, exercises, and alcohol and medication is used. The multilayer perceptron in Weka is used as the classification technique to uncover hidden patterns of the dataset. The main steps involves the import of the dataset which is broken into two parts: the first one is used for training the classifier and the second is used to test the accuracy of the classifier. The training set consists of 80% data while the testing set will contain 20% data. Once the model has been successfully created and tested, it is exported for use in the application.

3.3.1 Hypertension dataset pre-process

For the training of the model, the imported dataset has been pre-processed and attributes such as the waist length have been removed since BMI is already taking into consideration the overweight factor. Other redundant factors removed were the race and ethnicity, income level and residence that is either rural in urban. Removing these unnecessary factors improves the overall accuracy of the model. However, based on the research carried out other factors have been considered such as the occupation of the person, caffeine intake, medication that raise BP and more detailed factors leading to stress.

3.3.2 Training of the main model

The multilayer perceptron model is chosen in the classifier and functions category. There are different parameters to specify for the training purposes such as learning rate, the momentum and hidden layers. Fig. 9.2 illustrates the training process whereby the numbers of epoch, learning rate and momentum can be changed to test and train the model. The model yielded to a 98.79% accuracy.

After the model has been successfully constructed and tested, it needs to be imported and used for the diagnosis purpose. The model is stored on the server and is used for responding to patient's request coming from their mobile application. The result is then sent to the mobile application through the web service.

3.4 Decision Tree for PSL Model and PH Model

The decision tree is used to create the PSL model along with the PH model. The psychological factor is used as an input for the HD model since it predicts the stress level of the person. After predicting the stress level, the main diagnosis takes place since all the different factors are available. The PSL model is part of the HD model. A dataset, of over 2400 data records, was derived based on the factors considered to be important to predict psychological effect and stress level. This model was built using the decision tree J48 model whereby, it is yielding a 91.3% accuracy.

The PH model aims to predict whether a pregnant patient has hypertension based on factors described in Section 3.1.3. However, it is not a full prediction as the patient will have to carry out some medical tests to confirm her condition as several complications may occur during pregnancy. A decision tree J48 has been used to implement the prediction model and has yielded a 100% accuracy.

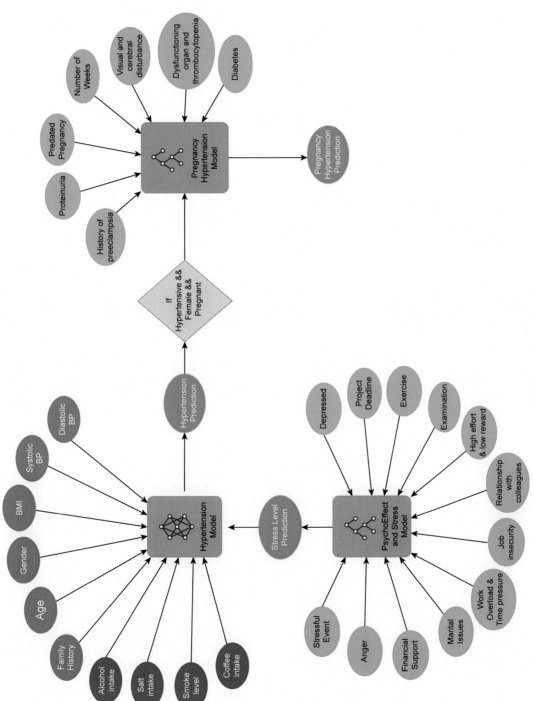

FIG. 9.1 Model workflow diagram.

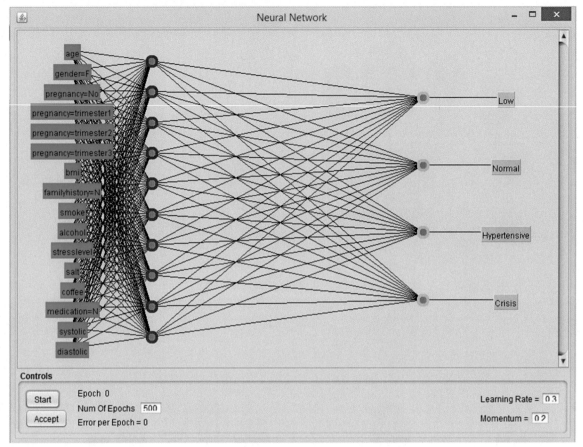

FIG. 9.2 Training of the artificial neural network.

4 EVALUATION

A mobile application has been developed based on the predictive model described in Section 3. The system is evaluated and compared to existing models found in literature.

A model from Polak and Mendyk [13] made use of the artificial neural network based internet hypertension. This model mainly considered factors such as the age, sex, diet, smoking, drinking habit and physical activity. It did not take into consideration psychological factors such as stress at work. Back propagation has been used for training the model along with tan_h activation and its accuracy was about 72.9% [13].

A second model from Ture et al. [11] made use of factors such as age, sex, family history, smoking habit, lipo-protein, triglyceride, uric acid, total cholesterol and the body and mass index. These factors are covered in our model as well. However, factors such as

lipoprotein and uric acid are not taken into consideration since the mobile application being developed is to be used by normal users who might not be aware of those data unless they have undergone medical test. This study has considered different intelligent techniques for the prediction and the most accurate result was that of the multilayer perceptron algorithm [11]. Table 9.1 illustrates the different techniques and their respective prediction rate.

Another study carried out by Su and Yang [14] made use of support vector machine to compute the hypertension prediction model and achieved and accuracy of 72%. Furthermore, a study carried out by Lafreniere [12] made use of a very large dataset of the Canadian Primary Care Sentinel Surveillance Network to train an artificial neural network using the back propagation algorithm, achieving an accuracy of 82%. The different factors used as input to the neural network are age, gender, BMI, obesity, stress, triglycerides, uric acid,

TABLE 9.1
Model Predictive Rates [11].

Model	TRAINING SET			TEST SET		
	Sensitivity (%)	Specificity (%)	Predictively Rate (%)	Sensitivity (%)	Specificity (%)	Predictively Rate (%)
Chi-squared Automatic Interaction Detector	93.00	60.60	80.95	88.24	70.00	83.10
Classification and Regression Tree	87.30	70.20	80.95	82.35	70.00	78.87
Quick, Unbiased, Efficient Statistical Tree	89.20	53.20	75.79	90.20	60.00	81.69
Logistic Regression Analysis	89.80	56.00	79.34	90.48	64.10	77.78
Flexible Discriminant Analysis	87.40	62.70	79.75	90.48	71.79	81.48
Flexible Discriminant Analysis/Multivariate Additive Regression Splines (degree = 1)	91.60	54.70	80.17	88.10	64.10	76.54
Flexible Discriminant Analysis/Multivariate Additive Regression Splines (degree = 2)	92.20	58.70	81.82	90.48	56.41	74.07
Multi-Layer Perceptron	94.00	66.70	85.54	90.48	71.79	81.48
Radial Basis Function	95.20	66.70	86.36	95.24	66.67	81.48

lipoproteins, story of smoking habit and hypertension family history [12].

From the above description of the performance of different models, our model has a higher accuracy of 98.79% and also makes use of a wider range of attributes that play an important role in the diagnosis of hypertension. The model of Lafreniere et al. [12] considers stress factor. However, it does not elaborate and consider the different factors that lead to stress such as work pressures and family issues. Furthermore, none of the above mentioned model consider hypertension prediction for pregnant women based on factors such as proteinuria, weeks reached, pre-eclampsia history among others. The other models also ask for factors that require medical check-up while our model mainly focuses on the different factors affecting a person on a daily basis which the user can provide.

5 USER ACCEPTANCE

For the purpose of the user acceptance test, a group of 20 users have been introduced to the application

and a brief overview of how to use the application was given to the users. The age of the group of people ranges from 22 to 54 years old. The group consisted of 9 men and 11 women, one of which was pregnant. After the usage of the application for a period of one week, a post-test survey questionnaire has been distributed to the participants to obtain feedback on the application. The details are summarized in Table 9.2.

Based on the ratings obtained in Table 9.2, it can be concluded that the application is helpful to users as most of the functionalities obtained a high rating from users.

6 CONCLUSION

Maintaining a proper blood pressure is quite challenging. People, nowadays, are living quite a hectic life in this fast developing world, whereby proper and healthy diet are not followed and finding some time for exercising is quite difficult. This results in more and more people are suffering from hypertension.

TABLE 9.2
User Response Based on Survey.

Features	Strongly Disagree	Disagree	Neutral	Agree	Strongly Agree
System helps to keep track of blood pressure reading			3	13	4
Reminder helps to adhere to diagnosis schedule			6	12	2
Application useful for estimating salt level		3	10	6	1
Application useful for estimating consumption details		1	6	12	1
Application provides proper diagnosis			2	11	7
Application provides meaningful recommendation			4	12	4
Application allows easy access to diagnosis details			5	12	3
Graph provides meaningful comparison and are informative			1	16	3
Application gives a proper estimate of stress level			5	12	3

Fortunately, advances in the mobile health (mHealth) facilities have helped a lot in monitoring the health condition of patients. The chapter therefore proposes a predictive model that can aid people in the diagnosis of hypertension. The predictive model is composed of the *Hypertension Diagnosis* model (HD), *Psychological and Stress Level* model (PSL) and the Pregnancy Hypertension (PH) model. A mobile application has been implemented based on the predictive model. The system integrates intelligent techniques such as Artificial Neural Network (ANN) by implementing the multilayer perceptron algorithm. Moreover, it also aids pregnant women in diagnosing the type of hypertension they are suffering from during pregnancy. The system has been evaluated and tested by a group of users whereby the latter claimed that the application provided proper diagnosis and meaningful recommendations.

REFERENCES

[1] O. Kingsley, A framework for intelligent remote blood pressure monitoring and control system for developing countries, J. Comput. Sci. Appl. 3 (1) (2015) 11–17.

[2] T.D. Giles, B.J. Materson, J.N. Cohn, J.B. Kostis, Definition and classification of hypertension: an update, J. Clin. Hypertens. 11 (11) (2009) 611–614.

[3] N.R. Poulter, D. Prabhakaran, M. Caulfield, Hypertension, 2015. Available from: http://www.sciencedirect.com/science/article/pii/S0140673614614689.

[4] J.M. Halimi, B. Giraudeau, E. Cacès, H. Nivet, J. Tichet, The risk of hypertension in men: direct and indirect effects of chronic smoking, J. Hypertens. 20 (2) (2002) 187–193.

[5] M.M. Ibrahim, A. Damasceno, Hypertension in developing countries, Lancet 380 (9841) (2012) 611–619.

[6] Preventing hypertension: a hopeless dream? Lancet 380 (2012).

[7] E. Falaschetti, J. Mindell, C. Knott, N. Poulter, Hypertension management in England: a serial cross-sectional study from 1994 to 2011, Lancet 383 (9932) (2014) 1912–1919.

[8] B.M. Sibai, Treatment of hypertension in pregnant women, N. Engl. J. Med. 335 (4) (1996) 257–265.

[9] A.R. Vest, L.S. Cho, Hypertension in pregnancy, Curr. Atheroscler. Rep. 16 (3) (2014) 395.

[10] I.P. Adebayo, Predictive model for the classification of hypertension risk using decision trees algorithm, Am. J. Math. Comput. Model. 2 (2) (2017) 48–59.

[11] M. Ture, I. Kurt, A.T. Kurum, K. Ozdamar, Comparing classification techniques for predicting essential hypertension, Expert Syst. Appl. 29 (3) (2005) 583–588.

[12] D. LaFreniere, F. Zulkernine, D. Barber, K. Martin, Using machine learning to predict hypertension from a clinical dataset, in: Computational Intelligence (SSCI), 2016 IEEE Symposium Series, IEEE, 2016, pp. 1–7.

[13] S. Polak, A. Mendyk, Artificial neural networks based Internet hypertension prediction tool development and validation, Appl. Soft Comput. 8 (1) (2008) 734–739.

[14] C.T. Su, C.H. Yang, Feature selection for the SVM: an application to hypertension diagnosis, Expert Syst. Appl. 34 (1) (2008) 754–763.

[15] S.A. Everson, D.E. Goldberg, G.A. Kaplan, J. Julkunen, J.T. Salonen, Anger expression and incident hypertension, Psychosom. Med. 60 (6) (1998) 730–735.

[16] S. Patel, H. Park, P. Bonato, L. Chan, M. Rodgers, A review of wearable sensors and systems with application in rehabilitation, J. NeuroEng. Rehabil. 9 (1) (2012) 21.

[17] R.G. Knight, B.J. Chisholm, J.M. Paulin, H.J. Waal-Manning, The Spielberger anger expression scale: some psychometric data, Br. J. Clin. Psychol. 27 (3) (1998) 279–281.

[18] M. Goebel, L. Gruenwald, A survey of data mining and knowledge discovery software tools, ACM SIGKDD Explor. Newsl. 1 (1) (1999) 20–33.

[19] O. Maimon, L. Rokach, Introduction to knowledge discovery and data mining, in: Data Mining and Knowledge Discovery Handbook, Springer US, 2009, pp. 1–15.

[20] P.R. Asha, S. Karpagavalli, Diabetic retinal exudates detection using machine learning techniques, in: International Conference in Advanced Computing and Communication Systems, 2015, IEEE, 2015, pp. 1–5.

[21] B.K. Bose, Expert system, fuzzy logic, and neural network applications in power electronics and motion control, Proc. IEEE 82 (8) (1994) 1303–1323.

[22] A. Benharref, M.A. Serhani, Novel cloud and SOA-based framework for E-Health monitoring using wireless biosensors, IEEE J. Biomed. Health Inf. 18 (1) (2014) 46–55.

[23] A. Abraham, Artificial Neural Networks. *Handbook of Measuring System Design*, 2005.

[24] P. Pereira, E. Duarte, F. Rebelo, P. Noriega, A review of gamification for health-related contexts, in: International Conference of Design, User Experience, and Usability, Springer, Cham, 2014, pp. 742–753.

[25] L. Meng, D. Chen, Y. Yang, Y. Zheng, R. Hui, Depression increases the risk of hypertension incidence: a meta-analysis of prospective cohort studies, J. Hypertens. 30 (5) (2012) 842–851.

[26] J.J. Hopfield, Artificial neural networks, IEEE Circuits Devices Mag. 4 (5) (1988) 3–10.

[27] J.D. Piette, H. Datwani, S. Gaudioso, S.M. Foster, J. Westphal, W. Perry, J. Rodríguez-Saldaña, M.O. Mendoza-Avelares, N. Marinec, Hypertension management using mobile technology and home blood pressure monitoring: results of a randomized trial in two low/middle-income countries, Telemed. e-Health 18 (8) (2012) 613–620.

[28] S. Michie, Causes and management of stress at work, Occup. Environ. Med. 59 (1) (2002) 67–72.

[29] C. Liu, Q. Zhu, K.A. Holroyd, E.K. Seng, Status and trends of mobile-health applications for iOS devices: a developer's perspective, J. Syst. Software 84 (11) (2011) 2022–2033.

[30] R. Pandya, J. Pandya, C5. 0 algorithm to improved decision tree with feature selection and reduced error pruning, Int. J. Comput. Appl. 117 (16) (2015).

[31] P.M. Kearney, M. Whelton, K. Reynolds, P. Muntner, P.K. Whelton, J. He, Global burden of hypertension: analysis of worldwide data, Lancet 365 (9455) (2005) 217–223.

[32] P.S. Vokonas, W.B. Kannel, L.A. Cupples, Epidemiology and risk of hypertension in the elderly: the Framingham Study, J. Hypertens. Suppl.: Official Journal of the International Society of Hypertension 6 (1) (1988) S3–S9.

[33] A.A. Aljumah, M.K. Siddiqui, Hypertension interventions using classification based data mining, Res. J. Appl. Sci. Eng. Technol. 7 (17) (2014) 3593–3602.

[34] NIH, National Heart, Lung, and Blood Institute, Your guide to lowering your blood pressure with DASH. https://www.nhlbi.nih.gov/files/docs/public/heart/new_dash.pdf.

[35] F.J. He, G.A. MacGregor, A comprehensive review on salt and health and current experience of worldwide salt reduction programmes, J. Hum. Hypertens. 23 (6) (2009) 363.

[36] A.V. Chobanina, G.L. Bakris, H.R. Black, W.C. Cushman, L.A. Green, J.L. Izzo Jr., D.W. Jones, B.J. Materson, S. Oparil, J.T. Wright Jr., E.J. Roccella, The seventh report of the joint national committee on prevention, detection, evaluation, and treatment of high blood pressure: the JNC7 report, JAMA 289 (19) (2003) 2560–2571.

[37] A.J. Moshfegh, J.M. Holden, M.E. Cogswell, E.V. Kuklina, S.M. Patel, J.P. Gunn, C. Gillespie, Y. Hong, R. Merritt, D.A. Galuska, Vital signs: food categories contributing the most to sodium consumption-United States, 2007–2008, MMWR Morb. Mortal. Wkly. Rep. 61 (5) (2012) 92–98.

[38] I.B. Puddey, L.J. Beilin, Alcohol is bad for blood pressure, Clin. Exp. Pharmacol. Physiol. 33 (9) (2006) 847–852.

[39] P.D. Arkwright, L.J. Beilin, I. Rouse, B.K. Armstrong, R. Vandongen, Effects of alcohol use and other aspects of lifestyle on blood pressure levels and prevalence of hypertension in a working population, Circulation 66 (1) (1982) 60–66.

[40] I.B. Puddey, L.J. Beilin, R. Vandongen, Regular alcohol use raises blood pressure in treated hypertensive subjects: a randomised controlled trial, Lancet 329 (8534) (1987) 647–651.

[41] Mayo clinic, Medications and Supplements That Can Raise Your Blood Pressure, 2016 [Online] Available from: https://www.mayoclinic.org/diseases-conditions/high-blood-pressure/in-depth/blood-pressure/art-20045245.

[42] R. Hurd, High Blood Pressure – Medicine – Related, 2016 [Online] Available from: https://medlineplus.gov/ency/article/000155.htm.

[43] A.G. Johnson, NSAIDS and increased blood pressure. What is the clinical experience, Drug Saf. 17 (1997) 277.

[44] K.C. Light, S.S. Girdler, A. Sherwood, E.E. Bragdon, K.A. Brownley, S.G. West, A.L. Hinderliter, High stress responsivity predicts later blood pressure only in combination with positive family history and high life stress, Hypertension 33 (6) (1999) 1458–1464.

[45] A. Groppelli, D.M. Giorgi, S. Omboni, G. Parati, G. Mancia, Persistent blood pressure increase induced by heavy smoking, J. Hypertens. 10 (5) (1992) 495−499.

[46] M.N. Kaplan, Smoking and Hypertension, 2015 [Online] Available from: https://www.uptodate.com/contents/smoking-and-hypertension.

[47] J.R. Mort, H.R. Kruse, Timing of blood pressure measurement related to caffeine consumption, Ann. Pharmacother. 42 (1) (2008) 105−110.

[48] J.E. James, Critical review of dietary caffeine and blood pressure: a relationship that should be taken more seriously, Psychosom. Med. 66 (1) (2004) 63−71.

[49] M.L. Nurminen, L. Niittynen, R. Korpela, H. Vapaatalo, Coffee, caffeine and blood pressure: a critical review, Eur. J. Clin. Nutr. 53 (11) (1999) 831.

[50] M. Reiff, S. Schwartz, M. Northridge, Relationship of depressive symptoms to hypertension in a household survey in Harlem, Psychosom. Med. 63 (5) (2001) 711−721.

[51] T. Rutledge, B.E. Hogan, A quantitative review of prospective evidence linking psychological factors with hypertension development, Psychosom. Med. 64 (5) (2002) 758−766.

[52] J.E. Dimsdale, Psychological stress and cardiovascular disease, J. Am. Coll. Cardiol. 51 (13) (2008) 1237−1246.

[53] D. Gasperin, G. Netuveli, J.S. Dias-da-Costa, M.P. Pattussi, Effect of psychological stress on blood pressure increase: a meta-analysis of cohort studies, Cad. Saúde Pública 25 (4) (2009) 715−726.

[54] R. Abouserie, Sources and levels of stress in relation to locus of control and self-esteem in university students, Educ. Psychol. 14 (3) (1994) 323−330.

[55] S. Cohen, D. Janicki-Deverts, G.E. Miller, Psychological stress and disease, JAMA 298 (14) (2007) 1685−1687.

[56] T.G. Vrijkotte, L.J. Van Doornen, E.J. De Geus, Effects of work stress on ambulatory blood pressure, heart rate, and heart rate variability, Hypertension 35 (4) (2000) 880−886.

[57] B. Gardell, Scandinavian research on stress in working life, Int. J. Health Serv. 12 (1) (1992) 31−41.

[58] B. Sharma, R. Wavare, Academic stress due to depression among medical and para-medical students in an Indian medical college: health initiatives cross sectional study, J. Health Sci. 3 (5) (2013) 029−038.

[59] W.Q. Chen, O.L. Siu, J.F. Lu, C.L. Cooper, D.R. Phillips, Work stress and depression: the direct and moderating effects of informal social support and coping, Stress Health 25 (5) (2009) 431−443.

[60] P. Salmon, Effects of physical exercise on anxiety, depression, and sensitivity to stress: a unifying theory, Clin. Psychol. Rev. 21 (1) (2001) 33−61.

[61] S.P. Whelton, A. Chin, X. Xin, J. He, Effect of aerobic exercise on blood pressure: a meta-analysis of randomized, controlled trials, Ann. Intern. Med. 136 (7) (2002) 493−503.

[62] American College of Obstetricians and Gynecologists, Hypertension in pregnancy. Report of the American College of Obstetricians and Gynecologists' task force on hypertension in pregnancy, Obstet. Gynecol. 122 (2013). Available on: https://www.acog.org/∼/media/Task%20Force%20and%20Work%20Group%20Reports/public/HypertensioninPregnancy.pdf.

[63] B.M. Sibai, Diagnosis and management of gestational hypertension and preeclampsia, Obstet. Gynecol. 102 (1) (2003) 181−192.

[64] B.M. Sibai, M. Lindheimer, J. Hauth, S. Caritis, P. VanDorsten, M. Klebanoff, C. MacPherson, M. Landon, M. Miodovnik, R. Paul, P. Meis, Risk factors for preeclampsia, abruptio placentae, and adverse neonatal outcomes among women with chronic hypertension, N. Engl. J. Med. 339 (10) (1998) 667−671.

[65] C.W. Yancy, M. Jessup, B. Bozkurt, J. Butler, D.E. Casey, M.H. Drazner, G.C. Fonarow, S.A. Geraci, T. Horwich, J.L. Januzzi, M.R. Johnson, ACCF/AHA guideline for the management of heart failure: a report of the American College of Cardiology Foundation/American Heart Association Task Force on Practice Guidelines, J. Am. Coll. Cardiol. 62 (16) (2013), 2013, e147−e239.

[66] Health Line, High blood pressure during pregnancy, Available on: https://www.healthline.com/health/high-blood-pressure-hypertension/during-pregnancy, 2013.

[67] X.Y. Djam, Y.H. Kimbi, Fuzzy expert system for the management of hypertension, Pac. J. Sci. Technol. 12 (1) (2011) 390−402.

[68] R.L. Campbell, L.W. Svenson, G.K. Jarvis, Perceived level of stress among university undergraduate students in Edmonton, Canada, Perceptual and motor skills 75 (2) (1992) 552−554.

[69] M. Mugo, G. Govindarajan, L.R. Kurukulasuriya, J.R. Sowers, S.I. McFarlane, Hypertension in pregnancy, Curr. Hypertens. Rep. 7 (5) (2005) 348−354.

CHAPTER 10

From *Telediagnosis* to *Teletreatment*: The Role of Computational Biology and Bioinformatics in Tele-Based Healthcare

SAHAR QAZI[a] • KHUSHNUMA TANVEER[a] • KHALID ELBAHNASY[b] • KHALID RAZA[a]

[a]Department of Computer Science, Jamia Millia Islamia, New Delhi, India; [b]Faculty of Computer and Information Sciences, Ain Shams University, Cairo, Egypt

1 INTRODUCTION

Telemedicine, a state-of-the-art concept, uses modern day electronic based technologies to accommodate healthcare services to the remotely residing public. Tele-based healthcare services were initially thought to be "fortuitous future" for the medical associations and federations, but with the incoming of vivacious and apex technologies it has now become a successful reality today. Telemedicine can be seen as a tool which is being utilized as a powerful source of reducing the long distance between remotely resided rural people and the healthcare services. It has myriad applications and has extended in many different medical care applications such as consultations, diagnosis, nursing, medications and treatment, psychiatry and psychology, rehabilitation and some specialized services. According to the WHO, telemedicine is simply the delivery of medical healthcare services by the medical professionals where distance is a belittling aspect utilizing information technology (IT) for providing a clear cut diagnosis, treatment and prevention of diseases. Moreover, it also helps to evaluate and work upon the misunderstood or least studied disorders and diseases [7,12].

1.1 Distinctness of Tele-Care From Traditional Care

Traditional medical treatment and healthcare refers to the common practice of providing healthcare services to the public where a patient either visits or carried to a nearby hospital or a clinic and gets diagnosed by a professional consultant [60]. One of the loopholes of traditional healthcare system is the fact that, people cannot always contact a consultant may be due to their tight schedules or any other reason [33]. Tele-healthcare is more or less similar to the traditional way of diagnosing and treatment, but differing at the few points. For instance, it uses information and communication technology (ICT) which connects the consultants with the patients making it healthcare services come at ease. Generally computers, mobile phones, radio, television etc., are the mode of delivering tele-based healthcare. It has helped people who lack the opportunities and advancements of medical care who come from a backward background. This system provides a new domain and opens up a layer of framework which dexterously describes the management of health related information across the computational platforms and its exchange from the consultant to the sufferer productively. It also ensures a good quality of diagnosis and prognosis and aids in giving an experience like a "live" consultation to the patients making it effortless in its delivery. Tele-based health care has improvised upon the following, and thus can be stated as an advantage over the traditional modus operandi of healthcare regime [9]:

i. Better experience and improved quality of consultation.
ii. Less of diagnosis/prognosis errors.
iii. Cheap healthcare.
iv. Healthcare at Home (no visits needed!)
v. Affordable healthcare regimen.
vi. Improved evaluation of the diseases.
vii. Early disclosure of epidemic outbreaks in countries.

Telemedicine Technologies. https://doi.org/10.1016/B978-0-12-816948-3.00010-6

1.2 Available Varieties of Tele-Care

Tele-health services come in many different forms as per the need of the public and are categorized as follows:

i. *Stock Storing & Stimulation:* This type of tele-based healthcare is revolved around taking in the clinical information such as medical images, patient's clinical profiles, etc. and then exciting this data to a consultant at an appropriate time for an adroit examination. It requires a proper well-structured format of medical record of a person and an audio/video communication of the patient which comes handy to the consultant for prescribing and diagnosis of the treatment [2].

ii. *Distant Supervision:* It is also known as remote monitoring or self assessment. It aids medical consultations to assess a person distantly based using many technological tools and devices. This service helps to provide an efficient, distinguishable, and an inexpensive mode of treatment [4,45].

iii. *Actual Time Interactions:* This is based on electronic communication of the patient with the consultant directly using a video-conferencing or phone call. This type of tele-based medical care is a more of an interactive healthcare which is composed of

diagnosis, prognosis, treatment, counseling and follow-ups of the patients. Here, the patients without any hassle can discuss issues and their conditions freely with the consultant without any time bound, unlike the traditional system of healthcare [16].

2 BIOINFORMATICS, TELEGENOMICS, TELEGENETICS AND TELEMEDICINE

Bioinformatics is an emerging interdisciplinary field of studies which has potential applications in biology and medicine (Fig. 10.1). It uses information technology to solve biological problems by in interpreting and analyzing data in a better way. It is the evolution of traditional sciences for obtaining an improvised version for achieving a better understanding of biological data. Many past projects, such as the Human Genome Project (HGP), paved a way for generating and handling huge amount of data for better understanding of disease mechanism and development of genome based personalized medicines.

Both genomics and telemedicine are two technologies driven and are rapidly growing areas which promise to deliver new configuration of healthcare without any

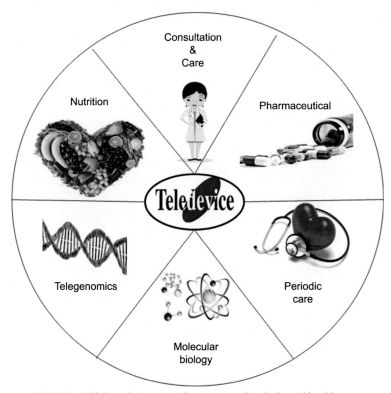

FIG. 10.1 Linkage between various aspects in tele-based healthcare.

TABLE 10.1
Summarizes the Most Important Genomic-Centric Databases Which are Commonly Used.

Database Name	Description	URL
dbSNP	A database encasing information on single nucleotide polymorphisms (SNPs) which allows its users to make queries and easy browsing.	https://www.ncbi.nlm.nih.gov/projects/SNP/
dbGaP	A database of both Genotypes and Phenotypes used to disseminate outcomes and data from human genotypic and phenotypic interaction studies respectively.	https://www.ncbi.nlm.nih.gov/gap
Direct-to-consumer genetic testing (Ancestry DNA, FTDNA, Geno 2.0 etc...)	A direct access to genetic testing without involving any practitioner or healthcare provider to the patients and provides information about the ancestry, common traits and the genetic composition.	https://ghr.nlm.nih.gov/primer/dtcgenetictesting/directtoconsumer
PharmaGKB	A freely accessible database which distributes information on how do the drug respond to human genetic variations, making researchers understand ones pharmacogenomics better.	https://www.pharmgkb.org/
PharmaGenEd	A program developed by UCSD Skaggs School of Pharmacy & Pharmaceutical Sciences which distributes information based on ones pharmacogenomics.	http://pharmacogenomics.ucsd.edu/

boundary of distance and time. The delivery methods and technologies used by telemedicine affect various areas of genomic practices. Telemedicine, apart from delivering various tele-based services, has been used to disseminate genomic information to both researchers and clinicians in order to make the genomic information quickly available [11]. Some of the important genomic information which has impact on healthcare and medicine include dbSNP, dbGaP, Direct-to-consumer genetic testing, PharmGKB, PharmGenEd, etc. [32]. Summary of important genomic-centric databases are presented in Table 10.1.

Several healthcare sectors started bringing telemedicine into precision medicine by integrating the technology to genetic and genomic research centers. It will bring together researchers, clinicians, and patients by using remote conferencing to expedite the impact of genetics and genomics on the diagnosis and treatment of diseases. As a result, genomic telemedicine can be introduced as tele-based healthcare. Further, due to reducing cost of DNA sequencing, personal genomic data may be easily available to both healthcare centers and patients which give patients access to genomically informed healthcare.

There are many research gaps which the research fraternity has to work upon in order to make *tele*-based healthcare a better and a reliable strategy, not only for patients and their consultants, but for the entire industrial and research domain as a whole. Currently, the only challenges that are present in the path of

electronic-healthcare system are as follows (https://munin.uit.no/bitstream/handle/10037/11275/thesis.pdf?sequence=2):

i. The shortage of human resources in medical infrastructures when compared to an apex patient flow.
ii. Lack of Electronic Patient Record (EPR) systems in medic facilities.
iii. Chaotic management of patient electronic records.
iv. Discrepancies in analyzing patient samples due to lack of Laboratory Information Systems (LIS).
v. Need to develop an electronic and an automated laboratory service for best experience.

2.1 Personalized Telehealthcare

Healthcare is not restricted to personalized medicines, but personalized telehealthcare as well. The existing situation of tele-based healthcare services have shown to be marvelously addressing to the health issues of the public, all this with a vision of providing personalization to every citizen [13]. Tele-based health services not only restrained to mild diseases but it is worthy for the treatment of chronic disorders as well.

According to Or and Karsh [36], personalization means that a telehealthcare unit must focus on the patient alone encapsulating healthcare regimen, patient's daily routine and lifestyle including their diet and social life by incorporating mobile phone applications and web based tele-medic programs. Personalization refers to defying the all time favorite — "one-size-fits-all" approach in medical well-being. For gaining more appreciations, telehealthcare systems must involve patients, family members, physicians and associated medic staff in enhancing the current healthcare organizations. In the earlier pretexts of this chapter, it has been stated that the consultants and fellow medical staff must be trained well enough on how to practice telehealthcare units, but, since telehealthcare concept is to provide a *"homey care"*, patients and their family members must also be trained before they actually start utilizing it. Therefore, tele-based healthcare systems must possess four properties in order to acquire an all-rounder edge to it [42]:

i. These must be *personalized*, by taking insights from the genomic information of the patients.

ii. These systems must be *predictive* enough so that risks of diseases can be determined beforehand.

iii. Telehealth systems must ensure *preventive* measures which can be used by the patients to cope up with a disease in time.

iv. Most importantly, it should be *participatory*, referring to it must be able to bridge up links between a patient and his/her respective consultant.

Developing awareness is something which can highlight the efficacies of personalized telehealthcare services. Some of the personalized tele-based healthcare platforms which aid in learning about telehealthcare services and technology are mentioned as follows:

Telegenomics (https://telegenomics.com/): A genomics based consulting company which gives an array for services for both the patients and the consultants and specializes in medical fields such as — neurology, cardiology, etc. It is world widely appreciated for its genetic counseling and genome-based medications. It has a user friendly interface which makes it even more easy to use.

CBInsights (https://www.cbinsights.com): Website unleashing medical and healthcare information to people. With its easy understanding, people get knowledge about personalized tele-based healthcare services and other related information just a click away.

Kanteron Systems (https://www.kanteron.com/): A private IT-based medical healthcare platform formed in 2005 helping practitioners in managing medical data for providing appropriate medications to their patients.

3 CATEGORIES OF TELE-BASED SERVICES

Telehealthcare has many facets which can serve various purposes for treatment and care of the patients. Some of them are as follows (Fig. 10.2):

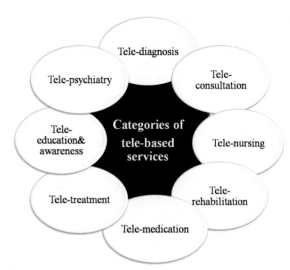

FIG. 10.2 Various tele-based services.

3.1 Telediagnosis

Telediagnosis is an assurance of the idea of an illness at a site distant from the patient based on Telehealth techniques for transmitted information. Tele-diagnostics, characterized as diagnosing from a separation, supposes remote access to test outcomes and medicinal archives of a patient who has no real physical contact with the doctor doing the examination remotely. This procedure can be done by utilizing different communication strategies, including telephonic discussions and instant messaging. In the last few years, many proposals have been developed for telediagnosis, two of them are explained as follows:

3.1.1 MTS application for telediagnosis

Mobile Telemedicine System (MTS) facilitates to obtain data from various instruments and its agreeable transmission and conveyance to social insurance and transfer to the healthcare provider using 3G/4G networks. MTS can be accessed in any geographical areas by using 3G/4G communication network [10] and web services to diagnose chronic diseases including diabetes, skin diseases and ECG [24,37]. An architecture of MTS is shown in Fig. 10.3, which comprises of two modules: client mobile and computer server.

3.1.2 Machine learning approach for telediagnosis

Machine learning, including deep neural networks, have been proved to be useful in various disease diagnosis. One of the examples of machine learning tele-based diagnosis is the diagnosis of Parkinson's

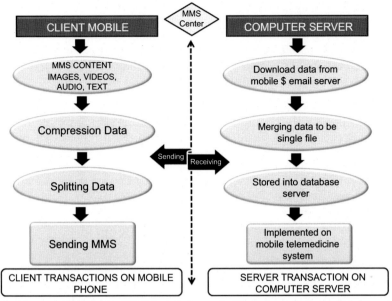

FIG. 10.3 Architecture of MTS using MMS for telediagnosis. (Credit by A. Setyono, M.J. Alam, C. Eswaran, Mobile telemedicine system application for telediagnosis using multimedia messaging service technology, Int. J. Wirel. Mob. Comput. 7 (4) (2014) 348−361.)

disease − a neurodegenerative disorder related to central nervous system and influences the motor system of the brain. Zhang [64] proposed smart-phone based approach for telediagnosis using deep neural network (Fig. 10.4). The idea of Parkinson's disease classification is based on text reading and speech recognition, following these steps:

i. The patient gives personal data such as age, gender health history and speech records as per given instruction in smart phone apps.
ii. The patient is requested to go through the data assuring that it is legitimate.
iii. Time recurrence based features are employed in order to analyze the speech record retrieved from the recorded voice samples
iv. Stacked Auto Encoder (SAE) and multiple layer encoders are used to encode and reduce the feature dimension, while k-nearest neighborhood (kNN) classifier is utilized for the training.
v. Diagnosis results can then be obtained by the patients.

3.2 Teleconsultation

Teleconsultation is a specific sort of tele-diagnostics which incorporates the period of discussion between the doctors, experts and patients where information isn't adequate for making an analysis. In spite of the conventional techniques, teleconsultations offer the

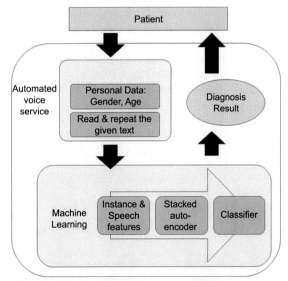

FIG. 10.4 Architecture of deep learning method for telediagnosis. (Credit by Y.N. Zhang, Can a smartphone diagnose Parkinson disease? A deep neural network method and telediagnosis system implementation, Parkinson's Dis. 2017 (2017).)

upside of taking out patient nearness at specific phases of the examination procedure. Serving heath care facilities in rural areas is a challenging task. Teleconsultation gives ease to people of rural areas through tele-communication by using web-based services [63]. A general framework for teleconsultation process between patients and doctors are as follows:

- Patient approach for the teleconsultation in the healthcare center through web-based facilities.
- Patient has to fill online form which is provided to him.
- Doctor reviews the online form of patients and start online real-time consultation [51].
- Outcomes are given to the patient and he has to pay electronic billing for consult.

Several web-based systems, called virtual polyclinic, were developed for teleconsultation that not only provides multimedia teleconsultation but also web-based healthcare records [30]. Patients can use these web-based healthcare records to exchange data among specialists and practitioners.

3.3 Telenursing

Telenursing facilitates nurses to examine and ascertain their patients by using web-based system and information of client, where doctors work as consultant (Fig. 10.5). It has been proved that telenursing raise the satisfaction in patients and increases the capability of nursing services [18]. In case of diabetes patient, it improves the effectiveness of HbA1c levels decrease the anxiety of diet therapy. Fig. 10.5 shows the cycle between patient and healthcare center. Patients communicate their health related issues to the doctor and after consultation, nurse provide the solutions to the patients. Patient gives feedback to the doctor after completing his treatment. Telenursing is provided to the patients in two ways: *Telephone nursing* and *telehomecare*. *Telephone nursing* helps in providing advice and information, appointments, symptoms managements and disease management to the patient according to the information of disease given by patient. *Telehomecare* provides supportive care, education, assessment, and disease management to the patient.

3.3.1 Telenursing system

A telenursingsystem framework utilizing web application were built up for overseeing health care utilizing the Internet services [23]. This web-based application was produced as a framework for nursing care utilizing bleeding edge data innovation. The highlights of Kawaguchi's telenursing framework incorporate having the capability to choose data innovation (on-request webcam discussions, email, and telephone calls) for intercession as per health require and care of nurse prerequisites. It has the facility to record understanding self-monitoring data as a feature of the wellbeing record, and diagramming of the self-observing data in the wellbeing record with the goal that patients can see their own particular information without any time limit. Telenursing consists of healthcare center server and mutual network which is connected to a nurse, a physician and a patient. A schematic overview of the Internet-based telenursing system proposed by Higano et al. [18] is depicted in Fig. 10.6.

3.4 Teletreatment and Telerehabilitation

Telerehabilitation is used to deliver rehabilitation services to distantly living people using ICT which gives sense of personal self-determination and authorization to the patient. Basically, they are turning into a dynamic accomplice as opposed to an inactive member in their care [55]. It empowers access to watch over people residing in remote regions or for the individuals who have portability issues related with physical disability, access to transport and financial components. It chops down the related travel expenses and time spent for both the social insurance supplier and the patient [20].

As telerehabilitation extends, persistent progression of care makes strides. It empowers clinicians to remotely connect with and convey understanding consideration outside of the restorative setting, along these lines dispensing with the issue of separation among clinician and patient [6]. The workflow for providing telerehabilitation services to the patient is shown in Fig. 10.7. Doctor first identify the patient's problem through the analysis and provide relevant solving factor to the patient. Treatment is started for recovery considering maintenance of the quality of life and set interposition settings. Treatment has provided

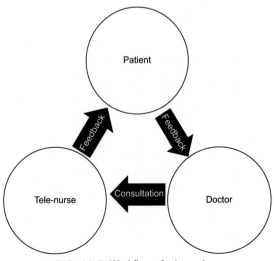

FIG. 10.5 Workflow of telenursing.

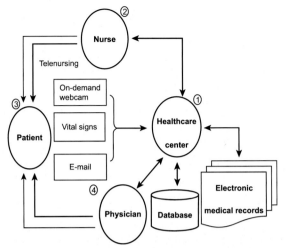

FIG. 10.6 Schematic outline of internet based telemedicine system. Telenursing consists of healthcare center server (1) and mutual network which is connected to a nurse (2), a physician (4) and a patient (3). The healthcare center mainly work for database in which graphing self monitoring and patient self monitoring information record are stored that patients are able to check on data by themselves from the house database at any time. Continue treatment and healthcare is applied through email by on-demand webcam, vital signs and telephone guidance. (Credit by K. Higano, T. Shibayama, M. Ichikawa, M. Motomura, H. Shimano, Y. Kawakami, et al., The effects of telenursing with goal attainment scaling in diabetic patients: a case report. Int. J. Nurs. Clin. Pract. 2 (2015) 117.)

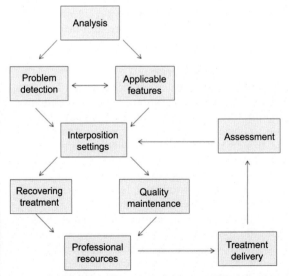

FIG. 10.7 Workflow for providing telerehabilitation services to patient. (Credit by V. Lubkina, G. Marzano, Building social telerehabilitation services, Procedia Comput. Sci. 77 (2015) 80–84.)

through the professional resources further that assessment is started.

Corriveau et al. [65] developed a multi-model telerehabilitation platform which led to the development of telerehabilitation software, called "Équipe Spécialisée en TéléRéadaptation À Domicile" (ESTRAD). Beside audio and video components, its different modules support an oximeter, an ECG, temperature, crucial signs, stationary bike, and speech therapy software. This platform is successfully used to treat various rehabilitation populations including orthopedics, pulmonary and neurologic.

3.5 Tele-Psychiatry

Tele-psychiatry is the act of conveying emotional well-being care remotely with the utilization of secure video conferencing innovation to take into account the assessment, determination, and treatment. Nowadays, there are several crisis of mental healthcare throughout the world with the need of administrations upgrading conventional practice techniques. Tele-psychiatry programming profits quality psychiatry care to almost everybody by utilizing ongoing progressions in correspondence innovation. Employing tele-psychiatry programming answers, a therapist can get best advices to any situation be it, the need for counseling with other medical practitioners or to get a moment supposition. Tele-psychiatry arrangement is a deliverer for the specialist situated in provincial or medicinally denied region, helping them to leave the box and treat their patients. On the basis of service delivery, tele-psychiatry introduces sub specialty which includes in-house tele-psychiatry, forensic tele-psychiatry, on-demand tele-psychiatry and scheduled tele-psychiatry [62].

4 BARRIERS, ETHICAL ISSUES AND AUTHORIZATION

4.1 Barriers

Telemedicine has proved its accomplishment in decreasing the geological and time snags caused in the receipt of care in conventional modalities with the same or more prominent adequacy; in any case, there are a few obstructions that should be tended to all together for telemedicine innovation to spread [27]. Some significant barriers are represented in Fig. 10.8.

4.1.1 Physical licensing

Telemedicine would permit to specialist for cross-state consultation that enables them to service maximum patients. Yet, doctors face several obstacles in the event that they want to practice in other state through the utilization of telemedicine, including controls that

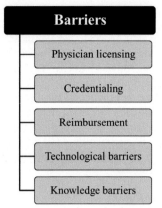

FIG. 10.8 Major barriers to telemedicine.

Several criteria have to meet with the end of goal for specialists to get reimbursement. State-by-state policy counselor must have the most forward information on these laws. When reimbursement is constrained, patients in the affected regions are underserved. High expenses, administrative and regulatory barriers which revolve around Tele-based healthcare services make the healthcare units experience angst, apprehensions and trepidation accordingly.

4.1.4 Technological barrier
The truancy of a broadband framework has severely affected tele-based services such as appeal video and store-and-forward administrations, which make an essential component of tele-based healthcare. Organization for Economic Cooperation and Development (OECD) gives a 15th rank to the U.S. broadband infiltration rate (26.4 associations for every 100 occupants) showing its debasement from a sixth rank back then in the year 2002. According to May 2013 survey, about 15% of American grownups (above 18 years of age) don't embrace internet services showcasing that homey-tele-based healthcare is not exhibited efficaciously.

As the incorporation and network between social insurance substances increase, in turn, allows a hearty innovation security approach to ensure classified patient money related and medicinal records to escalate as well. Security requires addition of classification of information and candor (amid transmission and maintenance), while improving its accessibility, convenience, and characterization of the measures for least necessities. An ongoing precise survey of the writing identified with telehealth security showed that there is a deficiency of institutionalization in telehealth security.

4.1.5 Knowledge barrier
Telemedicine, comprehensively characterized as the utilization of data innovation to convey restorative administrations over separations, is accountable for issues of openness, quality, and expenses of therapeutic care. In spite of the fact that telemedicine applications have multiplied as of late, their dissemination has stayed. This can be understood by viewing Attewell's hypothesis of information boundaries in order to clarify why dispersion of telehealth stays moderate [28]. A longitudinal, inserted numerous contextual analysis of telehealth programs in three restorative focuses in Boston and Massachusetts, uncovers that there are financial, authoritative, and social information hindrances that restrain the dissemination of telemedicine which are the specialized learning boundaries as

vary state-by-state. From time to time, experts may require a complete medical permit in both patient and specialist state where they live. Regulatory issues such as mal-practicing, authorization and attestation of the consultants and their instrumentations is one of the most crucial regulatory barriers in the efficient utilization of telehealthcare services. The only way out of this barrier is to allow the consultants to practice tele-consultations and related tele-services by giving them an authorized interstate license for which they might be required to compensate for the high expenditure. The rigorous nature of these principles beat advantages of telemedicine.

4.1.2 Credentialing
Credentialing procedure of telemedicine may get exceptionally sophisticated. For instance, in clinics that act from a cynosure and spoke model, all physicians must have telemedicine credential at all substitute clinics. The volume of printed material and administrating procedures are required for credentialing tedious through the rooftop. Instead of squandering time and high expenditure for finding a suitable licensed consultant for telehealthcare, it can be used to install a well equipped tele-based infrastructure, employing well trained and educated staff, according to the localities in a specific area of concern.

4.1.3 Reimbursement
In the field of telemedicine, there are two genuine financial issues: insufficient capital consumption and inadequacy of reimbursement. Variation in reimbursement models across the country is similar to licensing laws.

stated by Attewell. The abridgement of these boundaries involves serious learning endeavors by champions of utilizations inside adopter associations. They have to expand so that they imbibe attainability, therapeutic substantial, reimbursable, and institutionally bolstered applications with a specific end goal to legitimize the estimation of telemedicine and teleconsultations, respectively.

4.2 Ethical Challenges

Innovative advances in the public arena have brought huge changes and comfort to individuals. In any case, with innovation there is a level of depersonalization, as science is progressively supported over workmanship [22]. Some of the important ethical challenges which create obstacles in the development of telemedicine are depicted in Fig. 10.9 (Table 10.2).

For the mending experience to achieve its definitive objective of persistent advantage there must be an adjusted painstaking coalition of information, clinical expertize, and viable correspondence. Be it as it may, there is a fore fixing that must be observed in the intimacy and candor of a mutual individual association. That fixing is familiar to home mindfulness and contact that secures trust and reflects responsibility to a perfect consultation and healthcare. With expanding science and related technology, the present services which focus only on research and proof based solutions for healthcare problems don't budge an inch when it comes to

mutual understanding between the patient-consultant association [54]. Telehealth and telemedicine may energize depersonalization and curtail the trust and intimacy that characterizes the connection between

TABLE 10.2 Ethical Challenges, Legal and Ethical Issues of Telemedicines.		
S. No.	**Ethical Challenges**	**Legal and Ethical Issues**
1	Confidentiality	• Each patient has basic right not to have personal data disclosed to others without his/her permission. • It is required to make sure that the confidential information of patient can't be disclosed.
2	Avoiding harm	• Facilities should be provided to the patient to avoid any type of harm (mental or physical).
3	Informed Consent	• When the patient give permission to the doctor to disclose his/her personal information and physiological parameters then doctor should has to know the reasons of disclosing the information and what data will be revealed.
4	Establishing fees arrangements	• Each and every patient have right of healthcare so it should be provided keeping in view the financial condition of the patient. • Facilities of healthcare center should also provide to all patient who approach for it.
5	Service accessibility	• Service of healthcare center should be covered broad area of both rural and urban areas. • Provide suitable apparatus to the healthcare centers
6	Maintaining data	• Record history of patient that is taken before and after treatment should be stored.

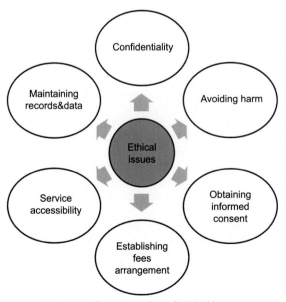

FIG. 10.9 Representation of ethical issues.

patients and their medical services suppliers. Telehealth innovation has existed for two decades, influencing wellbeing accessible to numerous patients in remote and generally underserved zones. Numerous patients have been profited by telehealth frameworks which would not have an access to medical services. Telemedicine has likewise empowered separation instruction and the quick, productive sharing of data, not just with patients but also between suppliers who are isolated either topographically or inside establishments. Using this type of innovation has shown to be practical and has brought about successful accomplishments in wellbeing for incessantly sick patients. From a mechanical and proof based point of view, it would create the impression that the virtual visit, similar to the vis-à-vis visit, is as solid and legitimate a sign of the mending relationship as an up close and a personal experience, however there are many moral concerns surrounding the same.

The telemedicine practice and the transmission of patient information are guided by the similar moral principles that have instructed all parts of specialist conduct. The ethical issues uncommon to telemedicine relate to the potential impact on the patching relationship, which go past the typical troubles of security and protection. The lack of physical check up touch, the depersonalization, the risk of mishandling and malpractice, possible unevenness during the conduct and management of telehealth organizations, are some of the ethical issues that should be considered.

4.3 Telemedicine Authorization

Telehealth authorization is a measurement that allows a person to practice telemedicine using a legitimate license. It enables the healthcare providers to use well planned healthcare services by initiating the system for the authorized use of telemedicine services. Prior authorization or approval is required by the physician from the health insurance companies so that specific medication can be prescribed to reduce medication cost. It also decides if the medication is suitable for the condition and medically necessary. Modern techniques for healthcare are used only when service providers have authorization, that is, they need its assessment before the proposal is covered. Drugs that require prior authorization include harmful drugs which often are misused or have dangerous side effects. Less expensive drugs prescribed by the physicians also needs authorization.

5 ENABLING BIOINFORMATICS AND COMPUTATIONAL TECHNIQUES TO IMPROVISE TELEHEALTH

Tele-based services encapsulate a complete range of all the medical services that one need in order to maintain a healthy life. Its umbrella is full of providence of services which have been mentioned in Section 3. Bioinformatics and Computational techniques have joined hands in order to improve the working of tele-based healthcare by giving better patient health care services. Telehealth approach is more of a "Connected Health Technology" (CHT) as it involves the amalgamation of technology and medical care [58]. Notwithstanding, the use of Bioinformatics and computational techniques into tele-based healthcare utilizes a bigger domain of population sizes, myriad geographical and topologies along with a diverse range of conditions which make it a better tool in this rat race of today [58].

5.1 Telegenomics and Telehealth

Telegenomics are a very useful tool for disseminating periodic counseling and has the capacity to assess patients who are suspicious to genetic conditions. It focuses to discern validated proof for consultations based on genetic of a person using video conferencing. This type of consultation is referred to as — "*Telegenetics-based consultations*" [17,34].

Hilgart et al. [19] have discerned in their work that patients are convinced with the use of tele-based healthcare in genomics services which includes genetics based counseling of patients and simultaneously diagnosis and prognosis. They have proved that telegenomics based consultations were more or less the same like the face-to-face consultations indicating that people are satisfied with the usage of telegenomics based medical deliveries and had no issues with the same. For instance, a research study proposed by Stalker et al. [52], determined that a geneticist counselor meticulously stated the pedigree of family and their ancestral history while a pediatrician conducted a physical diagnosis for genetic disorders. Their study concluded that tele-based genetic approach was way better in determining the genetic disorders when compared to the actual time diagnosis. Moreover, it also aids in providing rapid information about genetic disorders and malfunctioning of the body. Thus, telegenetics based counseling sessions had more benefits mainly reduced waiting consulting times and distances.

5.2 Biological–Omics Databases and Telehealth

Bioinformatics has opened up satisfactory opportunities in tele-based healthcare. One of the widely used bioinformatics application in tele-based medical care is BioWAP which stands for *Bioinformatics Wireless Application Protocol*, and is easily available and can be used on mobile devices. The BioWAP service is composed of all the major biological databases and tools which shows the implementation of the wireless tool [44]. Another personal digitized assistant (PDA) is a wireless ECG (Electro cardiogram) transmission system composed of an ECG signal module, a GPRS (General packet radio service) module and a pocket PC. Its main objective is to help in regular checkups and transferring the ECG signals of a cardiac patient. The system is capable of taking the ECG signals, projecting them on the LCD screen and simultaneously transmitting these signals to an ECG server [43].

Pavlopoulos et al. [39] designed an emergency telemedicine device which has the potential to escort the consultant at the emergency situation. This is based on the "*Store storing and stimulation*" type of tele-based healthcare where the information is exchanged during telehealth sessions which is very mandatory for administrative and clinical purposes. They developed a multimedia database which manages to collect data by the ambulance system and has a friendly interface which makes it a reliable and an easy toolkit. The multimedia database has features to display ECG's, X-ray, CAT scans and MRI imaging. Another database based on drug information and their potential targets is − DrugBank. This database amasses copacetic and beneficial information of various drugs and their protein targets and can be used in the practice of tele-based healthcare services. A person can him/herself check and gain knowledge about the medications which are prescribed to him/her by their consultants. This database has achieved appreciations for its usefulness in − pharmaceutical & medical, bioinformatics and cheminformatics fields of science (https://www.drugbank.ca/).

5.3 Manifolds of Computational Techniques and Telehealth

The advent of adroit tools and techniques of bioinformatics and computation archetypes has opened up a new path in medical systems and their mechanisms. Computational techniques, such as Artificial Neural Network (ANN), Fuzzy Logic Systems (FLS), Evolutionary Algorithms like Genetic Algorithm (GA) and Genetic Programming (GP), etc., makes it easy to provide a sensible justification when it comes to medical systems. These techniques can actually mimic the biological phenomenon and in turn help in assembling the information which eventually helps in making an appropriate conclusion in healthcare systems [3]. Modus operandi such as Kernel Density Estimation (KDE), Apriori's algorithm and Text Compression Analysis (TCA) are used for developing the decision making of a medical system. Several studies have utilized the same concept for identification of breast cancer in females which employs Bayesian networks as the main decision maker [56]. It has again been observed in the wireless capsule endoscopy, discerned as a pattern recognition system for characterizing the composition and the consistency and allocation of the endoscopic pictures [26]. Furthermore, it is also used in the selection of antibiotics for cardiac surgeries [8]. A multi-agent based healthcare system is focused to aid in tele-based services, regular patient check-ups and diagnosis [21].

5.4 Growing Data in Tele-Based Healthcare

It is obvious, with the growth of computational power the data has also grown tremendously over the past years. In terms of medical data, gigantic data has been and is still being collected using sensory tools, and is expanding with every passing moment. Quiescent and tabulated steps must be ensured so that this huge data is not lost and unequivocal interpretation and analysis can be done easily. For that, one has to look up to the concept of Big Data and Fog Data for the same. Data contraction, data mining and data assessment are the need of the hour which must be taken up by medical fraternity. One such motivation has been stated by Dubey et al. [15]. The authors have tried to put forth an exclusive architecture for the efficient working of telehealthcare services. The antecedent unit of the stated fog-computing based architecture is a low power enclosed machine that channels methods of data mining and data analytics completely on crude data which is extracted from many dressable sensors used for tele-based health services. The enclosed computer congregates the sensed data as time series, evaluates it, then on, searches for matching patterns existing. Common patterns are stored while emblematic patterns are transmitted. Furthermore, the enclosed computer system takes out clinically germane information which is eventually sent to the cloud.

5.5 Personal Sensors in Telehealthcare

With an expansion of computational intelligence, the society has made a good move in incorporating the IT intelligence with healthcare. We have become quite health conscious and every individual tries to keep up well with their health assessments. Many personal sensing devices have been launched lately which help every being to keep a track of their health. One can view this as tele-personal healthcare (TPH) systems. In the TPH systems, a person can watch out for his/her well-being without disturbing their complacencies. There is no need of visiting the local clinicians, one can simply purchase these easy to wear personal sensors and keep a track of their medical regimen. TPH systems are nothing but alliance between the personal area networks (PAN) and body sensing networks (BSN). Some of the TPH systems are mentioned as follows:

i. *Vital Connect:* It is an auto-schediastic healthcare strip which aims to congregate continuous important signs in remote settings [49].

ii. *Agglutinant Strips:* The famous company, Philips discharged a sticky patch to keep an eye on Lung functioning, especially Chronic Obstructive Pulmonary Disease (COPD) [40,41].

iii. *Fitbit:* These have actually become a part and parcel of people today. These are wristwatch-like sensors composed of a locomotion sensor for reckoning health indicators for instance, distance trotted, calorie intake and suspension of consciousness (http://www.fitbit.com/).

iv. *Blood-Glucose Meters:* Most commonly known meters are AccuChek which instantly gives a reading of the blood-glucose level of an individual. These are easy to use tools and are inexpensive (https://www.accu-chek.in/meter-systems/instant-s).

v. *Echo Wear:* It is a watch which gives a home-based speech tele-therapy which produces around 100 Mb data per day per person [14].

5.6 ITAREPS: Data Mining Approach to Telehealth

It is one of the best approaches to prevent schizophrenic episodes and was started up in Prague, Czech of Republic in 2006. The program was matured by first employing it for 400 people. With its success, it has now been utilized in many countries [35]. ITAREPS is a short questionnaire composed of questions on problems in sleeping - could either be insomnia or hypersomnia (over sleeping), restlessness, appetite changes, experiencing dysphoric symptoms, changes in behavior, problems in cognitive skills, hearing voices or hallucinations, etc. These questions are answered using a SMS message which is sent on the patient's phone. A persons schizophrenic state is assessed in the following way, if the answers range from 0 to 4, 0 means neither change nor improvement in the state, and 1 to 4 means mild to extreme cases of schizophrenia accordingly. If the total sum of answers increases a given threshold for a due period of time, ITAREPS forwards an alert message to the outpatient psychiatrist. The respective consultant then decides to either increase or decrease the prescribed medicament as per the requirement of the patient. Data mining approaches such as Naive Bayes, Support Vector Machines (SVM) and AdaBoost were employed to improve the performance and henceforth, improve its overall efficiency [35].

6 PRESENT OPPORTUNITIES AND FUTURE PROSPECTS

Tele-based healthcare is new to general public today, but it was initially specific for the armed forces and space administration stringently. With time and its tremendous efficacy, it expanded over and everywhere. Harking down the memory lane, one learns that it was in the mid 1800s when telegraph messages were used to know about the American soldiers who were wounded in the American Civil War and their treatment [5]. Polishing novelty to obtain better outcomes surely takes time and is a challenging task, but the newer form of e-healthcare comprises of the following components [48]:

i. The treasury aspect comprising *e-Commerce*.

ii. The learning aspect comprising of *e-Education*, such as — the training, awareness etc…

iii. The informatics aspect composed of *e-Informatics*, data segregation, its analysis, distribution etc…

iv. The *telehealth* aspect (Stock Storing & Stimulation, Distant Supervision & Actual time Interactions).

6.1 Potential of Telehealth and the World

The word '*potential*' refers to the capability of connected healthcare technology (CHT), which must be equal everywhere, which is quite a disputable talk. Developing countries face much of the barriers as mentioned in Section 4 of this chapter. One can take up the example of Africa to understand the lackadaisical approach of tele-based healthcare in the continent. Africa makes up a 14% of the world's population and is a hub of diseases. The disheartening fact of Africa is that it has a smaller section of health workers (3%) which have admittance to only 1% of the world healthcare expenses [61]. Moreover, it has a total of 320 consultants per 100,000 individuals, the ratio being arguable still [59]. In addition to Africa, Asia is another continent which faces more or less similar problems.

6.2 Challenges to Tele-Based Healthcare

The mostly observed loopholes in the path of efficient employment of telehealthcare services are mentioned as follows [1,46−48]:

i. *Lack of Medic Pupils*: To deal with the shortage of medical fraternity related to telehealthcare, more human resources can be trained and utilized at different levels.

ii. *Probabilistic Populace*: There is a common drift toward growing older population and has been materializing for the past many years now. Thus, it puts more grain tension on current healthcare systems. More attention needs to be paid to telehealth services to eliminate the tension.

iii. *Inadequacy of Awareness*: There is an erroneous understanding of connected healthcare technology (CHT) among people. The need of the hour is to expand a greater foundation of adequate knowledge among common people and professionals so that stupendous expertize and fortuitous contingencies can be worked on [47].

iv. *Destitution*: The five countries which were tagged to be "extremely poor" in 2010 were: India (33%), Nigeria (9%), People's Republic China (13%), Bangladesh (5%) and Democratic Republic of Congo (5%). Tele-based healthcare services which mainly pay heed to "proof-based medical needs" recognizes the indigence [1].

v. *Abominable Management*: With technology and medical practices, there is also the need of "being ready" to accept the revived healthcare management systems. It is all human psychology, for an advantageous functioning of tele-based healthcare systems, there must not be any resistance, which can only be brought in when there is proper awareness and willingness within the people to accept and adopt the change in the human well-being. Management of such telehealth systems is for sure a challenging task, but with its sophisticated learning and practice, its influential impact would open up new avenues [48].

If tele-based services start to face the real scenario and answer such desperate problems it will have to adopt a generic strategy which merges both the global and the local health systems. It makes e-health a more convenient and adaptable in developing countries which are in high need of transparency of medical healthcare organizations [66,67]. This is the only way tele-based health technology can win hearts and can provide a broad range of medical healthcare services through teleconsultations, telediagnosis, teletreatments, by expanding greater knowledge and expertize in this field through appropriate learning and training of individuals involved with healthcare. Furthermore, developing well-equipped infrastructure for its practice, which abstain all the physical barriers, has been explained in Section 4. It must be stated here that it is not the penultimate solution to the problem but, can help in the betterment of the current situation [48]. When tele-medical systems will be implemented in a sophisticated manner, it will help the underdeveloped countries to access a better and healthy medical care [38]. With all this, tele-based healthcare should be able to save on time and expenses furthermore it improves the suitability of admission to emergencies and hospital departments accordingly [57].

6.3 Future Prospects

Telehealthcare technologies have already set a benchmark for its counterparts. Some of the future prospects are as follows [25,29,53,57]:

i. Increment in the providence of enhanced delivery of medical care and elimination of regular follow ups to nearby clinicians.

ii. Can be utilized and installed in disease managements units which can reduce the risks of panic and strenuous attacks during calamitous situations.

iii. Also can be employed in early detections of disease risks in patients.

iv. Casual checkups and analysis can help in eliminating rates of disability, morbidity and mortality.

v. In the near future, with help of AI based techniques, telehealthcare will become a more united milieu for e-health for the general public.

vi. An intelligent decision making support systems for telehealthcare services if devised can do wonders and give substantial outcomes in the treatment and diagnosis of patients remotely based.

vii. For setting up a more reliable and private healthcare, more secured software and hardware are needed to protect patient's demanded policies.

viii. Need for well structured data management of medical records.

ix. With more and more awareness, it will need to develop and install telehealthcare units which in turn would make sophisticated public healthcare protection systems.

x. Nonetheless, thus it will contribute in both propagating medical healthcare in every region of the country and flourishing the nations employment rates.

7 CONCLUSION, DISCUSSION AND FUTURE DIRECTIONS

Telemedicine uses modern ICT to provide healthcare services to the remotely residing public and it can be seen as a powerful tool which is being utilized to reduce the long distance between remotely residing rural people and the healthcare services. It has myriad applications including consultations, diagnosis, nursing, medications and treatment, psychiatry and psychology, rehabilitation and some specialized services. Generally, computers, smart phones, radio, television etc., are the mode of delivering tele-based healthcare services. Bioinformatics has several potential applications in biology and medicine. It uses information technology to solve biological problems by interpreting and analyzing data to extract useful knowledge. Both genomics and telemedicine are two technologies driven and rapidly growing area which promise to deliver new configuration of healthcare without any boundary of distance and time. Several healthcare sectors started bringing telemedicine into precision medicine by integrating the technology to genetic and genomic research centers. It will bring together researchers, clinicians, and patients by using remote conferencing to expedite the impact of genetics and genomics on the diagnosis and treatment of diseases. As a result, genomic telemedicine can be introduced as tele-based healthcare. Further, due to reduced cost of DNA sequencing, personal genomic data may be easily available to both healthcare center and patients which give patients access to genomically informed healthcare.

Telemedicine has proved its accomplishment in decreasing the geological and time snags, there are a few obstructions that should be tended to all together for telemedicine innovation to spread. Some significant barriers and ethical challenges are maintaining patient's records and data, service accessibility, confidentiality, obtaining informed consent and so on. Telehealth authorization allows a person to practice telemedicine using a legitimate license which enables the healthcare

workers to use well planned healthcare services. Prior authorization or approval is required by the physician before prescribing any harmful drugs or less expensive drugs. Bioinformatics and Computational techniques have joined hands in order to improve the working of tele-based healthcare by giving better patient healthcare services. The use of Bioinformatics and computational techniques into tele-based healthcare utilizes a bigger domain of population sizes, myriad geographical and topologies along with a diverse range of conditions which make it a better tool in this rat race of today. From bioinformatics perspective, telegenomics are a very useful tool for disseminating periodic counseling and has the capacity to assess patients who are suspicious to genetic conditions, referred to as — *"Telegenetics-based consultations"*. It has opened up several opportunities in tele-based healthcare including BioWAP (Bioinformatics Wireless Application Protocol), PDA (personal digitized assistant) for ECG transmission, multimedia databases for data collection by ambulance, DrugBank, computational intelligence application for critical decision making, application of big data and fog data, and so on. In addition, several personal sensing have been launched which help to keep a track of our health, such as tele-personal healthcare systems which is a personal area networks and body sensing networks, blood-glucose meters, echo wears. ITAREPS — a data mining approach to telehealth, has been used treat insomnia, hypersomnia, restlessness, appetite changes, changes in behavior, hallucinations, etc.

Some of the challenges to telehealthcare services are lack of medic pupils, probabilistic populace, inadequacy of awareness, destitution, and abominable management. Telehealthcare technologies have already set a benchmark for its counterparts. Some of the future prospects are (i) enhanced delivery of medical care and elimination of regular follow ups, (ii) installation of disease managements units to reduce risks of panic attach of diseases, (iii) application of more artificial intelligence based techniques for critical decision making, semi-automated surgery, better design of drugs, (iv) secure software and hardware for more reliable and private healthcare, need for well structured medical record management, and (v) propagation of telehealth services would flourish the nations employment rates.

The future is open to changes when it talks about *Tele*-based healthcare systems. This diagnostic and prognostic strategy has already gained many positives since its arrival and the governments worldwide are also stringently paying heed to its dynamics. The government funding in this medical domain has gained drastic momentum and is still going very high and will surely make it one of the best remedial approach in case of emergencies or resident crucial cases for sure. Another aspect is that this e-based healthcare management will introduce tele-based milieu to consultants helping them to interact with their patients without stepping out of their houses or workplaces across the country. With this, great reliability will be achieved in managing patient results and their follow-ups by their particular physicians. Furthermore, with lightening fast growth of electronic-based healthcare systems, it will also be encouraged, accepted and supported by the common public. Once demands of people increase for tele-healthcare industry, there won't be any looking back!

REFERENCES

[1] A. Alwan, Global Status Report on Noncommunicable Diseases 2010, World Health Organization, Geneva, 2011.

[2] American Telemedicine Association, *What Is Telemedicine?* 2013. Washington, DC.

[3] E.R. Andersson, et al., Recent Advances in Decision Making, SCI 222, 2009, pp. 145−159.

[4] P. Andreas, Nocturnal hemodialysis: dialysis for the new millennium, Can. Med. Assoc. J. 161 (9) (1991).

[5] R. Bashshur, G.W. Shannon, History of telemedicine: evolution, context, and transformation, Healthc. Inform. Res. 16 (2009) 65−66.

[6] D.M. Brennan, S. Mawson, S. Brownsell, Telerehabilitation: enabling the remote delivery of healthcare, rehabilitation, and self management, Stud. Health Technol. Inf. 145 (2009) 231−248.

[7] N. Brown, A brief history of telemedicine, Telemed. Inf. Exchange 105 (1995) 833−835.

[8] P.B. Cerrito, Choice of antibiotic in open heart surgery, Intell. Decis. Technol. 1 (1−2) (2007) 63−69.

[9] Chaudhry, et al., Systematic review: impact of health information technology on quality, efficiency, and costs of medical care, Ann. Intern. Med. 144 (10) (2006) 742−752.

[10] Y. Chu, A. Ganz, Mobile Telemedicine Systems Using 3G Wireless Networks, 2005. Retrieved November, 20, 2007.

[11] A. Darkins, P. Ryan, R. Kobb, et al., Care coordination/home telehealth: the systematic implementation of health informatics, home telehealth, and disease management to support the care of veteran patients with chronic conditions, Telemed. J. E Health 14 (10) (2008) 1118−1126.

[12] A. Dasgupta, S. Deb, Telemedicine: a new horizon in public health in India, Indian J. Community Med. 33 (1) (2008) 3−8.

[13] B. Dinesen, B. Nonnecke, D. Lindeman, E. Toft, K. Kidholm, K. Jethwani, et al., Personalized telehealth in the future: a global research agenda, J. Med. Internet Res. 18 (3) (2016).

[14] H. Dubey, J.C. Goldberg, et al., Echo Wear: smartwatch technology for voice and speech treatments of patients with Parkinson's disease, in: Proceedings of the Wireless Health 2015 on National Institutes of Health, ACM, 2015.

[15] H. Dubey, J. Yang, et al., Fog data: enhancing telehealth big data through Fog computing, in: Proceedings of the ASE Big Data & Social Informatics, ACM, 2015, p. 14.

[16] F. Fatehi, N.R. Armfield, et al., Clinical applications of videoconferencing: a scoping review of the literature for the period 2002-2012, J. Telemed. Telecare 20 (7) (2014) 377−383.

[17] J. Gray, K. Brain, et al., A pilot study of telegenetics, J. Telemed. Telecare 6 (2000) 245−247.

[18] K. Higano, T. Shibayama, M. Ichikawa, M. Motomura, H. Shimano, Y. Kawakami, et al., The effects of telenursing with goal attainment scaling in diabetic patients: a case report, Int. J. Nurs. Clin. Pract. 2 (2015) 117.

[19] J.S. Hilgart, J.A. Hayward, et al., Telegenetics: a systematic review of telemedicine in genetics services, Genet. Med. 14 (9) (2012) 765−776.

[20] D. Kairy, P. Lehoux, C. Vincent, M. Visintin, A systematic review of clinical outcomes, clinical process, healthcare utilization and costs associated with telerehabilitation, Disabil. Rehabil. 31 (6) (2009) 427−447.

[21] E. Kang, et al., Remote control multi-agent system for u-healthcare service, in: N.T. Nguyen, A. Grzech, R.J. Howlett, L.C. Jain (Eds.), KES-AMSTA 2007. LNCS (LNAI), vol. 4496, Springer, Heidelberg, 2007, pp. 636−644.

[22] B. Kaplan, S. Litewka, Ethical challenges of telemedicine and telehealth, Camb. Q. Healthc. Ethics 17 (4) (2008) 401−416.

[23] T. Kawaguchi, M. Azuma, K.I. Ohta, Development of a telenursing system for patients with chronic conditions, J. Telemed. Telecare 10 (4) (2004) 239−244.

[24] A. Shaikh, M. Misbahuddin, M.S. Memon, A system design for a telemedicine health care system, in: International Multi Topic Conference, 2008, pp. 295−305.

[25] B.A. Kobrinskii, E-health and telemedicine: current status and future steps, E-Health Telecommun. Syst. Netw. 3 (2014) 50−56.

[26] V.S. Kodogiannis, Decision support systems in wireless capsule endoscopy, Intell. Decis. Technol. 1 (1−2) (2007) 17−32.

[27] C.S. Kruse, P. Karem, K. Shifflett, L. Vegi, K. Ravi, M. Brooks, Evaluating barriers to adopting Telemedicine worldwide: a systematic review, J. Telemed. Telecare 24 (1) (2018) 4.

[28] C. LeRouge, M.J. Garfield, Crossing the telemedicine chasm: have the US barriers to widespread adoption of telemedicine been significantly reduced? Int. J. Environ. Res. Public Health 10 (12) (2013) 6472−6484.

[29] F. Lindén, EpSOS local data providers, Acta Inf. Med. (3) (2009) 142–145.

[30] S. Lončarić, E. Pek, A. Margan, M. Šilović, Virtual poly-clinic: a web-based system for medical teleconsultation, in: Telemedicine Symposium Emerging Technologies and Life Sciences: Medicine and Communication, January 2000.

[31] V. Lubkina, G. Marzano, Building social telerehabilitation services, Procedia Comput. Sci. 77 (2015) 80–84.

[32] G.M. Kuo, J.D. Ma, K.C. Lee, P.E. Bourne, Telemedicine, genomics and personalized medicine: synergies and challenges, Curr. Pharmacogenomics Personalized Med. (CPPM) 9 (1) (2011) 6–13.

[33] J. Matusitz, G.M. Breen, Telemedicine: its effects on health communication, Health Commun. 21 (1) (2007) 73–83.

[34] J.A. Mitchell, G. Demiris, Telegenetics: the next phase in the provision of genetics services? Genet. Med. 7 (2005) 1–2.

[35] P. Nalveka, V. Svatek, Improvement efficiency of telemedical prevention programs through data mining on diagnostic data, in: 4th International Conference on Bioinformatics and Biomedical Technology, IPCBEE, vol. 29, 2012.

[36] C.K. Or, B. Karsh, A systematic review of patient acceptance of consumer health information technology, J. Am. Med. Inf. Assoc. 16 (4) (2009) 550–560.

[37] A.A. Tahat, Mobile personal electrocardiogram monitoring system and transmission using MMS, in: International Caribbean Conference on Devices, Circuits and Systems, 7th, 1–5, IEEE, 2008.

[38] S.K. Pal, G.S. Pandey, et al., Telemedicine: E-Health and hospital of the future, J. Sci. Ind. Res. 61 (2002) 414–422.

[39] S. Pavloupos, A. Berler, et al., Design and development of a multimedia database for emergency telemedicine, Technol. Health Care 6 (2,3) (1998) 101–110.

[40] Philips, COPD Sensor, 2014. https://www.engadget.com/2014/10/13/philips-copd-sensor/.

[41] J. Polisena, K. Tran, et al., Home telehealth for chronic obstructive pulmonary disease: a systematic review and meta-analysis, J. Telemed. Telecare 16 (3) (2016) 120–127.

[42] S. Qazi, Personalized medicines in psychiatry: promises and challenges, J. Appl. Comput. 2 (2) (2017) 50–55.

[43] Z. Qiang, W. Mingshi, A wireless PDA-based electrocardiogram transmission system for telemedicine, in: Engineering in Medicine and Biology Society, 2005. IEEE-EMBS 2005. 27th Annual International Conference of the, IEEE, January 2006, pp. 3807–3809.

[44] P. Riikonen, J. Boberg, T. Salakoski, M. Vihinen, Mobile access to biological databases on the Internet, IEEE Trans. Biomed. Eng. 49 (12) (2002) 1477–1479.

[45] Z. Salehahmadi, F. Hajialiasghari, Telemedicine in Iran: chances and challenges, World J. Plast. Surg. 2 (1) (2017) 18–25.

[46] R.E. Scott, M. Mars, The spectrum of needed e-Health capacity building – towards a conceptual framework for e-Health 'training', Stud. Health Technol. Inf. 206 (2014) 70–77.

[47] R.E. Scott, M. Palacios, A. Saeed, Foundations in global e-health: a global capacity-building course, Stud. Health Technol. Inf. 161 (2010) 122–131.

[48] R.E. Scott, M. Mars, Telehealth in the developing world: current status and future prospects, Smart Homecare Technol. TeleHealth 3 (2015) 25–27.

[49] N. Selvaraj, Long term remote monitoring of vital signs using a wireless patch sensor, in: Healthcare Innovation Conference (HIC), IEEE, 2014, pp. 83–86.

[50] A. Setyono, M.J. Alam, C. Eswaran, Mobile telemedicine system application for telediagnosis using multimedia messaging service technology, Int. J. Wirel. Mob. Comput. 7 (4) (2014) 348–361.

[51] A.D. Sorknaes, M. Bech, et al., The effect of real-time tele-consultations between hospital-based nurses and patients with severe COPD discharged after an exacerbation, J. Telemed. Telecare 19 (8) (2013) 466–474.

[52] H.J. Stalker, R. Wilson, et al., Telegenetic medicine: improved access to services in an underserved area, J. Telemed. Telecare 12 (2006) 182–185.

[53] S. Tachakra, X.S. Wang, et al., Mobile e-health: the unwired evolution of telemedicine, Telemed. J. E Health 9 (3) (2003) 1–11.

[54] H. Tanriverdi, C.S. Iacono, Diffusion of telemedicine: a knowledge barrier perspective, Telemed. J. 5 (3) (1999) 223–244.

[55] D. Theodoros, T. Russell, R. Latifi, Telerehabilitation: current perspectives, Stud. Health Technol. Inf. 131 (2008) 191–210.

[56] M. Velikova, A decision support system for Breast cancer detection in screening Programs, in: Proceedings of the 18th European Conference on Artificial Intelligence, IOS Press, Amsterdam, 2008, pp. 658–662.

[57] M. Vitacca, S. Scalvini, et al., Telemedicine and home care: controversies and opportunities, Breathe 3 (2) (2006) 149–158.

[58] S.M. Walker, Medical informatics and Telemedicine how digital technology is transforming healthcare and telemedicine, J. Health Med. Inf. 8 (4) (2017) 70.

[59] D. Watson, K. McGrail, More doctors or better care? Healthc. Policy 5 (10) (2009) 26–31.

[60] World Health Organization, Traditional Medicine: Definitions, 2008.

[61] World Health Organization, World Health Report 2006: Working Together for Health, WHO Press, Geneva, 2006, pp. 1–209.

[62] P. Yellowlees, M. Burke Parish, Á. González, S. Chan, D. Hilty, A.M. Iosif, et al., Asynchronous telepsychiatry: a component of stepped integrated care, Telemed. E Health 24 (5) (2018) 375–378.

[63] P. Zanaboni, S. Scalvini, P. Bernocchi, G. Borghi, C. Tridico, C. Masella, Teleconsultation service to improve healthcare in rural areas: acceptance, organizational impact and appropriateness, BMC Health Serv. Res. 9 (1) (2009) 238.

[64] Y.N. Zhang, Can a smartphone diagnose Parkinson disease? A deep neural network method and telediagnosis system implementation, Parkinson's Dis. 2017 (2017).

[65] H. Corriveau, M. Tousignant, S. Gosselin, P. Boissy, L. Azevedo, G.J. Gelderblom, Patients satisfaction with an in-home telerehabilitation exercise program and physiotherapists' satisfaction toward technology for an acute stroke population: a pilot study, Assistive Technology: from research to practice 33 (2013) 753−757.

[66] R.E. Scott, M.F. Palacios, E-Health challenges for going Global, in: C.M. Scott, W.E. Thurston (Eds.), Collaboration in context. Calgary: Institute for Gender Research & Health Promotion Research Group, University of Calgary Press, 2003, p. 45.

[67] R.E. Scott, M. Mars, Principles & Framework for e-Health Strategy development, J. Med. Internet Res 15 (7) (2013) e-155.

FURTHER READING

[1] M. Akhavansaffar, A. Nakhaei, A security architecture for use in mobile medical and electronic health, J. Comput. 13 (7) (2018) 794−805.

[2] A.G. Ekeland, A. Bowes, S. Flottorp, Effectiveness of telemedicine: a systematic review of reviews, Int. J. Med. Inform. 79 (11) (2010) 736−771.

[3] A. Galletta, L. Carnevale, A. Bramanti, M. Fazio, An innovative methodology for Big Data Visualization for telemedicine, IEEE Trans. Industr. Inf. (2018).

[4] N.M. Hjelm, Benefits and drawbacks of telemedicine, J. Telemed. Telecare 11 (2) (2005) 60−70.

[5] W.Y. Hsu, Clustering-based compression connected to cloud databases in telemedicine and long-term care applications, Telematics Inf. 34 (1) (2017) 299−310.

[6] Z. Jin, Y. Chen, Telemedicine in the cloud era: prospects and challenges, IEEE Pervasive Comput. 14 (1) (2015) 54−61.

[7] E.A. Krupinski, History of telemedicine: evolution, context, and transformation, Telemed. E Health 15 (8) (2009) 804−805.

[8] A.C. Norris, A.C. Norris, Essentials of Telemedicine and Telecare, Wiley, Chichester, 2002, p. 106.

[9] D.M.M. Pacis, E.D. Subido Jr., N.T. Bugtai, Trends in telemedicine utilizing artificial intelligence, in: AIP Conference Proceedings, Vol. 1933(1), AIP Publishing, February 2018, p. 040009.

[10] F. Rezaeibagha, Y. Mu, Practical and secure telemedicine systems for user mobility, J. Biomed. Inf. 78 (2018) 24−32.

[11] F. Stradolini, N. Tamburrano, T. Modoux, A. Tuoheti, D. Demarchi, S. Carrara, IoT for telemedicine practices enabled by an Android™ application with cloud system integration, in: Circuits and Systems (ISCAS), 2018 IEEE International Symposium on, IEEE, May 2018, pp. 1−5.

m-Health in Public Health Practice: A Constellation of Current Evidence

SURAJUDEEN ABIOLA ABDULRAHMAN[a] • KURUBARAN GANASEGERAN[b]

[a]Emergency Medicine Department, James Paget University Hospital, Great Yarmouth, Norfolk, United Kingdom;
[b]Clinical Research Center, Seberang Jaya Hospital, Ministry of Health Malaysia, Seberang Jaya, Penang, Malaysia

1 INTRODUCTION

Public health practice and research has benefitted tremendously from the expansion and growth of m-Health technologies. Not only has m-Health revolutionized service delivery in key areas related to family health, it has also improved approaches to prevention and control of non-communicable diseases (NCDs), including mental health. The capacity for epidemic or pandemic preparedness and response has improved disease surveillance, outbreak detection and management with the advancement of better communication platforms and monitoring systems through m-Health. More than ever before, m-Health has expanded the capacity for knowledge dissemination, health behavior change and better self-health practices at the individual, community and population level matrices. This has undoubtedly improved community participation in disease prevention programs as well as advocacy and policy dialogues for effective health promotion. Table 11.1 summarizes the context in which m-Health has been successfully applied in public health practice.

The availability of local infrastructure such as reliable networks (cellular, broadband and wireless) and advances in technology (mobile devices and apps, network solutions, interoperability of platforms and health information systems, diversity of contents etc) will continue to enable public health practitioners to leverage m-Health solutions through advanced mobile functions.

2 M-HEALTH BENEFITS IN PUBLIC HEALTH PRACTICE

2.1 Community-Related Approach to Service Delivery

2.1.1 Maternal and child health

2.1.1.1 Antenatal and postnatal care attendance, skilled midwifery and antenatal health.
According to 2016 UNICEF data, only about 50% of women worldwide receive the recommended level of care during pregnancy. The report revealed a global disparity between the proportion of pregnant women who received antenatal care with a skilled health personnel at least once (86%) and those who attended the recommended minimum of four visits (62%), with wider disparities observed in sub-Saharan Africa (52%) and South Asia (46%) where maternal mortality rates are highest [1]. Literature has highlighted successful applications of m-Health technologies in improving antenatal, midwifery and post-natal care in low-to-middle income countries (LMICs) [2]. Evidence from RCTs and other studies conducted in Zanzibar, Thailand, Malawi, Sierra Leone and Nigeria [3–7] suggests that the use of text-messaging and/or voice messaging (singly or in combination with ancillary interventions such as provision of call vouchers, toll-free case management hotline, finance vouchers, micro health insurance etc) to provide health education and appointment reminders to pregnant mothers significantly improved antenatal care visits, skilled attendance at birth,

Telemedicine Technologies. https://doi.org/10.1016/B978-0-12-816948-3.00011-8

TABLE 11.1
Common Contexts of m-Health Application in Public Health Practice.

Context/Purpose	BASIC MOBILE PHONES		SMART PHONES			Others[b]
	Basic Text Message	Voice Calls	Special Apps	Wearable Devices[a]	Smart Text/Instant Messaging/Chats	
MATERNAL HEALTH						
ANC attendance	✔	✔			✔	
PNC attendance	✔	✔			✔	
Skilled birth attendance	✔	✔			✔	
Family planning	✔	✔	✔		✔	
Health education	✔	✔			✔	
CHILD HEALTH						
Immunization	✔	✔	✔		✔	
Nutrition	✔	✔			✔	
NON-COMMUNICABLE DISEASES						
Physical inactivity	✔	✔	✔	✔	✔	✔
Diet control	✔	✔	✔		✔	✔
Smoking	✔	✔	✔		✔	✔
Alcohol use	✔	✔	✔		✔	✔
Sleep patterns	✔		✔	✔	✔	✔
Health education, promotion and disease prevention	✔	✔	✔	✔	✔	✔
COMMUNICABLE DISEASES						
Surveillance	✔	✔	✔		✔	✔
Outbreak detection and management	✔	✔	✔		✔	✔
Diagnosis	✔	✔	✔		✔	✔
Monitoring and treatment	✔	✔	✔		✔	✔
Occupational Health	✔	✔	✔		✔	✔
Other areas[c]	✔	✔	✔		✔	✔

NOTE: Some of the above interventions are used singly or in combination with other range of non m-Health related interventions, and with varying degrees of success.
[a] These include watches, wristbands (accelerometers, pedometers, etc.), cutaneous patches, and clothing with smartphone-enhanced connectivity.
[b] Includes digital workplace interventions, virtual reality, desktop-based applications, web-coaching, video messaging and other behavioral therapies.
[c] Includes mobile job aids for healthcare workers, drugs and commodity logistics and supply chain management system.

post-natal care and health facility utilization rates. In addition, these interventions enhanced communication between healthcare workers and pregnant women, and between healthcare workers and traditional birth attendants [6]. Variable results regarding the effectiveness of these interventions in improving antenatal care knowledge [8,9] or breast-and infant feeding practices [10,11] have also been reported.

These evidences have underpinned the conceptualization and implementation of several health intervention programs that addresses a wide range of barriers to maternal and health care services such as lack of information at the community level about locally available services, forgetting clinic appointment schedules, difficulty in reaching health facilities (large distance to facilities and lack of affordable transport), and ineffective communication between patients and community-based or facility-based healthcare workers [12−16]. For example, the United Nations Foundation-funded WAHA International's m-Health program implemented in the Tambacounda region in eastern Senegal employed a comprehensive communication and transportation support strategy in which a mass communication campaign via SMS was provided to the general community to sensitize them on the locally available maternal and neonatal health services, while targeted SMS communication campaign was provided to pregnant women and new mothers to inform them of the importance of ANC visits, delivery with a skilled birth attendant, and postnatal care. The project addressed transportation barriers through the implementation of a motorcycle ambulance system, a call center to efficiently manage the referral system, and provision of mobile phones to ambulance drivers to coordinate the referral process. Additionally, women who attended ANC consultations were provided with a mobile phone SIM card for receiving informational messages or for contacting a healthcare worker at the call center, in addition to receiving additional credit as an incentive for attending their antenatal care consultation at the health care facility [12]. Where such interventions are fully integrated with the existing health care system, the potential to achieve project outcomes and sustainability while also meeting clients' expectations are further enhanced.

2.1.1.2 Family planning.
As at 2010, a worldwide contraceptive prevalence of 63.3% and an unmet need for family planning of 12.3% was reported among women aged 15−49 years who were married or in a union [17]. By 2015, the global contraceptive prevalence had marginally increased to 63.6%, while the unmet need for family planning dropped to 11.9% [18]. Unmet need is especially higher among adolescents, migrants, urban slum dwellers, refugees and women in the postpartum period. These less than desirable statistics on family planning, the projected growth in the number of contraceptive users worldwide worsened by severe shortages in skilled providers, has catalyzed the need for increased investments and more innovative solutions to address the gaps, with m-Health offering promising results in this regard.

The adoption of m-Health technologies in providing the information that couples need to make informed family planning decisions or address their concerns regarding same, has been largely successful [19]. For example, the United Nations foundation-funded Mobile for Reproductive Health (m4RH) [20] and Mobile Alliance for Maternal Action (MAMA) [15] projects provide evidence-based results in this area. FHI360's m4RH − implemented in Kenya and Tanzania − developed family planning text messages based on best practices from health communication programs, guided by international and context-specific standards. The SMS were delivered to men and women through an automated, interactive and on-demand text message (SMS) system that provides essential facts about contraception while addressing common misconceptions. In addition, the system allowed users to locate nearby family planning clinics. Similarly, MAMA developed and disseminated adaptable messages based on WHO and UNICEF guidelines, to extend hospital-based support provided to pregnant women by providing them with health information and promotional messages regarding pregnancy, postnatal and child care.

Beyond these, m-Health technologies have been leveraged to improve accessibility and quality of family planning services, addressing commodity logistics, and monitoring and evaluation of family planning services in several parts of the world including Tanzania and India. Results from a project implemented in Shinyanga region of Tanzania by D-tree International and Pathfinder International indicate that not only does m-Health (specific mobile phone family planning app with or without financial incentives) improve workflow and quality of family planning services provided by community health workers, it also improved client registration and follow-up, as well as clients' adherence to and satisfaction with their chosen contraceptive method [21]. In India, the United Nation Foundation-funded Institute for Reproductive Health's CycleTel offers Standard Days Method of family planning service to eligible women through text messages [22]. Through this service, women are alerted of their unsafe (fertile) days each month through SMS and encouraged to use a barrier method or abstain from sex during this period. In addition, reproductive health information is provided to the women through this platform.

Despite these evidences, not much is known regarding the potential efficacy of m-Health interventions on post-abortion family planning (PAFB) [23], an area requiring more systematic approaches and studies to establish its efficacy or otherwise [24]. Further, literature suggests that interventions comprising a variety of daily educational text-messages and oral contraceptive (OC) reminders [25] were more effective compared to simple text message interventions (contraception reminders) alone [26,27]. It is imperative for program managers, public health practitioners and policy makers to conduct social diagnosis and formative research to understand the unique socio-cultural determinants of family planning coverage and service utilization among their population, in order to successfully tailor m-Health strategies that may benefit them.

2.1.1.3 Childhood immunization.
The World Health Organization (WHO) 2017 report estimates that about 2−3 million annual deaths from vaccine-preventable diseases (VPDs) are saved by immunization [28]. Despite the obvious benefits of vaccination, global vaccination rates have stalled at 86% for the past few years, a situation requiring concerted efforts by all stakeholders to address the gaps in coverage by reaching an estimated 19.5 million infants who currently miss out on basic immunization services. As with under-5 mortality rate, the challenges facing expansion of childhood immunization coverage are more evident in developing countries.

Like earlier systematic reviews [29], a recent study by Abdulrahman and Olaosebikan [30] which examined literature evidences regarding the efficacy of m-Health interventions in improving childhood immunization coverage, suggested a global consensus from high quality studies (RCTs) regarding the efficacy of m-Health interventions − used singly or in combination with other interventions − in improving childhood immunization coverage (uptake and completion rates) in many parts of the world.

Most notably within the scope of public health programming, m-Health has been deployed in many parts of the developing world to improve immunization coverage (uptake and completion rates), timeliness of receipt, commodity logistics and monitoring by digitizing the entire cascade using mobile phones or computers [31−34]. While majority of these innovations have explored the ubiquitous availability of mobile phones to target pregnant women and new mothers with simple SMS reminders on immunization clinic appointments, others have employed specialized mobile apps that enhance real-time update and reporting of vaccination records, as well as vaccination schedule and stocks monitoring. For example, the United Nations and WHO worked in partnership with stakeholders in Vietnam [31], Nepal [32] and Pakistan [33] to address the issue of poor vaccination coverage using variety of m-Health solutions described above. In one of such robust systems implemented by PATH in Ben Tre Province, Southern Vietnam, the project sought to enhance real time access to immunization data, and also allow timely generation of reports at the community health center and district level. The project implemented a digital immunization registry system comprising of a web-based application accessible via smartphone or computer, allowing mother-infant pairs to be registered and receive SMS reminders on immunization clinic appointments and also track the vaccines received. The system also improves the quality and efficiency of services provided by the healthcare workers by generating reports on the number and types of vaccine they need to administer every month as well as the list of individuals due for vaccination, thereby reducing the chances of stock-out of essential vaccines and commodities and minimizing drop-outs or loss to follow up. In the long-term, the project aimed to replace the existing paper-based immunization registry system with a more efficient and reliable digital system that improves performance and confidence in the health care system.

Today, in many parts of Asia, Africa and South America, the use of m-Health (particularly SMS reminders for immunization clinic appointments) is gaining popularity as a quick win, yet sustainable and cost-effective approach to improving immunization coverage, timeliness of receipt, commodity logistics and monitoring.

2.1.1.4 Maternal and child nutrition.
In many developing countries, undernutrition remains a very important cause of mortality among children under 5 years, accounting for nearly 50% of all deaths in this age group (about 3 million annual deaths) [35]. Not only does undernutrition increase the risk of children dying from common infectious diseases, it also worsens the frequency, severity and chances of recovery from these diseases, with increased potential for worse outcomes among undernourished children less than 1000 days old. Child undernutrition has a very important association with nutritional status of their mothers. Children whose mothers are underweight (with a body mass index less than 18.5 kg/m^2) are much more likely to be stunted, wasted and underweight than other children [36].

Whereas poor nutrition among mothers and children are largely preventable through concerted multisectoral efforts that include poverty alleviation, food security and women empowerment, public health approaches aimed at providing nutritional education, early detection and referral, nutrition and growth monitoring of children have been shown to yield better outcomes at population level. The emerging use of m-Health in addressing the challenge of maternal and child undernutrition has shown promising results and offers hope to public health programmers in this regard. One of such innovative use of m-Health is the United Nations Foundation-funded Society for the Elimination of Rural Poverty's (SERP) mobile Nutritional Day Care Centers (mNDCC) project implemented in rural Andhra Pradesh India [37] to improve monitoring and reporting between local and state-level actors, enhance tracking of nutrition, health status and behavior of clients enrolled in the program. The project uses the mNDCC device — a mobile handset with several modules — to survey, enroll, track and report the health (hemoglobin level, immunization status, antenatal care and prenatal care), nutrition (feeding practices, growth monitoring) and behavior (daily attendance at the NDCCs) of enrollees. Based on the information inputted into the system, relevant health-related action can be taken based on software-generated alerts as well as decision-making on other program-related matters including financial information. The mNDCC provides real-time data to guide timely decision-making on the functionality and performance of the women's self-help groups (NDCC).

2.2 Non-Communicable Diseases Prevention and Risk Factor Modification

In the last decade, most LMICs have witnessed a progressive increase in the prevalence of NCDs and its risk factors believed to result from increasing life expectancy, industrialization and westernization of lifestyle. Two-thirds of all NCD deaths (most importantly from cancer, cardiovascular and respiratory disease as well as diabetes) occur in these countries and coupled with the high prevalence of communicable diseases such as TB, this "dual burden of disease' exerts significant negative effect on the health care system and economy of these countries [38]. Experts have called for innovative approaches of curbing and reversing this trend.

A range of digital interventions have been developed to support these efforts [39]. Widely described as digital health, these tools exist in various forms including text messaging, mobile and desktop-based applications, wearables (watches, wristbands, cutaneous patches, and clothing), digital workplace interventions, and virtual reality. The beneficial effects of these tools are believed to include improved access, efficient communication, affordability as well as an increase in efficiency and value of health care [40]. Although the evidence-base to support the efficacy of these tools is still emerging, however, there appear to be a consistency of evidence suggesting beneficial effects in reducing cardiovascular risk [39] and potentially CVD outcomes [41].

2.2.1 Physical inactivity

About 1 in three adults worldwide were physically inactive, with proportions ranging from 17% in Southeast Asia to about 43% in the Americas and the Eastern Mediterranean [42]. A 2015 RCT conducted in Baltimore USA utilized a wearable triaxial accelerometer with a Bluetooth-enabled connectivity to participants' smartphones to track their daily physical activity levels and also deliver smart text prescriptions by physicians to motivate respondents toward achieving the 10,000 steps/day goal. The study findings suggest that a fully automated tracking-texting intervention increased physical activity with, but not without, the texting component [43].

2.2.2 Diet control

According to a 2016 Global Burden of Disease report [44], poor/sub-optimal diet is the second leading risk factor for deaths and DALYs globally, accounting for nearly one-fifths of all deaths. Other diet-related risk factors include high blood glucose, high blood pressure, high BMI and high total cholesterol. The risk associated with sub-optimal diet exists in two extremes of malnutrition among poor people and unhealthy diet among the rich. The study reports the biggest driving factor of dietary risk is that millions of people are eating a diet which consists of too much salt and saturated fat, and not enough fruit, vegetables, nuts, seeds, omega 3, and whole grains.

Beyond the traditional approaches of in-person or telephone counseling on dietary habits, recent advances regarding the use of m-Health technologies in improving dietary risk awareness and driving behavioral change toward healthy eating have shown promising results [45]. For example, the use of supportive mobile phone text-messaging to encourage healthy eating is gaining wider acceptability because of its relative inexpensiveness and capacity for wider reach even among populations with limited access to smartphones or internet connectivity. In one prospective cohort study (mDiabetes study) [46] conducted among 1925 mobile phone subscribers in India, the authors sent 56 text messages (twice-weekly messages containing facts and behavioral modification content regarding diabetes, healthy eating and physical activity) in one of 12 languages to 982 randomly selected Nokia subscribers over 6 months, and compared the results against 943 controls (non-Nokia subscribers who received no text messages) with respect to changes in 4 health behaviors that lessen diabetes risk: engagement in exercise, avoidance of fat foods, fruit intake of 2 servings a day or more, and vegetable intake of 2 servings a day or more.

The authors reported a differential positive improvement in diabetes preventive behaviors among intervention compared to control group, especially with regards to improved fruit and vegetable intake and reduced fat intake.

Current trends tend toward the use of m-Health technologies for multiple behavior change interventions including dietary education, promoting healthy eating and physical activity simultaneously [47]. Although an emerging area of research, early findings are encouraging and point toward beneficial effects.

2.2.3 Smoking

Smoking (all forms) is a leading cause of death and DALYs worldwide, accounting for 7.1 million deaths and 177.3 million DALYs in 2016 [44]. m-health technologies offer a glimpse of hope in combating the burden of tobacco. Of available m-Health interventions, text-messaging for smoking cessation appear to be the most popular approach, and also yields comparable results to other types of smoking cessation interventions such as telephone quit lines, web-coaching, video messaging, counseling and behavioral therapies or nicotine replacement therapy (NRT) [48]. A recent meta-analysis on the efficacy of SMS text messaging for smoking cessation found that quit rates was 35% higher in the intervention group as compared to the control group [48]. While this area of research has witnessed significant evolution and yielded promising results, important determinants/moderators of the efficacy of this approach are believed to include frequency of messaging (fixed vs. decreasing/variable schedules), length of follow-up (≤ 3 months or ≥ 6 months), intervention design (text alone vs. text and additional interventions) and variations in content. Other factors include message direction and initiation, extent of message tailoring and availability of on-demand messaging support features [49,50].

2.2.4 Alcohol use

Alcohol was estimated to be the seventh leading cause of death and DALYs in 2016 [44]. A variety of m-Health approaches are available for addressing alcohol misuse or dependence, many of which rely on the different functionalities of mobile phones available to clients [51], the most basic and popular approach being text messaging. While the evidence regarding the efficacy of text messaging in achieving alcohol abstinence have been largely variable [51–54], newer approaches of using smartphone-based alcohol abstinence apps have been shown to be more effective [55]. Many such apps exist, and while systematic studies evaluating

their comparative efficacy in promoting reductions in alcohol consumption or treating Alcohol Use Disorders (AUD) are relatively sparse, initial evidence suggest beneficial effects in this regard [56,57]. For example, the Alcohol − Comprehensive Health Enhancement Support System (A-CHESS) which is a relapse prevention app for individuals in recovery for alcohol dependence who have been recently discharged from residential care, is reported to provide the strongest evidence of efficacy [58] compared to other apps for AUD (such as Location-Based Monitoring and Intervention system for Alcohol use disorders (LBMI-A), Promille-koll, PartyPlanner, HealthCall-S, and Chimpshop), with participants in a recent RCT who received A-CHESS reporting significantly fewer risky drinking days over a 12-month period as well as higher likelihood of abstinence than controls [56]. The A-CHESS is GPS-enabled and has a number of inbuilt support features to encourage sobriety, such as a warning noise when users get too close to a bar, and a panic button that the user can press if they're thinking about drinking or buying alcohol. The button contacts the user's closest peers and medical professionals who can reach out and discourage the person from relapsing (Fig. 11.1). While the patient is waiting for a response, it offers them links to potentially helpful materials such as tutorials on relaxation exercises.

2.3 Communicable Diseases

2.3.1 Surveillance, outbreak detection and management

In the last decade, the world has witnessed several pandemics/epidemics from emerging and re-emerging infectious diseases such as Avian Influenza, MERS-CoV, Ebola, Zika, etc causing morbidity and mortality of devastating proportions especially in developing countries. Emerging zoonoses underscores the growing need for all countries to develop capacities to rapidly detect, contain, respond to, and cope with such pandemics. Central to achieving this is the development and maintenance of a robust surveillance, outbreak detection and management capacity. Traditional approaches to surveillance, outbreak detection and management are often too cumbersome and faced with a lot of challenges and inefficiencies in communication, contact tracing, skilled manpower and other infrastructural issues, that militate against rapid containment of outbreaks, hence resulting in devastating outcomes.

Lessons learnt from the 2014/2015 Ebola Viral Disease (EVD) outbreak in West Africa suggest that more robust surveillance and outbreak management infrastructure are needed [59]. Incidentally, the burden of

FIG. 11.1 A-CHESS app main menu and panic button. (Courtesy GoWireless, 2017. Available from: https://www.gowireless.com/blog/study-smartphone-app-may-help-people-overcome-alcoholism#more-4284. Reproduced with permission from original work of D.H. Gustafson, F.M. McTavish, M.Y. Chih, A.K. Atwood, R.A. Johnson, M.G. Boyle, M.S. Levy, H. Driscoll, S.M. Chisholm, L. Dillenburg, A. Isham, D. Shah, A smartphone application to support recovery from alcoholism a randomized clinical trial, JAMA Psychiatry 71 (5) (2014) 566–572, https://doi.org/10.1001/jamapsychiatry.2013.4642.)

these infectious diseases is highest in developing countries where capacity for rapid containment is often limited. The high penetration of mobile phones in these countries offer a significant opportunity to leverage m-Health technologies in overcoming many of the communication, transportation and outbreak management challenges in these settings [60,61]. A variety of m-Health tools and apps with diverse functionalities have been developed and tested for many diseases such as EVD, H1N1, polio, dengue fever, SARS, MERS-CoV, malaria, measles etc, including with general disease surveillance functionalities. For example, the open source "SORMAS" [62], "Ebola Sense Followup" [63] and "CommCare" [64] tools were tested during the EVD outbreak and included functionalities such as surveillance notification, contact tracing, case management, automated GPS tracking, tailored health-specific user profiles with variable functionalities, laboratory function and rumor management, enabled on smartphones and/or tablets.

The development and deployment of the "Ebola Sense Followup" tool (Fig. 11.2) contributed significantly to the timely containment and control of the EVD outbreak in Nigeria by providing valuable platform for early warning and timely action. More of such tools are in development, and initial evidence of

their successful application serves as a reference point of their efficacy in infectious disease surveillance and management. Future areas of improvement include the need to fully integrate these tools with the existing Health Management Information System (HMIS) infrastructure in these countries, which enhances overall sustainability.

2.3.2 Enhancing infectious disease diagnosis, monitoring and treatment

m-Health technologies have been exploited to overcome challenges in timely diagnosis and initiation of treatment arising from delays in receipt of laboratory results and poor continuity of care, especially in remote settings with limited communication and transportation infrastructure. m-Health technologies have been successfully deployed to enhance efficiency across the continuum of care in many African countries with basic telecommunication networks. One of such applications is in facilitating early infant diagnosis of HIV and follow-up and retention of mothers and their exposed infants in Nigeria [65], Kenya [66] and Malawi [67]. For example, the United Nations Foundation-funded Clinton Health Access Initiative's (CHAI) "SMS Printers to Accelerate Return of Test Results for Early Infant Diagnosis of HIV/AIDS' (SMART) project implemented

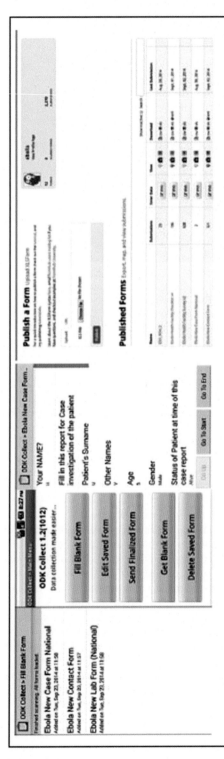

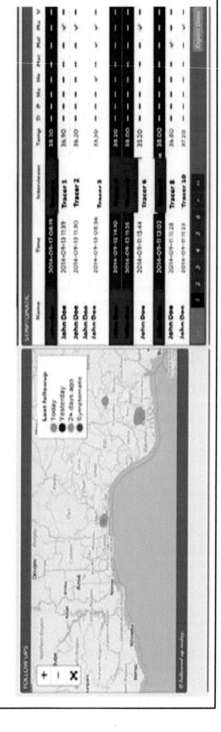

Dashboard Technology

FIG. 11.2 Screen shots from Ebola Sense Followup app and web dashboard. (Reproduced with permission from the original work of D. Tom-Aba, A. Olaleye, A.T. Olayinka, P. Nguku, N. Waziri, P. Adewuyi, O. Adeoye, S. Oladele, A. Adeseye, O. Oguntimehin, F. Shuaib, Innovative technological approach to Ebola virus disease outbreak response in Nigeria using the open data Kit and form Hub technology, PLoS One 10 (6) (2015) e0131000, https://doi.org/10.1371/journal.pone.0131000. Available from: https://journals.plos.org/plosone/article?id=10.1371/journal.pone.0131000.)

in Nigeria utilized text messaging to rapidly communicate infants' HIV test results from laboratories to the health facilities, especially those in very remote hard-to-reach locations with irregular power supply, limited skilled manpower and internet access. This approach has facilitated instant transfer (from the laboratory to health facility) of test results which can be printed out from the battery-operated SMS printers to enhance timely initiation of antiretroviral therapy (ART) and limit loss-to-follow up [68].

2.4 Environmental and Occupational Health

To date, little is known about the prospects and potential application of m-Health technologies to improve occupational and environmental health, apparently because research in this area has been very sparse. However, to stay in conformity with Labor laws and Occupational Safety and Health regulations, employers are increasingly leveraging mobile technologies in form of text messaging and apps to track and report employee vaccination status, report incidents, monitor employee safety and training, keep OSHA records as well as leave and medical management [69]. For employees whose work schedule involves a lot of traveling, employers are now able to use a journey management software program to track employees' movements and receive instant alerts when an employee has not checked-in to their destination at their scheduled time [70]. These tools help to minimize risks and improve employee safety.

2.5 Other Areas

m-Health technologies are increasingly being adapted for other purposes that impact the overall health of populations such as providing mobile job aids to improve efficiency and effectiveness of community health workers and nurses [71,72], improving surveillance, reporting and decision-making by health care workers [73], and improving drugs and commodities logistics and supply chain management system [74]. Whereas m-Health has the potential to improve access to mental health care, the evidence-base to support effectiveness of mobile apps in reducing depression, anxiety, stress and substance use symptoms is still emerging and remains a controversial subject [75].

3 CONCLUSION

m-Health has positively impacted the global public health landscape. The increasing penetrance of mobile devices even among the most vulnerable and lower income populations around the world, has continued to ease most of the challenges faced in "traditional' public health programming – which had hitherto militated against the timely delivery of essential services, implementation of robust surveillance and disease prevention efforts, and health promotion activities using traditional approaches – by offering an acceptable, cost-effective platform and alternative for knowledge dissemination, behavior change communication/interventions, program monitoring, evaluation and research. Not only has m-Health adoption revolutionized awareness creation about health and disease at population-level, it has also improved self-consciousness and encouraged better self-health practices at individual level. Notwithstanding some documented drawbacks of its use, the overarching implication of its application in public health practice is a drive toward an informed, healthier global population and the possibility of achieving the global health targets through coordinated advocacies, policy dialogues and community participation in health programs. m-Health must continue to evolve in its responsiveness to stakeholders' (communities, care providers, health systems, policy-makers) needs and functionalities that address its current drawbacks such as the development of cryptographic technologies to address security and privacy of information.

REFERENCES

[1] UNICEF Data: Monitoring the Situation of Women and Children, 2016. Available at: https://data.unicef.org/topic/maternal-health/antenatal-care/.

[2] J.L. Watterson, J. Walsh, I. Madeka, Using m-Health to improve usage of antenatal care, postnatal care, and immunization: a systematic review of the literature, BioMed Res. Int. 2015 (2015) 153402, https://doi.org/10.1155/2015/153402.

[3] S. Lund, B.B. Nielsen, M. Hemed, I.M. Boas, A. Said, K. Said, M.H. Makungu, V. Rasch, Mobile phones improve antenatal care attendance in Zanzibar: a cluster randomized controlled trial, BMC Pregnancy Childbirth 14 (1) (2014) 29, https://doi.org/10.1186/1471-2393-14-29.

[4] J. Kaewkungwal, P. Singhasivanon, A. Khamsiriwatchara, S. Sawang, P. Meankaew, A. Wechsart, Application of smart phone in "Better Border Healthcare Program": a module for mother and child care, BMC Med. Inf. Decis. Mak. 10 (2010) 69, https://doi.org/10.1186/1472-6947-10-69.

[5] S.C. Watkins, A. Robinson, M. Dalious, Evaluation of the Information and Communications Technology for Maternal, Newborn and Child Health Project Known Locally as 'Chipatala Cha Pa Foni' (Health Center by Phone), Invest in Knowledge Initiative, Balaka District, Malawi, 2013.

[6] H. Jalloh-Vos, H. Ormel, K. de Koning, A.M. Jalloh, K. Herschderfer, R. Khadduri, et al., Mobile Health: Connecting Managers, Service Providers and Clients in Bombali District, Sierra Leone, 2014. Amsterdam.

[7] S.O. Oyeyemi, R. Wynn, Giving cell phones to pregnant women and improving services may increase primary health facility utilization: a case-control study of a Nigerian project, Reprod. Health 11 (2014) 8, https://doi.org/10.1186/1742-4755-11-8.

[8] S.S. Datta, P. Ranganathan, K.S. Sivakumar, A study to assess the feasibility of Text Messaging Service in delivering maternal and child healthcare messages in a rural area of Tamil Nadu, India, Australas. Med. J. 7 (2014) 175–180, https://doi.org/10.4066/AMJ.2014.1916.

[9] Y.K. Lau, T. Cassidy, D. Hacking, K. Brittain, H.J. Haricharan, M. Heap, Antenatal health promotion via short message service at a Midwife Obstetrics Unit in South Africa: a mixed methods study, BMC Pregnancy Childbirth 14 (2014) 284, https://doi.org/10.1186/1471-2393-14-284.

[10] N.M. Tahir, N. Al-Sadat, Does telephone lactation counselling improve breastfeeding practices? A randomised controlled trial, Int. J. Nurs. Stud. 50 (2013) 16–25, https://doi.org/10.1016/j.ijnurstu.2012.09.006.

[11] H. Jiang, M. Li, L.M. Wen, Q. Hu, D. Yang, G. He, L.A. Baur, M.J. Dibley, X. Qian, Effect of short message service on infant feeding practice: findings from a community-based study in Shanghai, China, JAMA Pediatr 168 (2014) 471–478, https://doi.org/10.1001/jamapediatrics.2014.58.

[12] World Health Organization, Improving Access to Maternal Health Care in Senegal: WAHA International's M-Health Programme, 2014.

[13] World Health Organization, MAMA South Africa: Putting the Power of Health in Every Mama's Hand, 2013.

[14] World Health Organization, Increasing Facility Deliveries in Western Kenya: Changamka's Mobile e-Vouchers, 2013.

[15] World Health Organization, Supporting Pregnant Women and New Mothers in South Africa: Cell-Life's MAMA SMS, 2013.

[16] World Health Organization, Improving Maternal and Newborn Access to Services in Ghana: Grameen Foundations's MOTECH, 2013.

[17] L. Alkema, V. Kantorova, C. Menozzi, A. Biddlecom, National, regional, and global rates and trends in contraceptive prevalence and unmet need for family planning between 1990 and 2015: a systematic and comprehensive analysis, Lancet 381 (9878) (2013) 1642–1652, https://doi.org/10.1016/S0140-6736(12)62204-1.

[18] United Nations Department of Economic and Social Affairs, Population Division, Trends in Contraceptive Use Worldwide 2015 Report. Available at: http://www.un.org/en/development/desa/population/theme/family-planning/index.shtml.

[19] K. L'engle, H. Vahdat, E. Ndakidemi, C. Lasway, T. Zan, Evaluating feasibility, reach and potential impact of a text message family planning information service in Tanzania, Contraception 87 (2) (2013) 251–256, https://doi.org/10.1016/j.contraception.

[20] World Health Organization, SMS-based Family Planning in Kenya and Tanzania: FHI 360's m4RH, 2014.

[21] The David, Lucile Packard Foundation, Using Mobile Health Applications to Improve Family Planning Services, 2015. Available at: https://www.packard.org/awardee/m-health-improve-quality-family-planning-services-tanzania/.

[22] World Health Organization, Family Planning through Mobile Phones in India, Institute for Reproductive Health's CycleTel™, 2014.

[23] J. Tripney, I. Kwan, K.S. Bird, Post-abortion family planning counseling and services for women in low-income countries: a systematic review, Contraception 87 (1) (2013) 17–25, https://doi.org/10.1016/j.contraception.

[24] C. Smith, U. Vannak, L. Sokhey, T.D. Ngo, J. Gold, C. Free, Mobile Technology for Improved Family Planning (MOTIF): the development of a mobile phone-based (mHealth) intervention to support post-abortion family planning (PAFP) in Cambodia, Reprod. Health 13 (1) (2016) 1, https://doi.org/10.1186/s12978-015-0112-x.

[25] P. Castano, J. Bynum, R. Andres, M. Lara, C. Westhoff, Effect of daily text messages on oral contraceptive continuation; a randomised controlled trial, Obstet. Gynecol. 119 (1) (2012) 14–20, https://doi.org/10.1097/AOG.0b013e31823d4167.

[26] M. Hou, S. Hurwitz, E. Kavanagh, J. Fortin, A. Goldberg, Using daily text-message reminders to improve adherence with oral contraceptives: a randomized controlled trial, Obstet. Gynecol. 116 (3) (2010) 633–640, https://doi.org/10.1097/AOG.0b013e3181eb6b0f.

[27] L. Tsur, E. Kozer, M. Berkovitch, The effect of drug consultation center guidance on contraceptive use among women using isotretinoin: a randomized, controlled study, J. Wom. Health 17 (4) (2008) 579–584, https://doi.org/10.1089/jwh.2007.0623.

[28] World Health Organization, Immunization Coverage Fact Sheet. Available at: http://www.who.int/mediacentre/factsheets/fs378/en/.

[29] A. Odone, A. Ferrari, F. Spagnoli, S. Visciarelli, A. Shefer, C. Pasquarella, C. Signorelli, Effectiveness of interventions that apply new media to improve vaccine uptake and vaccine coverage: a systematic review, Hum. Vaccines Immunother. 11 (1) (2015) 72–82, https://doi.org/10.4161/hv.34313.

[30] S.A. Abdulrahman, M.O. Olaosebikan, mHealth: a narrative synthesis of evidence of its application in improving childhood immunization coverage, J. Hosp. Manag. Health Policy 1 (2017) 6. https://doi.org/10.21037/jhmhp.2017.10.01.

[31] World Health Organization, Improving Immunization Registration, Coverage and Monitoring in Viet Nam–PATH's Digital Immunization Registry (IR) System, 2014.

[32] World Health Organization, An Automated Mobile Vaccination Reminder System in South Asia: Tika Tracker, 2013.

[33] World Health Organization, Small Incentives Improve Vaccine Coverage in Pakistan: IRD's Interactive Alerts, 2013.

[34] World Health Organization, Better Access to Life-Saving Medicines through Interactive SMS in Malawi: JSI's cStock, 2014.

[35] UNICEF Data: Monitoring the Situation of Women and Children, 2017. Available at: https://data.unicef.org/ topic/nutrition/malnutrition/.

[36] F. Arnold, et al., Nutrition in India: National Family Health Survey (NFHS 3), 2005–2006, International Institute for Population Sciences, Mumbai, India, 2007.

[37] World Health Organization, Improving Maternal and Child Health in India: SERP's Nutrition Day Care Centers Mobile App, 2013.

[38] N. Probst-Hensch, M. Tanner, C. Kessler, C. Burri, N. Kunzli, Prevention – a cost-effective way to fight the non-communicable disease epidemic: an academic perspective of the United Nations High-level NCD Meeting, Swiss Med. Wkly. 141 (2011) w13266, https:// doi.org/10.4414/smw.2011.13266.

[39] L.E. Burke, J. Ma, K.M.J. Azar, G.G. Bennett, E.D. Peterson, Y. Zheng, W. Riley, J. Stephens, S.H. Shah, B. Suffoletto, T.N. Turan, B. Spring, J. Steinberger, C.C. Quinn, Current science on consumer use of mobile health for cardiovascular disease prevention: a scientific statement from the American Heart Association, Circulation 132 (2015) 1157–1213, https://doi.org/10.1161/CIR.00000000000 00232.

[40] C.K. Chow, N. Ariyarathna, S.M. Islam, A. Thiagalingam, J. Redfern, m-Health in cardiovascular health care, Heart Lung Circ. 25 (2016) 802–807, https://doi.org/ 10.1016/j.hlc.2016.04.009.

[41] R.J. Widmer, N.M. Collins, C.S. Collins, C.P. West, L.O. Lerman, A. Lerman, Digital health interventions for the prevention of cardiovascular disease: a systematic review and meta-analysis, Mayo Clin. Proc. 90 (2015) 469–480, https://doi.org/10.1016/j.mayocp. 2014.12.026.

[42] P.C. Hallal, L.B. Andersen, F.C. Bull, R. Guthold, W. Haskell, U. Ekelund, Lancet Physical Activity Series Working Group. Global physical activity levels: surveillance progress, pitfalls, and prospects, Lancet 380 (9838) (2012) 247–257, https://doi.org/10.1016/ S0140-6736(12)60646-1.

[43] S.S. Martin, D.I. Feldman, R.S. Blumenthal, S.R. Jones, W.S. Post, R.A. McKibben, E.D. Michos, C.E. Ndumele, E.V. Ratchford, J. Coresh, M.J. Blaha, m-Active: a randomized clinical trial of an automated m-Health intervention for physical activity promotion, J. Am. Heart Assoc. 4 (11) (2015) e002239, https://doi.org/10.1161/JAHA.115. 002239.

[44] E. Gakidou, GBD 2016 Risk Factors Collaborators. Global, regional, and national comparative risk assessment of 84 behavioral, environmental and occupational, and metabolic risks or clusters of risks, 1990–2016: a systematic analysis for the Global Burden of Disease Study 2016, Lancet 390 (10100) (2017) 1345–1422, https:// doi.org/10.1016/S0140-6736(17)32366-8.

[45] B. Spring, K. Schneider, H.G. McFadden, J. Vaughn, A.T. Kozak, M. Smith, A.C. Moller, L.H. Epstein, A. DeMott, D. Hedeker, J. Siddique, D.M. Lloyd-Jones, Multiple behavior change in diet and activity: a randomized controlled trial using mobile technology, Arch. Intern. Med. 172 (10) (2012) 789–796, https://doi.org/ 10.1001/archinternmed.2012.1044.

[46] A. Pfammatter, B. Spring, N. Saligram, R. Dave, A. Gowda, L. Blais, M. Arora, H. Ranjani, O. Ganda, D. Hedeker, S. Reddy, S. Ramalingam, m-Health intervention to improve diabetes risk behaviors in India: a prospective, parallel group cohort study, J. Med. Internet Res. 18 (8) (2016) e207, https://doi.org/10.2196/jmir.5712.

[47] B. Spring, K. Schneider, H.G. McFadden, J. Vaughn, A.T. Kozak, M. Smith, A.C. Moller, L. Epstein, S.W. Russell, A. DeMott, D. Hedeker, Make Better Choices (MBC): study design of a randomized controlled trial testing optimal technology-supported change in multiple diet and physical activity risk behaviors, BMC Public Health 10 (2010) 586, https://doi.org/10.1186/1471- 2458-10-586.

[48] S.A. Spohr, R. Nandy, D. Gandhiraj, A. Vemulapalli, S. Anne, S.T. Walters, Efficacy of SMS text message interventions for smoking cessation: a meta-analysis, J. Subst. Abuse Treat. 56 (2015) 1–10, https://doi.org/ 10.1016/j.jsat.2015.01.011.

[49] M.L. Ybarra, J.S. Holtrop, A.T. Bosi, S. Emri, Design considerations in developing a text messaging program aimed at smoking cessation, J. Med. Internet Res. 14 (4) (2012) e103, https://doi.org/10.2196/jmir.2061.

[50] R. Whittaker, H. McRobbie, C. Bullen, R. Borland, A. Rodgers, Y. Gu, Mobile phone-based interventions for smoking cessation, Cochrane Database Syst. Rev. 11 (2012) CD006611.

[51] A. Quanbeck, M. Chih, A. Isham, R. Johnson, D. Gustafson, Mobile delivery of treatment for alcohol use disorders, Alcohol Res. 36 (1) (2014) 111–112.

[52] V.I. Agyapong, S. Ahern, D.M. McLoughlin, C.K. Farren, Supportive text messaging for depression and comorbid alcohol use disorder: single-blind randomized trial, J. Affect. Disord. 141 (2–3) (2012) 168–176, https:// doi.org/10.1016/j.jad.2012.02.040.

[53] V.I. Agyapong, D.M. McLoughlin, C.K. Farren, Six-months outcomes of a randomised trial of supportive text messaging for depression and comorbid alcohol use disorder, J. Affect. Disord. 151 (1) (2013) 100–104, https://doi.org/10.1016/j.jad.2012.02.040.

[54] V. Keoleian, D. Polcin, G.P. Galloway, Text messaging for addiction: a review, J. Psychoact. Drugs 47 (2) (2015) 158–176, https://doi.org/10.1080/02791072. 2015.1009200.

[55] S.M. Alessi, N.M. Petry, A randomized study of cellphone technology to reinforce alcohol abstinence in the natural environment, Addiction 108 (5) (2013) 900–909, https://doi.org/10.1111/add.12093.

[56] D.H. Gustafson, F.M. McTavish, M.Y. Chih, A.K. Atwood, R.A. Johnson, M.G. Boyle, M.S. Levy, H. Driscoll, S.M. Chisholm, L. Dillenburg, A. Isham, D. Shah, A smartphone application to support recovery from alcoholism a randomized clinical trial, JAMA Psychiatry 71 (5) (2014) 566–572, https://doi.org/10.1001/jamapsychiatry.2013.4642.

[57] V.M. Gonzalez, P.L. Dulin, Comparison of a smartphone app for alcohol use disorders with an internet-based intervention plus bibliotherapy: a pilot study, J. Consult. Clin. Psychol. 83 (2) (2015) 335–345, https://doi.org/10.1037/a0038620.

[58] S.E. Meredith, S.M. Alessi, N.M. Petry, Smartphone applications to reduce alcohol consumption and help patients with alcohol use disorder: a state-of-the-art review, Adv. Health Care Technol. 1 (2015) 47.

[59] World Health Organization, Situation Report 2016: Ebola Virus Disease in West Africa. Available at: http://apps.who.int/iris/bitstream/10665/206536/1/ebolasitrep_19May 2016_eng.pdf?ua=1.

[60] G. Nasi, M. Cucciniello, C. Guerrazzi, The role of mobile technologies in health care processes: the case of cancer supportive care, J. Med. Internet Res. 17 (2) (2015) e26, https://doi.org/10.2196/jmir.3757.

[61] C. Fahnrich, K. Denecke, O.O. Adeoye, J. Benzler, H. Claus, G. Kirchner, S. Mall, R. Richter, M.P. Schapranow, N. Schwarz, D. Tom-Aba, Surveillance and outbreak response management system (SORMAS) to support the control of the Ebola virus disease outbreak in West Africa, Euro Surveill 20 (2015).

[62] SORMAS, Surveillance Outbreak Response Management Analysis System, 2016. Available at: http://www.sormas.org.

[63] D. Tom-Aba, A. Olaleye, A.T. Olayinka, P. Nguku, N. Waziri, P. Adewuyi, O. Adeoye, S. Oladele, A. Adeseye, O. Oguntimehin, F. Shuaib, Innovative technological approach to Ebola virus disease outbreak response in Nigeria using the open data Kit and form Hub technology, PLoS One 10 (6) (2015) e0131000, https://doi.org/10.1371/journal.pone.0131000.

[64] COMMCARE, Commcare Ebola Contact Tracking App 2015 Feb 01, 2015. Available at: http://www.dimagi.com/wp-content/uploads/2015/02/Ebola-Response-Final.pdf.

[65] World Health Organization, SMS Printers Aid Early Infant Diagnosis of HIV/AIDS in Nigeria. CHAI's SMART, 2013.

[66] World Health Organization, Text Messaging to Improve Early Infant Testing for HIV in Kenya: KEMRI's TextIT, 2014.

[67] World Health Organization, Facilitating Follow-Up and Retention Among HIV+ Mothers and Exposed Infants in Malawi – CHAI's Mother-Infant Pair Clinic, 2013.

[68] Early Infant Diagnosis. Interagency Task Team (IATT) on the Prevention and Treatment of HIV Infection in Pregnant Women, Mothers and Children, IATT Laboratory & Child Survival Working Group, 2012. Available at: www.emtct-iatt.org/wp-content/uploads/2012/12/EID-GSG.pdf.

[69] E. Wicklund, Occupational Health Executives Find New Uses for M-Health Tools. From the mHealthNews Archive, 2013. Available at: http://www.mobihealthnews.com/news/occupational-health-executives-find-new-uses-mhealth-tools.

[70] J. Hoolahan, Using Technology to Improve Safety in the Workplace, 2016. Available at: https://jesi.io/using-technology-improve-safety-workplace/.

[71] World Health Organization, Assisting Community Health Workers in India, Dimagi's CommCare, 2013.

[72] World Health Organization, Supporting Treatment of Childhood Malnutrition in Zanzibar: D-Tree International's eNUT, 2013.

[73] World Health Organization, Assisting Community Health Workers in Rwanda: MOH's RapidSMS and mUbuzima, 2013.

[74] World Health Organization, Preventing Stock-Outs of Antimalarial Drugs in Sub-saharan Africa: Novartis's SMS for Life, 2013.

[75] T. Donker, K. Petrie, J. Proudfoot, J. Clarke, M.R. Birch, H. Christensen, Smartphones for Smarter Delivery of Mental Health Programs: A Systematic Review, J Med Internet Res 15 (11) (2013) e247. https://www.jmir.org/2013/11/e247/.

CHAPTER 12

The Egyptian-African Telemedicine Network: The Treat and Teach Comprehensive Model

HODA MF WAHBA[a] • TAMER EMARA[b] • AHMED ELBOKL[b]
[a]Geriatrics Department, Ain Shams University, Cairo, Egypt; [b]Neurology Department, Ain Shams University, Cairo, Egypt

1 INTRODUCTION

Africa is the second largest and second most populous continent on earth with an estimated population of more than 1.25 billion expected to reach 2.4 billion in 2050 if left unchecked [1]. The heart of the world consists of 54 countries that are diverse in everything with the exception of inadequate education and healthcare services. The African continent constitutes 15% of the world's population. It has 24% of the disease burden of the world, surprisingly managed by only 3% of the world's healthcare workers. Moreover, its health expenses constitute less than 1% of global health expenditure [2].

Africa faces an imbalance between needs and resources with the upper hand being the needs. There is a deficit in quality and availability of healthcare services. Furthermore, there is a geographical maldistribution of healthcare providers being more clustered in urban areas given that 60% of Africa's people are living in rural areas [3]. It is outrageous to state the fact that patient doctor ratio (general physicians and specialists) in Africa ranges from 750 in Tunisia, 1900 in Egypt to 50,000 in Tanzania [4], compared to 142 in Monaco, 232 in Russia, 358 in the UK and 408 in the USA. The WHO estimates a maximum of 400 patients per physician, nurse or midwife is needed to provide adequate primary health care [2].

Current solutions like sending medical convoys and short term visits from international experts lack sustainability. Moreover, students who travel for training in international centers, with no clear plan for integrating them in their local healthcare services, usually do not come back. They prefer pursuing the future of their careers or better socioeconomic communities.

The continent needs to overcome the discrepancy between resources and needs by benefiting from resources, namely manpower. It needs to invest more in empowering humans. Moreover, with the rapid technological advances, new inventive models and approaches are sought to compensate for the discrepancy between needs and resources in healthcare.

It would be quite ridiculous in 2018 to debate how telemedicine considerably decreases mortality, decreases number of patients in waiting rooms, decreases ER visits, brings healthcare services to patients in distant locations, considerably decreases mortality, and decreases hospital admissions and hence healthcare costs (I am even not going to put a reference). Yet it is worth mentioning it is an indispensable innovative tool that can definitely overcome many of the challenges faced in healthcare in Africa.

Telemedicine was actually first used via telegraph to give advice on the management of the wounded following an attack on the Barrow Creek Telegraph Station in Australia in 1874; more than a century ago [5]. Africa isn't a stranger either to telemedicine; the first reported use of modern telemedicine was in 1984 to diagnosis Crouzon's syndrome via a satellite link using slowscan television transmission between Swaziland and London [6]. There was even a "first Ethiopian telemedicine project" [7] and a "National Telemedicine Project in South Africa" [8]. Evidently, these and many other telemedicine projects have either failed or only managed to solve a minor problem in healthcare service providing. There is no comprehensive model that is incorporated in and complementary to the standard healthcare services provided.

Telemedicine Technologies. https://doi.org/10.1016/B978-0-12-816948-3.00012-X

183

Egypt the third highest populated country in the continent and alone accommodates five of the top twenty universities in Africa according to the 2017 African University Ranking by the Spanish National Research Council [9]. ASU is one of the leading universities in Egypt and is the home of many talented and innovational cadres in all areas of expertize. It has the only comprehensive Geriatrics hospital in the continent. It has specialized outpatient clinics, inpatient wards, geriatrics ICU, long-term care, adult daycare and undergraduate and postgraduate studies in Geriatrics and Gerontology. The Neurology department has been a regional leader in stroke care with a functioning unit that had started since 1991.

Stroke is the number one cause of disability and one of the leading causes of death in the world. It has a dramatic mode of onset that can happen at anytime during day or night. Stroke care needs a team functioning 24/7 with technical and logistic experiences that is usually headed by a neurology consultant. This team should take prompt, sophisticated and aggressive treatment measures in the first few hours after the onset of the symptoms. The timely guideline interventions are the key to effective management.

There is a general deficiency of neurologists in the world; this problem is most pronounced in Africa due to general lack of neurology training programs in Subsaharan African countries. For these reasons telestroke services have been established as an efficient solution for stroke management, saving lives, time and effort. Telestroke is the number one application of hospital to hospital telemedicine solutions (hub-spoke model). Its role has been established by rigorous research over the past two decades [10].

In the next few pages we will display the telemedicine model founded with the ultimate goal of equitable healthcare for the continent; The Egyptian-African telemedicine network. The network operates as a hub-spoke\s model. The hub being Ain Shams University Virtual Hospital (ASUVH), an institutional telemedicine provider affiliated to Ain Shams University (ASU), Cairo, Egypt.

2 THE TEAM

The Egyptian-African telemedicine network was the yield of joint efforts of a team of consultants at ASU. The vision of the team was to benefit from the expertize of ASU in medical education and medical care to revolutionize healthcare in the area hence improving disease outcome, decreasing healthcare cost and improving socioeconomic wellbeing. We envisioned the "Treat and Teach initiative" to connect more than 3000 consultants and academics in 38 departments at the Faculty of Medicine, ASU to those who need their services in underserved areas in Egypt and Africa.

The specialties of the founders were a fundamental basis for the network. The team constituting of Dr. Tamer Emara; assistant professor of Neurology, Dr. Hoda MF Wahba; assistant professor of Geriatrics and Gerontology and Dr. Ahmad Elbokl; lecturer of Neurology were the nucleus for "Ain Shams University Virtual Hospital". Both neurology and geriatrics consultants are quite scarce. According to a study by Bower and Zenebe in 2005 [11] in 50 countries of the 54 African countries, only 11 nations had more than 10 neurologists, 23 nations had 1−4 neurologists and 12 nations with a total population above 5 million had no neurologists at all. As for geriatrics the situation was worse. There are only seven registered geriatricians in Africa excluding Egypt, three of whom are in private practice and only 3.1 % medical students had been exposed to any post-graduate training in geriatric medicine [12].

One of the crucial steps in the planning and instigating ASUVH was the corporate hierarchy. It had to be a combination of medical service provider personnel and ICT service provider personnel in a frame that suites the governmental origin yet satisfies the business model. None of ongoing national and international telemedicine and e-learning providers had the same criteria to benefit from their experience in team and hierarchy assembly.

Currently, the hierarchy is made up of a managerial team, IT team, quality control team, legal and financial team, medical coordinators, research and development team and a marketing team. A board of members was formed headed by the dean of the Faculty of Medicine, ASU and consists of one or more representatives from the teams as well as experts from the civil society.

The managerial crew; the founders, constitutes the director and executive directors for the teach and treat limbs. They established ASUVH as a legal body endorsed by the Faculty of Medicine and Ain Shams University. They contributed to drafting the operational legislations and board assembly. They envisioned a central hub in Cairo linked to a network of spokes efficiently covering the continent of Africa with quality, safe and up to date healthcare. They are responsible for initiating new spokes and maintaining current spokes and coordinating the operation of the hub. They continually seek new technologies and methodologies to improve and revolutionize healthcare provision in the continent.

The IT team has the responsibility of initiating the technical links between the spokes and the hub and maintaining their smooth operation. Another task includes managing the social media page, developing curricula using novel ICT methods and updating and upgrading the software platform. Digital health is a rapidly progressing sector that needs continual updating and upgrading both to maintain quality and to be able to compete with emerging similar services.

The quality control team essentially maintains quality of the service given the highly variable packages offered. Satisfaction surveys are used with customers, consultants and medical coordinators.

Legal and financial representatives are personnel already working at ASU hence aware of legal and financial policies of the facility and have modified them to suite the operational requirements of ASUVH. Since the Egyptian law doesn't have laws that specifically govern the instigation and operation of telemedicine, it was important to investigate already present laws that may apply. Laws governing data transfer electronically, electronic signatures, privacy and anonymity of data, trans-border data transfer and general laws governing the medical practice were found.

Medical coordinators are physicians representing their specialty departments. They are responsible for the content of health promotion messages and medical courses. They have also contributed to the matrix of the software platform to suite each medical specialty requirements. Consent adjustments were also necessary especially for specialties like psychiatry. Coordinating medical teleconsults within the department and ensuring best management are part of their activities.

Marketing is indispensable to any venture especially a service provider that offers a relatively novel solution in the national and regional market. At the present time it consists of social media material, orientation sessions at universities, social clubs, social networks, conferences and workshops and brochures.

3 CONCEPT AND ACTION PLAN

ASUVH "Treat and Teach" initiative aims to transfer scientific and clinical experiences in different medical fields from centers of excellence to where they are needed in a sustainable model.

Learning from previous telemedicine projects in the area that failed to sustain themselves we noticed that telemedicine projects had been either teleconsultations, usually in one or a few specialties, or tele-mentoring/tele-education. Both provided great momentarily aid that failed to progress into long-term improvement in the healthcare delivery system.

Many initiatives have appreciated the role of capacity building through medical education to attain sustainability. This win-win partnership has been adopted by several initiatives in the past few years. Among those, the Global Health Service Partnership that combines a consortium of American and African Universities with Seed Global Health (SEED), US Presidential Emergency Plan for AIDS Relief (PEPFAR) and Peace Corps. GHSP would deploy qualified doctors and nurses in host institutions based on the priorities of the collaborating African governments and universities. The main objective of the visiting faculty team is to improve the capacity of a team of local physicians and nurses to tackle the pressing healthcare issues stated by governmental and professional bodies [13].

Similarly, the Medical/Nursing Education Partnership Initiative (MEPI, NEPI) focuses on upgrading the educational capacity of medical institutions by facilitating joint work to increase the availability of digital educational resources. This facilitates the building of an environment of collaboration and exchange of information between institutions in the same country, or across borders [14].

ASUVH adopts a blended, pragmatic, and priority based approach to deliver safe, efficient, timely, equitable and cost-effective healthcare through innovative telemedicine techniques. It is done through a mix of telemedicine and e-learning, on one hand, and clinical education at the bedside on the other (to deliver direct and effective communication and transfer of clinical skills). Mixing online and onsite education and health care services in the hub-spoke model ensures economic viability and sustainability.

This is achieved through health development packages that are tailored solution packages that satisfy the health needs, through an operating healthcare service, within a specific timeline. The packages are multidisciplinary solutions that are highly customizable (Fig. 12.1).

The health development package is offered to individuals, healthcare service providers and to medical higher education institutes (spokes). ASUVH offers a package that is tailored to the needs of the spoke according to a needs assessment survey done. The survey puts into consideration the demands of the spoke as well as an objective needs assessment onsite visit. The assessment is on-line and onsite shedding light on the medical needs, site, infrastructure and skills required to launch the healthcare service within a certain timeline. Surveys and inspection checklists are used to collect data and different stakeholders are involved.

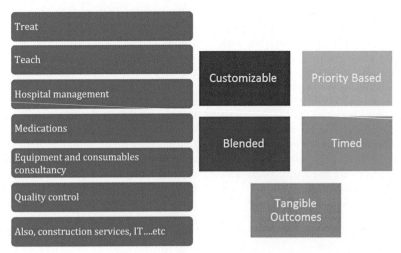

FIG. 12.1 The components and characteristics of the healthcare development package solution.

The ASUVH team analyzes data collected and a task force is assigned. Accordingly a multidisciplinary educational course is designed to help fulfill the medical knowledge and skills. Advice regarding infrastructure is available through ASU's building consultancy office. Equipment specs can also be provided when needed to help establish the service. Furthermore, ASUVH has succeeded in manufacturing some telemedicine equipment in the local market at a very competitive price and that can be introduced commercially to the African market. Hospital management courses are also offered at ASU.

The courses provided target the whole health professional team that will operate the health care service; physicians, nurses, etc. The educational course is in the form of lectures, clinical rounds, case discussions and quizzes. The courses are offered part online and part onsite at the hub (ASU teaching hospitals) by ASU staff members in English; the formal language used in medical education in Egypt. Regular evaluation is performed and minor adjustments to the plan are done to suite the requirements. This has partly discounted French speaking African countries. But due to huge resource requirements it has been delayed to a later phase.

The course is ensued by the team returning to their local healthcare service. They remain attached to the hub for medical and technical consultations until the service is running at the potential planned. These champions are now competent healthcare professionals that have both the knowledge and skills to improve healthcare in the geographical area they serve. Local health professionals give the added value of social and linguistic comprehension that gives the healthcare plan realistic quality.

The local health professionals in the course of their medical education are supplemented with the fundamentals of telemedicine. We encourage and mentor the team to lead their own telemedicine networks hence increasing their outreach and probably reaching local healthcare providers that may otherwise never be identified. They may be non-licensed or non-health professionals providing medical care to natives. These groups have the advantage of trust, credibility, social knowledge and communication skills that the hub team lack. They comprehend ingrained false medically harmful beliefs and have a better chance in modifying these attitudes.

One of the very important missions of ASUVH is health education as part of preventive health measures. The practice offers its spokes (Egyptian governorates at the time being) educational messages in simple nonmedical terms with practical culturally adjusted advice. It is a future plan to offer such advice to foreign spokes but putting in consideration the language barrier and social and cultural ignorance the information supplied may be of similar quality to that available on searching the internet. Therefore, for quality material that will have true impact in preventing disease and maintaining the best medical and functional levels in the population, it should be adjusted by the local champions to suite the population it targets (Fig. 12.2).

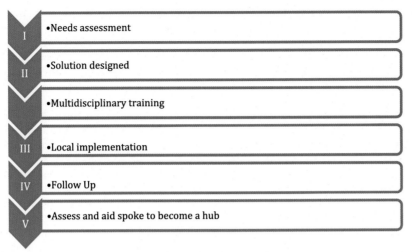

I • Needs assessment

II • Solution designed

• Multidisciplinary training

III • Local implementation

IV • Follow Up

V • Assess and aid spoke to become a hub

FIG. 12.2 Stages for spoke development.

4 SWOT ANALYSIS

The strength of the Ain Shams University Virtual Hospital lies in its affiliation and in its "Treat and Teach" initiative.

Ain Shams University was founded in July 1950, the third-oldest native public Egyptian university. The university's academic structure includes 14 faculties, 1 college and 2 high institutes plus 12 centers and special units [15]. According to the 2014 Webometrics World Universities rankings, Ain Shams University is ranked 3rd in Egypt, 25th in Arab World and 15th in Africa [16,17].

Moreover, the "Treat and Teach" initiative boosts the service as it empowers healthcare service providers allowing them to competently operate their organizations. As the ancient Chinese proverb says "Give a man a fish and you feed him for a day; teach a man to fish and you feed him for a lifetime". Continued medical education (CME) is teaching the healthcare professionals to fish, it is how they can be weaned from senior staff consultations and the best way to proficient case management. It is a necessity to maintain competency of the healthcare team. It provides the healthcare professionals with updated guidelines. It allows transmission of experiences and display of practical methods to manage real life situations. Unfortunately CME opportunities and material may be unavailable to many health professionals resulting in incompetent, distressed and sometimes lost personnel who crave for guidance in managing presenting cases. The initiative is the golden opportunity to facilitate and recuperate quality healthcare services in the continent.

E-learning through any of its portals e.g. m-learning (mobile learning) bypasses the geographical barriers and time limitedness and saves effort hence making CME more available to all healthcare practitioners.

The quality of material offered is ensured considering it is offered by an eminent university whose staff members have pronounced experience in teaching local and foreign medical students. CME is offered in English suiting all Egyptian governorates and many African countries. Many of the courses offered are already being provided in a blended form part online and part onsite, while others are provided only onsite. Many of the onsite courses are being adjusted to become blended to better suite the international students subsequently spending less time away from their local communities.

Online courses are comparable in nature to similar available courses; consisting of lecture material that may be accompanied by audio and video content, assignments, discussion forums and quizzes. Online courses are available for academic and clinical programs. Part of these courses have been designed as part of available programs offered by the Faculty of Medicine, ASU, while others have been personalized to specific needs of health professionals. Novel e-learning methods are continuously sought to improve quality of material offered and to benefit from the added value of these new technologies in simplifying and enlivening educational material to stimulate students.

Lack of appropriate funding represents one of the major challenges facing the project. Most funding bodies prefer providing operational cost rather than funding infrastructure and equipment. Another obstacle to funding lies in the fact that the project is affiliated to a governmental university putting a lot of restraints on funding bodies eligible.

Yet, a couple of funds were raised from governmental funding organizations allowing the kick off. They provided a few inexpensive equipment and the running cost required to be able to recruit spokes and render them sustainable.

Language was another confronting factor as more than one language is required to provide the service (Medical and medical education) in Egypt and Africa. It was a glitch that required enrolling team members fluent in Arabic, English and French.

Another quite challenging factor was ignorance; illiteracy and/or technology. The illiteracy rate in Egypt is around 25% which is similar to many other African countries. Unfortunately, technology ignorance from experience exceeds this number greatly. We have tried to overcome this obstacle through posting orientation videos and health promotion videos on social media. Onsite orientation visits were done in social clubs and clinics to verbally explain the steps required to perform teleconsults. Performing doctor-doctor teleconsults also provided a good alternative that bypassed the illiteracy and technology barrier.

Another major barrier is the availability of network connectivity and electricity availability which limit expansion to certain geographical areas. Collaboration with primary health care centers became the best option to ensure coverage of all geographical areas.

Social acceptance and cultural restraints have been a challenge for the Egyptian-African telemedicine network. Within Egypt some cultural beliefs have prevented patients to seek medical advice without physical face to face meetings. Similarly, cultural differences between the Egyptian and African and intra-African also poses a threat to the continuity of the network. This necessitated the assistance of a local middle man who would eliminate miscommunication and cultural ignorance. This would not just improve better management of cases it bestows credibility and trust.

Establishing this regional health service provider provided a golden opportunity to regain the political and commercial relationships within the continent with the added value of cultural exchange and continental tourism.

Strict legislations binding the service as it is affiliated to a governmental university threaten the rapid upgrade and innovative solutions that are required for sustainability and maintenance of the service.

5 THE KICK OFF

The kickoff of the project was a small scale package delivered to a private hospital at Kafr El Sheik. Kafr El Sheik is an Egyptian governorate about 134 km north of Cairo. According to the Central Agency for Public Mobilization and Statistics, Egypten, 2006 [18] it had a population of 147,380 inhabitants. Unfortunately, the governorate didn't have any neurologists.

In 2015 an insightful neurology postgraduate student at ASU agreed to implement the "Treat and Teach" initiative at his home town. Along with his formal neurology program courses he broadcasted to ASU neurology consultants his neurology cases, mainly stroke patients. Assessment of cases and timely management were performed using open source videoconferencing applications, mobile data internet service and a tablet at the Kafr El Sheikh side and a desktop on the ASU side. The quality of transmission wasn't optimal but sufficient for input of data and output of decisions clearly. Quality upgrading was attained using a laptop, at Kafr El Sheikh that was mounted on an ECG table which gave stability and quality to the picture. A landline internet service was used providing real time conversation with minimal lag.

Case management was a perfect way to consolidate knowledge provided in the doctorate neurology program. The postgraduate candidate gained confidence, knowledge and skills without losing the favorable outcome in his patients. Teleneurology consults were scheduled twice per week and on demand teleconsults were welcomed.

Today, a competent neurology consultant is available at Kafr El Sheikh and several neurology postgraduate students are enrolled at ASU masters and doctorate neurology programs. Teleconsults are still available for difficult cases and specialized cases that require highly experienced consultants as for muscle dystrophy cases. Preliminary assessment is performed by the staff at Kafr El Sheikh, followed by a teleconsult with possible referral of the patient to ASU muscle dystrophy clinic and laboratory. The Kafr El Sheikh consultant has been encouraged to become a hub and serve spokes to expand the geographical area served.

The next project was invigorating; the East African University health services. The ASUVH team was introduced to Somalian health professionals through a neurology conference where the initiative was being presented. We were then invited to help them start a neurology service as the country had no certified neurologists.

The needs assessment discovered the absence of a general ICU that should serve a population of 700,000. Discussions lead to the decision of instigation a multidisciplinary ICU that serves the population in neurology, cardiology an acute care. The assessment was on-line and onsite regarding the medical needs, site, infrastructure and skills required to launch the

ICU within one year. Surveys and inspection checklists were used to collect data and different stakeholders were involved.

The ASUVH team analyzed data collected and designed a course to help fulfill the medical requirement as knowledge and skills. A multidisciplinary course was designed and a taskforce assigned. They were also given advice regarding infrastructure and equipment to launch their ICU. An online course was started to give them the essential academic knowledge for four months. Then the team of nine doctors and nurses, males and females received a three month course at ASU. The multidisciplinary course involved five specialties with a condensed program of lectures, clinical rounds and onsite case management at Ain Shams University Hospital. Regular evaluation was performed and minor adjustments to the plan were done to suite the requirements.

The team returned to Somalia and started cardiology and neurology outpatient clinics and started accepting inpatients at their wards for the first time. Difficult cases are seen by ASU consultants. Currently they are preparing the ICU which is scheduled to operate in a few months. Ain Shams University Virtual Hospital is also aiding East African University to seek funding for the infrastructure of the ICU construction.

The service will be up and running within 1 year. The patients are already benefiting from the service. The health professionals will be able to competently run the maturing service and they will be leaders of change in their communities decreasing the brain drain.

Work will be done to facilitate East Africa to transform from a dependent spoke to become an independent hub with its own network of spokes. This contributes to the economic sustainability and scalability of the project. Startup expenses are minimal as open source software and basic hardware are used. Revenue shares will be subsidized from the running service. The results of this pilot attracted funding bodies, other potential spokes (Nigeria, Comoros) and other academic partners.

6 SUCCESS DYNAMICS

Why hasn't telemedicine been incorporated in healthcare systems in the region despite its existence for decades? Many telemedicine ventures never proceed beyond the pilot phase. What made Ain Shams University Virtual Hospital develop beyond the pilot stage?

A review by Chen and colleagues in 2013 [19] on telemedicine business models studied eight successful (longstanding minimum five years) ventures, three in developed countries and five in developing countries of which three were in India. They compared them using Osterwalder's "Business Model Canvas". The ventures were heterogeneous regarding geographical coverage, socioeconomic contexts, services offered, healthcare fields, general or specific demographics served, value propositions and revenue streams.

One of the interesting findings is that ventures in developed countries targeted organizations while those in developing countries mainly targeted individuals in underserved areas. The authors postulate that the ICT maturity in advanced countries allows the large scale, economically more profitable targeting of organizations.

The advancement of ICT and cultural preferences have also allowed the provision of the service in front-end primary customer interface in developed countries while in rural areas in developing countries a rural care center acts as the interface.

The authors also postulate that non-dedicated assistance, where the patient doesn't necessarily see the same doctor each time is less expensive but of course has the obvious disadvantage of losing, personal connection with the patient and medical consistency.

When comparing the above efficacious ventures with Ain Shams University Virtual Hospital it seems that even though the hub is centered in a developing country its business model has merged features from both models described by Chen and colleagues. Although it started by offering services to individuals it has expanded to offer service packages to universities, hospitals, ministry of health and medical societies among others. The ministry of health through the primary care centers provides the rural care reliable interface.

As for the type of assistance and for technical reasons non-dedicated assistance is the rule unless dedicated assistance is a necessity as in psychiatric cases or if requested by the patient.

ASUVH also adopts a model where some customers, as health insurance companies and private hospitals, help cover the expenses required to serve underprivileged customers usually individuals.

Mixing both the standard medical management hierarchy with corporate hierarchy has provided strong managerial team and associate teams that run the service by the business model planned. A successful business model is one of the pillars of a sustainable model.

Similarly, sustainability can be ensured by maintaining quality of the service; teleconsults and e-learning.

Collaborations and partnerships can provide stability, expertize, outreach, networking and resources.

7 PARTNERSHIPS AND COLLABORATIONS

In sustainable business models usually partnerships are a mainstay. Partnerships are continually sought with technical partners, medical and social societies, Universities, Ministries, and others. Some examples of these partnerships will be displayed to explain the unique added value of each.

One of the earliest collaborations was with the East African University. The cooperation was mainly targeting the teach limb of the initiative ensuring a whole competent generation of neurologists. The collaboration provided facilitation of postgraduate studies to Somalian students at ASU. The blended courses allowed practice at their local communities and timely application and open discussions. Facilitation of accommodation and other logistics were provided. It concentrated its efforts on capacity building rather than only temporary patient consults. Challenges faced helped adjust and upgrade the model. It provided a model that can be replicated in similar communities. It was also a proof of concept when the project was initiated.

Collaboration with ministries gave credibility and power to Ain Shams University Virtual Hospital. It also expedited many governmental procedures and paperwork.

Joint projects with national and international societies and organizations helped target areas of need and disadvantaged populations and sometimes provided funding for initiating spokes.

Ain Shams University Virtual Hospital was introduced to partners through national and international conferences, personal acquaintances, or approached by the team.

Ain Shams University Virtual Hospital has held two international conferences in the past three years. As it is not restricted to national services it was important to introduce to the international community the new service, its providers and the packages offered. "The First Arab African Teleneurology conference: A Treat and Teach initiative" was held at the League of Arab States under patronage of the President of Ain Shams University and several neurology societies, January 19–20, 2016 [20].

The conference established numerous constructive measures toward better stroke care in disadvantaged areas. Several agreements with Egyptian community hospitals and African universities have been signed to start a proof of concept phase of hospital to hospital acute care teleneurology service that would be complemented with bilateral mobility to facilitate service development in remote areas.

After the success of the first conference and the discussion of several potential collaborative projects in the continent it was important to expand the activities and hence the team. An African sector was instigated at the International scientific office at the Faculty of Medicine, ASU. It was responsible for connecting with African countries, receiving their requests and analyzing them. Several formal receptions were held with the President of ASU, Dean of Faculty of Medicine, ASU, African Ambassadors and the ASUVH team. These meetings concluded slow steady steps toward establishing either new healthcare services in African countries under patronage of ASU or initiating telemedicine consultancy services.

The second Arab African Telemedicine Conference: Toward Equitable Healthcare, held as part of the 39th Faculty of Medicine, ASU conference which had several parallel conferences. The conference was under patronage of Minster of higher education, Minster of telecommunication, and president of ASU. The conference displayed the vision of ASUVH in developing and implementing new comprehensive medical services in low resource settings. It displayed the use of telemedicine to provide safe, equitable, cost-effective healthcare for the continent through innovative telemedicine techniques. To urge policymakers to assist in propelling the African initiative beyond restraints and challenges, to patronage and collaborate with our leadership of African telemedicine in a step forward toward the ultimate African healthcare network.

The event highlighted strategies and experiences in improving medical training and practice in low resource settings. It demonstrated the inevitable growing role of technology and how to wisely use it. It hosted a consortium of policymakers, service providers, health professionals, IT professionals and many other stakeholders. Many subjects were discussed including legal and ethical issues binding local and trans-border consults. It discussed the available laws that govern the telemedicine practice even though there are no recent laws for telemedicine in Egypt. It also discussed consents in available telemedicine practices and how they have been adjusted to suite the Egyptian and African telemedicine service. Details of how to maintain quality and practices implemented internationally and nationally were compared. The Egyptian, Tunisian, Irish and international societies of telemedicine protocols have been displayed. It was also important to show how the ministry of health was incorporating telemedicine in the new insurance system to ensure provision of the service to all the population. Previous small scale telemedicine experiences were also demonstrated especially in rural areas. Availability and scope of novel methods in e-learning as virtual reality, augmented reality among others and their feasibility were one of the interesting content of the program. Representatives from the Arab League of States and the WHO revealed the health needs in the region and how for years digital health has aided to overcome the challenges in healthcare.

ASUVH recently contributed in drafting the Global Digital health Guidelines for eight m-health interventions during a meeting convened at the WHO, Geneva, Switzerland, June 6–8, 2018. The panel consisted of WHO staff and their team of methodologists and eminent digital health experts from all over the world. The discussions that lead to the recommendations took in consideration available evidence and the various experiences of the experts that represented societies, institutes, organizations and universities. The recommendations are expected to be published later this year.

8 SERVICES

Ain Shams University Virtual Hospital offers its services as part of its comprehensive package described earlier and also provides individual educational courses and teleconsults.

Teleconsults for individuals are available through the website www.asuvh.com at which the patient uses the customized booking system where he provides a short history and can choose a specialty or ask for help choosing the appropriate consultant. He then meets his consultant at the chosen date and time via a videoconsult platform and can upload laboratory studies, pictures or imaging. The patient receives a report with the case, recommendations and electronically signed. For follow up consults the patient has the option of visiting the same specialist or choosing another although he is primarily directed to the one who saw him first. Some consultations are store and forward where the patient sends his complain via a message and receives a reply within 24 h. Sometimes the patient may require a videoconsult or to be physically seen. Most patients are second opinion patients or follow up cases.

Cases are seen in almost all specialties; neurology, geriatrics and dermatology being the most frequent. Despite the importance of the service to rural areas in Egypt only around quarter of consults present from Egyptian governorates while the rest are from Cairo, the capital.

The patients were introduced to the service via social media where specialists posted short attractive health promotion messages of prevalent problems as vitamin D deficiency, the importance of water, headache, and others. Virtual clinics were announced and eminent staff members assigned days which attracted patients especially the service for individuals is for free.

9 CONCLUSION

Despite the available literature on telemedicine and its beneficial effects in providing equitable healthcare especially in developing countries there are no solid guidelines or models to direct new telemedicine providers to initiate their hubs.

The diversity of services that could be offered through telemedicine with or without e-learning along with mixed evidence in literature warrants rigorous research and innovative models that capture the added value of telemedicine and avert from its negative consequences.

The Egyptian-African network has provided several packages with a similar outer frame yet adapted content proposing that no one model suites all and that flexible designs are important in such a novel intervention.

REFERENCES

[1] United Nations, World Population Prospects-Population Division, United Nations, 2017. http://www.un.org/en/development/desa/population/.

[2] WHO, World Health Report: Working Together for Health, WHO Press, Geneva, 2006. http://www.who.int/whr/2006/en/.

[3] United Nations, World Urbanization Prospects, Department of Economic and Social Affairs, Population Division, New york, 2012. www.un.org/.../development/.../population/publications/.../urbanization/WUP2011_Re....

[4] EuroRSCG. http://bigthink.com/strange-maps/185-the-patients-per-doctor-map-of-the-world.

[5] R.H. Eikelboom, The telegraph and the beginnings of telemedicine in Australia, Stud. Health Technol. Inf. 182 (2012) 67–72.

[6] Telecommunication Development Bureau, Impact of Telecommunications in Health-Care and Other Social Services, International Telecommunication Union, Geneva, 2013.

[7] F. Shiferaw, M. Zolfo, The role of information communication technology (ICT) towards universal health coverage: the first steps of a telemedicine project in Ethiopia, Glob. Health Action 5 (2012), https://doi.org/10.3402/gha.v5i0.15638. Addis Ababa, Ethiopia: Department of Medical Services.

[8] M. Gulube, S. Wynchank, The national telemedicine system in South Africa — an overview and progress report, S. Afr. Med. J. 92 (7) (2002) 513–515.

[9] Webometrics. Available from: http://www.webometrics.info/top100_continent.asp?cont=africa.

[10] M.N. Rubin, K.E. Wellik, D.D. Channer, B.M. Demaerschalk, A systematic review of telestroke, Postgrad. Med. 125 (1) (2013) 45–50, https://doi.org/10.3810/pgm.2013.01.2623.

[11] J.H. Bower, G. Zenebe, Neurologic services in the nations of Africa, Neurology 64 (2005) 412–415.

[12] M. Ferreira, Geriatric medicine in South Africa — a Cinderella subspecialty? SA Fam Pract 48 (5) (2006).

[13] M. Eileen, E.C. Stuart-Shor, The global health service partnership: an academic–clinical partnership to build nursing and medical capacity in Africa, Front Public Health 5 (2017) 174.

[14] E.P. Goosby, The medical and nursing education partnership initiatives, Acad. Med. 89 (8 Suppl. l) (2014) s5–7.

[15] Ain Shams University. http://www.asu.edu.eg/article.php?action=show&id=255#.WvLC8YiFOUk.

[16] Cybermetrics Lab, Webometrics Ranking of the World Universities. Available from: www.africaranking.com/top-25-universities-in-africa.

[17] timeshighereducation. https://www.timeshighereducation.com/student/best-universities/best-universities-arab-world-2017.

[18] Central Agency for Public Mobilization and Statistics, Egypten, 2006. http://capmas.gov.eg.

[19] S. Chen, A. Cheng, K. Mehta, A review of telemedicine business models, Telemed. J. E Health 19 (4) (2013) 287–297.

[20] T. Emara, M. S. Al-Kotb, M. Nawara, H. Farouk, A. Elbokl, World Neurology. http://www.worldneurologyonline.com/article/the-first-arab-africanteleneurology-conference-a-treat-and-teach-initiative/.

CHAPTER 13

An Extended Views Based Big Data Model Toward Facilitating Electronic Health Record Analytics

VIVEK TIWARI[a] • RAMJEEVAN SINGH THAKUR[b]
[a]IIIT Naya Raipur, Naya Raipur, India; [b]MA-NIT Bhopal, Bhopal, India

1 INTRODUCTION AND BACKGROUND

The studies say that speed of data birth growing exponentially and around 88% of the entire till date data of universe were generated during back two years [1]. It could possible due to development in information technology [2] i.e., cheap and efficient data storage, fine network, reliable communication, ultra-speed network etc. One of the consequences of such powerful infrastructure availability is a public social site where peoples uploads around 2 million snaps per day [3] and if consider snap size is 3 MB, total storages is need around 5.4 TB per day. Traditionally, these pics include social gathering, events, parties, marriages, and birthdays but, the scenario is changing rapidly and now pics bucket includes every moment's pics, smileys, selfies, traffic, and all 360° pics. The aforesaid picture make more clearly the Big data applications and must notice the growth of data collection where it went beyond the perimeter of standard available software in any domain. As data analytics perspective, the utmost complexity with Big data is to knowledge generation form such high volume heterogeneous data. Moreover, in critical data analytics application, knowledge generation should be dynamic and considerable fast [4,5].

Xindong et al. [2] has taken a very real life example to make clear understanding on how identifiable signs of Big data make it considerable complex to apply standard data mining techniques. With the influence of presented example [2], we are describing the same issue with taking different scenario and example form real life as depicted in Fig. 13.1. A crow and fox story is graphically illustrated [6] and easily understood if carefully observe images sequentially from image 1 to image 8. Most importantly, it

is feasible (interpretation for story) if and only if (Scenario-1):

- all the images must be visited in order
- all images are accessible, stored in single place
- Image position is stationary, i.e., images are not changing their position/order

Now, try to take this scenario-1 in traditional centralized data context where data (story images) are stored in singly repository system and easily accessible. Such data are relatively small and static (data (images) position and order never change).

Scenario-2: All stuff depicted in Fig.13.1 could be seen in Big data context where the situation are redefined, but goal is intact (interpretation of the story). There was need to re-adjust the underlying parameters of Fig. 13.1 to make it more resemblance with Big data context as below:

- Images are scattered and stored in difference places
- Images access in order in not possible
- Image position is dynamic, i.e., images are changing their position/order

In the same line, we try to make this picture by add torches where it represent the accessibility. Assume, There are four person having torches A, B, C, D respectively and can visit respective image such as Image 1, Image 3, Image 6, and Image 7 respectively. Because each person's view is limited to his local region, it is not surprising that the peoples will conclude independently depending on the region each of them is limited to, such as: (A) Crow is flying, (B) Fox is walking, (C) Crow is sitting on the tree branch and (D) Crow and fox in fighting. In this scenario, none on the person can understand the whole story since:

- Accessibility of single torch is limited to single picture

Telemedicine Technologies. https://doi.org/10.1016/B978-0-12-816948-3.00013-1

FIG. 13.1 A story and the torches: concise and restricted (localize) view of torches may interpret the story wrongly.

- By seeing the small part of whole story, difficult to guess the story correctly
- Situation become more complex as images position become dynamic

In the same view, knowledge discovery (i.e., applicability of present data mining techniques) is complex, difficult, close to infeasible since data are huge, scattered, heterogeneous, continuously growing and moving. Exploring the Big Data in this scenario is equivalent to aggregating heterogeneous information from different sources (torches) to help draw a best possible picture to reveal the genuine gesture story in a real-time fashion. They may even have privacy concerns about the messages they deliberate in the information exchange process.

2 SEMATIC VIEWS IN CONTEXT OF PROPOSED WORK

In this work, the meaning of semantic view is significantly differ from the conventional literature. A semantic view is readable and presentable structure of knowledge. Furthermore, knowledge is pattern/artifact/unknown-information which is generally extracted through data mining techniques [7]. A brief process to extract semantic view from raw data is depicted in Fig. 13.2.

In other words, semantic views are nothing but a semi-structural way of presentation patterns which can be further processed (manipulate/aggregate/update/retrieve etc.). The patterns are represented with triple (Pattern_Type, Pattern, and Context) in the proposed schema [8,9]:

Pattern_Type: A pattern type *'pt' is a quintuple -*
pt = (n, ss, ds, ms, f)
Where:
"n" is the name of pattern type.
"ss" (structure schema) is a definition of pattern space.
"ds" (source schema) define related raw data space.
"ms" (measure schema) quantify the quality
"f" is a formula that describes the relationship between context space and pattern space.

Example (Association rule): Pattern type for association rule is defined as-
"n": Association rule
"ss": TUPLE (head: SET(STRING), body: SET(STRING))
"ds": BAG (transaction: SET(STRING))
"ms": TUPLE (confidence: REAL, support: REAL)
"f": $\forall x$ ($x \in$ transaction and $x \in$ context source, i.e., transaction $\in$ context source)

Example (Classification): Pattern type for classification is defined as-
"n": Classification
"ss": Class Label ("ABC", "PQR", "STU")
"ds": Bank (transaction: SET(STRING))
"ms": CLASS (Entropy: Min-x, Information Gain: Min-y, Error: Max-z)
"f": $\forall x$ ($x \in$ transaction and $x \in$ context source, i.e., transaction $\in$ context source)

Pattern: Let pt = (n, ss, ds, ms, f) is a pattern type. A pattern "p" instance of "pt" is a quintuple:
P = (pid, s, d, m, e)
Where,
"pid" − pattern identifier
"s" − is a value for type "ss"
"d" − dataset
"m" − is a value of type 'ms'
"e" − region of the source space
Context: It is defined as:
c = (cid, cn, cs, pattern-type, pc)
Where,
"cid" − context identifier
"cn" − context name
"cs" − context source
Pattern-type ("pt")
pc-collection of pattern of type "pt"

It clearly depicts the complexity of pattern representation (able to complete define through three level of information: Pattern_Type, Pattern, and Context) [8,9]. In this regards, a definition of semantic view of complete patterns consists of all its underlying depending terms as presented in Fig. 13.3. Furthermore, Data representation format toward structure and knowledge carrying capabilities is depicted in Fig. 13.4.

3 BIG DATA MINING WITH EXTENDED SEMANTIC VIEWS

A generic distributed data storage architecture (bigdata) is depicted in Fig. 13.5 where data are scattered in various locations and linked through the cloud. Such highly data distributed environment is the most

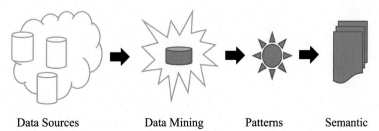

Data Sources   Data Mining   Patterns   Semantic

FIG. 13.2 Process to extract semantic view from raw data.

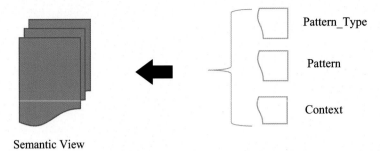

Semantic View

FIG. 13.3 Semantic view structure.

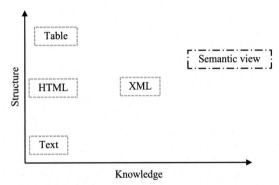

FIG. 13.4 Data representation format toward structure and knowledge carrying capabilities.

infeasible to apply conventional data mining techniques (i.e., data is not available centrally).

One of the solution from data warehouse concept can be thought as gather data centrally and then apply data mining but it is again infeasible (i.e., data is huge, volatile, continues). In this problem domain, semantic views based solution comes in picture where distributed location wise semantics views are maintained (detailed process is depicted in Fig. 13.5). Extended sematic view follow pre-defined flexible styles (more readily configurable) and it depends upon kind of knowledge are being carried. These semantic view are very light in size (need very less memory space) and hence easy to relocated somewhere centrally and get final view after proper integration through underlying operations [9]. Benefits with semantic view based big data mining are:

- Improve structure and visualization of knowledge
- Carry actionable information
- Data independence
- Fast processing
- Less memory consumption
- User friendly

- Raw data protection
 Challenges:
- Does not explain meaning (inference of new information)
- Critical information
- Similarity measures (same sematic view identification)
- Complex operations (post operation interpretation, semantic views aggregation and integration)
- Linking of semantic view (cannot be easily automated)
- Refreshment

4 PROPOSED SOLUTION TO FACILITATING ELECTRONIC HEALTH RECORD PROCESSING

In the last one decade, the significant advancements in broadcast communications have been experienced. Additionally, high end utilization and development of cell phones (3G, 4G and Internet of Things (IoT)) make the things more automotive [10−12]. The population of aging people with illnesses such as diabetes, heart disease and much more is increasing. In addition, youngsters are also experiencing diseases like asthma and obesity due to a lavish lifestyle [13]. The Republic of India; a youthful nation with the world's most populous community, has battled for making emergency medical care safe [14]. Nowadays, Government is thinking about its socioeconomic expansion through Digital India and Skilled India, but it also plagues many challenges [14]. The private medical sector of India is holding more or less 81% medical services and 88% medical professionals [10,14]. India started its health reforms journey in the most recent decade. Although, healthcare workforce in India remains inadequate and underutilized. This journey now needs to pick up momentum. It seems feasible to analyses health records to find out patterns and hidden trends which can be further used for policy making, budgeting,

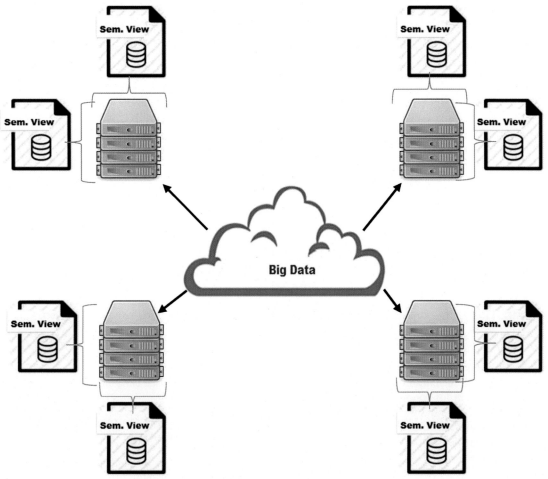

FIG. 13.5 Semantic views for distributed data cluster.

disease investigation, continuous observation of patients and elderly individuals for their wellbeing [15,16]. Such arrangements enable medical experts to analyses current patient status, minimize reaction time, increase livelihood, scalability, and availability.

Data warehouse helps to manage huge amount of data efficiently and allow data mining techniques to extract hidden patterns [17–19]. The pattern is useful because it describes a recurrent behavior [20–22]. The volume of extracting patterns from various knowledge discovery applications is increasing exponentially, so there requires an effective pattern management system [23–25]. A tool is that will permit us to compare, query and store the patterns [26–28]. In this regards, a concepts is found in literature (pattern warehouse) which will serve our purpose. Pattern warehouse give the capability to store, manipulate, update various kind of

patterns [29]. In the proposed work, semantics views which are actually patterns are generated in different locations (Health Record Cluster) send to pattern warehouse (central pattern repository system) as depicted in Fig. 13.6.

4.1 Health Record Cluster (HRC)

Considering the view of large demographic area (i.e., country) where health related data is being generated at bottom level of line (village). Such data is forwarded toward root of the hierarchy and stores somewhere in district level. Similarly, other district maintain their health related data. We are considering little light view to make the discussion simple and feasible to understand the inclusion of proposed concept. Suppose, we have divided whole large demographic area into four big region to store the health records of concerns district

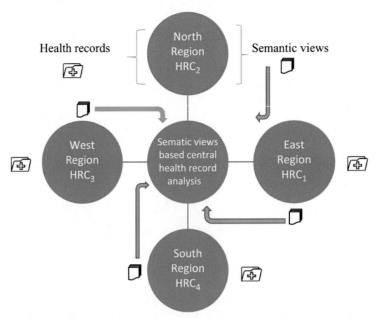

FIG. 13.6 An extended views based big data model toward facilitating electronic health record analysis.

and we are using term "Health Record Cluster" for those region. In other word, such health records data of whole country is maintain in for regions/clusters (east, west, north, and south). This kind of data arrangement are prevalent and as taken as it is. Data distribution and its big size are the major concern of such arrangement and big data technology is there to tackle it [30]. The primary objective of the chapter is to propose an analytical model which give holistic view of patterns/hidden trends as whole, fast responsive and effective in term of memory consumption. A semantic view based analytics system is proposed and discussed in Sections 3 and 4, which is suitable to implement health data analytics under distributed environment. The proposed system will work as:

1. Health records are stored in HRC.
2. Health records are heterogeneous data and so need attention while integrated into HRC.
3. Various data analysis and data mining methods are available and can be chosen as per requirement.
4. Patterns and trends are extracted.
5. Transform patterns into semantic form (as discussed in Section 2)
6. Send semantic view to centrally repository system
7. Intergrade the related pattern and analyses to get holistic view.
8. Update the old patterns as new pattern arrive.

REFERENCES

[1] IBM What Is Big Data: Bring Big Data to the Enterprise, IBM, 2012. http://www-01.ibm.com/software/data/bigdata/.

[2] X. Wu, X. Zhu, G.Q. Wu, W. Ding, Data mining with big data, IEEE Trans. Knowl. Data Eng. 26 (1) (2014) 97–107.

[3] F. Michel, How Many Photos Are Uploaded to Flickr Every Day and Month?, 2012. http://www.flickr.com/photos/franckmichel/6855169886/.

[4] A. Rajaraman, J. Ullman, Mining of Massive Data Sets, Cambridge Univ. Press, 2011.

[5] E. Chondrogiannis, V. Andronikou, A. Tagaris, E. Karanastasis, T. Varvarigou, M. Tsuji, A novel semantic representation for eligibility criteria in clinical trials, J. Biomed. Inf. 69 (2017) 10–23. Elsevier.

[6] https://hubpages.com/literature/Short-stories-for-kids-with-pictures.

[7] V. Tiwari, S. Gupta, R. Tiwari, Association rule mining: a graph based approach for mining frequent itemsets, in: 2010 International Conference on Networking and Information Technology (ICNIT), IEEE, 2010, pp. 309–313.

[8] V. Tiwari, R.S. Thakur, Contextual snowflake modeling for pattern warehouse logical design, Sadhana Acad. Proc. Eng. Sci. 40 (1) (2015) 15–33.

[9] V. Tiwari, R.S. Thakur, Towards elementary algebra for on-line knowledge processing of pattern cube, Natl. Acad. Sci. Lett. 41 (1) (2018) 35–39.

[10] Y.P. Yang, T.Y. Lai, The application of intelligent keywords to patient Adherence management system,

Int. J. E-Health Med. Commun. 4 (4) (2013) 102–119. IGI-Global.

[11] M. Kuhn, K. Prettner, Growth and welfare effects of health care in knowledge-based economies, J. Health Econ. 46 (2016) 100–119. Elsevier.

[12] R. Mirani, A. Harpalani, Business benefits or incentive maximization? impacts of the medicare EHR incentive program at acute care hospitals, ACM Trans. Manage. Inf. Syst. 4 (4) (2013) 20.

[13] P.B. Jensen, L.J. Jensen, S. Brunak, Mining electronic health records: towards better research applications and clinical care, Nat. Rev. Genet. 13 (6) (2012) 395.

[14] R. Srinivisan, Health Care in India-Vision 2020, Government of India, Planning Commission of India, New Delhi, India, 2012.

[15] H. Banaee, M.U. Ahmed, A. Loutfi, Data mining for wearable sensors in health monitoring systems: a review of recent trends and challenges, Sensors 13 (12) (2013) 17472–17500.

[16] B. Tiwari, A. Kumar, Role-based access control through on-demand classification of electronic health record, Int. J. Electron. Healthc. 8 (1) (2015) 9–24. Inderscience.

[17] M. Eldridge, Enterprise Data Warehouse: A Patterns Approach to Data Integration, Microsoft IT Showcase, 2010 Microsoft Corporation, 2010. http://www.microsoft.com/technet/itshowcase.

[18] C. Barbara, M. Anna, PSYCHO: a prototype system for pattern management, in: Proceeding of the 31st International Conference on Very Large Data Bases (VLDB), ACM, Trondheim, Norway, 2005, pp. 1346–1349.

[19] S. Rizzi, Conceptual modeling solutions for the data warehouse, in: R. Wrembel, C. Koncilia (Eds.), Data Warehouses and OLAP: Concepts, Architectures and Solutions, IGI-Global, 2007, pp. 1–26.

[20] D. Batra, Conceptual data modeling patterns: representation and validation, J. Database Manag. 16 (2) (2005) 84–106.

[21] A. Nandi, C. Yu, P. Bohannon, R. Ramakrishnan, Data cube materialization and mining over MapReduce, IEEE Trans. Knowl. Data Eng. 24 (10) (2012) 1747–1759.

[22] J. Pardillo, J.N. Mazón, J. Trujillo, Bridging the semantic gap in OLAP models: platform-independent queries, in: Proceedings of the ACM 11th International Workshop on Data Warehousing and OLAP, ACM, 2008, pp. 89–96.

[23] T. Zdenka, Data modeling and ontological semantics, Int. J. Data Anal. Tech. Strateg. 4 (3) (2012) 237–255.

[24] S. Rizzi, E. Bertino, B. Catania, M. Golfarelli, M. Halkidi, M. Terrovitis, P. Vassiliadis, M. Vazirgiannis, E. Vrahnos, Towards a logical model for patterns, in: Proceeding of ER Conference, Chicago, IL, USA, Springer, Berlin, Heidelberg, 2003, pp. 77–90.

[25] V. Tiwari, R.S. Thakur, P2MS: a phase-wise pattern management system for pattern warehouse, Int. J. Data Min. Model. Manag. 7 (4) (2014) 331–350.

[26] T. Manolis, P. Vassiliadis, Architecture for Pattern Base Management Systems, Department of Electrical and Computer Engineering, 2003. PANDA Workshop, National Technical University of Athens.

[27] M. Golfarelli, S. Rizzi, I. Cella, Beyond data warehousing: what's next in business intelligence?, in: Proceedings of the 7th ACM International Workshop on Data Warehousing and OLAP ACM, Washington, DC, USA, 2004, pp. 1–6.

[28] T. Manolis, P. Vassiliadis, S. Spiros, Modeling and language support for the management of pattern-bases, Data Knowl. Eng. 62 (2) (2007) 368–397.

[29] V. Tiwari, R.S. Thakur, Pattern warehouse: context based modeling and quality issues, Proc. Natl. Acad. Sci. India Sect. A Phys. Sci. 86 (3) (2016) 417–431.

[30] J. Andreu-Perez, C.C. Poon, R.D. Merrifield, S.T. Wong, G.Z. Yang, Big data for health, IEEE J. Biomed. Health Inform. 19 (4) (2015) 1193–1208.

CHAPTER 14

Security and Privacy in Remote Healthcare: Issues, Solutions, and Standards

PIJUSH KANTI DUTTA PRAMANIK[a] • GAURAV PAREEK[b] • ANAND NAYYAR[c]
[a]National Institute of Technology, Durgapur, India; [b]National Institute of Technology, Goa, India; [c]Duy Tan University, Da Nang, Vietnam

1 INTRODUCTION

As a child, we remember, when we used to get sick, our family doctor was called upon at home. We rarely had to visit the doctor's place unless the family doctor was out of town or it was a critical case required consultation with other physicians. But the scenario is reversed today. It is a very rare case that doctors visit the patients' house. We have to go to the clinic whenever we need to visit a doctor. Besides other social and economic reasons, the major reason behind this, probably, the alarmingly decreasing ratio of physicians and the general population. For instance, as per National Health Profile's latest data, released by the Central Bureau of Health Intelligence, India, across the country, for approximately 11,082 people only one allopathic government doctor is available [1].

This changed health care scenario has affected patients from urban and rural areas differently. For urban patients, it is an economic problem. People have to take time off their work, drive or travel to the clinic, and spend a significant amount of time there. This causes productivity loss and a decrease in GDP and revenue. If the visit to the doctor is frequent and regular, as in the case of chronic patients, the loss will be more. But for rural people, the degree of suffering is much more mainly due to the absence of proper healthcare infrastructure in rural areas. If we consider the Indian scenario, out of 1.33 billion people, 70% are living in villages and a majority of which do not even have the basic and minimum healthcare facility. In many villages, even if there is some government medical center, they have no qualified physicians. Getting specialist doctors is out of a dream. In India, nearly 80% of doctors, 75% of pharmacies, and 60% of hospitals are in urban areas [2]. Of all the medical visits in India, nearly 86% are made by rural peoples, and the majority of them have to travel more than 100 km to avail healthcare facility [3], often spending a month's wage. Because of the inaccessibility of primary healthcare, these people sought for medical help at the later stage of the disease-cycle. This increases the medical expenses by nearly 1.5 times compared to those from urban areas. And approximately 80% of this cost is borne by themselves [4], most often by selling their assets, farmlands, and homes. This pushes them below the poverty line further.

Remote healthcare has come out as a promising way out of this challenge in the healthcare scenario. In this method of health care service, patients do not have to move to the clinics to visit a doctor in case of need of medical help. They can communicate with a remote city-based physician with the help of telecommunication and advanced features of ICT. Physicians diagnose the patients remotely and prescribe treatments and medicines accordingly. It is expected that with proper implementation of remote healthcare, 76% of rural patients need not have to go beyond their village for medical care [2]. Remote healthcare offers several benefits (see Section 2.2) and is capable of changing the healthcare crisis, especially for those people who are underserved and do not have easy access to urban healthcare facilities as well as for elderly peoples.

Remote health care is achieved through telemedicine and remote monitoring. Telemedicine refers to a system that allows patients from remote and rural areas to get a consultation from distance doctors. Whereas remote health monitoring allows doctors to continuously monitor patients remotely using health monitoring systems that are connected to the Internet. Though

Telemedicine Technologies. https://doi.org/10.1016/B978-0-12-816948-3.00014-3

telemedicine and remote monitoring seem to be different, in the waking of RFID [5], Internet of Things (IoT) [6], pervasive systems [7], healthcare data analytics [8], etc. and successful use of these technologies in healthcare, the differences between the two are getting blurred. The popularity and usability of nano-sensors and IoT along with the advancement of high-bandwidth, energy-efficient wireless technologies, remote health monitoring has become more practical and affordable. More often, telemedicine is getting blended with remote monitoring for precision diagnosis and medication. In this text, we refer remote health care as both to remote health monitoring and telemedicine.

Though it has promising potentials, successful implementation of remote healthcare is not free of concerns. Several challenges need to be addressed to get the full utilization of remote healthcare (see Section 2.3). But, being a networked system and since healthcare data are sensitive to mishandling, maintaining security, privacy, and confidentiality are the primary concerns and of foremost priority. Doctors, in remote healthcare, heavily depends on the digital health record for diagnosis. Hence information integrity and reliability are essential for error-free diagnosis. These systems are vulnerable to a wide range of security threats. The threats can emerge at each level of a remote healthcare system, for instance, at data collection, data transmission, data storage, and data access levels [9]. In some cases, inadequate awareness and training lead to these threats. But, most of them have been surfaced because security threats are inherent in multi-attributed distributed networked systems and implementing end-to-end security in such a complex and the evolving system is really challenging [10]. The security threats can be mitigated by opting measures such as user authentication, access control, audit control, enforcing information integrity, confidentiality, and privacy, non-repudiation, secured data transmission, etc. Furthermore, considering the evolving nature and proliferating coverage of the threats, predictive and proactive measures are required [10].

Besides the aforementioned threats, another issue that holds back remote healthcare is that the users' trust in remote healthcare and its acceptability to the masses. It is necessary to infuse confidence to the patients in adopting the remote healthcare system. In this direction, patients' privacy and confidentiality need to be protected. In doing so, the obvious challenge faced is that having the balance between confidentiality and availability. Doctors will be able to provide better quality healthcare service if the patient's medical history and clinical records are accessible from other sources as well.

More the patients' data are integrated and available across different medical service applications the patient can obtain more quality, time-saving, and economical healthcare services, and importantly, from anywhere because sharing medical data helps in realizing pervasive healthcare [7]. But again, all the data cannot be permitted to be accessed by all because of security and privacy reasons. It is to be determined wisely, which part of the patients' data should be confidential and which part should be pervasively accessible. The optimal balance between these two will result in delivering the best possible healthcare to the patients [9].

The rest of the chapter is organized as follows. Section 2 covers the basics of remote healthcare, mentioning its benefits and challenges. It differentiates between "health care" and "healthcare". Telemedicine and remote monitoring, the two approaches of remote health care, and their components and architecture are discussed briefly. Section 3 is devoted to security and privacy issues of remote healthcare. It presents the roles of various entities and their impact on the security of the health care system. It then highlights vital security and privacy issues in remote healthcare, formulates the concrete security requirements and describes a range of solutions to satisfy these requirements. The security standards for local and wide-range network transmission are described. The application-level security solutions like access control, privacy and reliability are also presented. It also presents important trade-offs and the challenges facing the implementation of security solutions in remote healthcare. Future of remote healthcare security is also discussed briefly. And finally, Section 4 concludes the chapter.

2 REMOTE HEALTHCARE

The traditional approaches of medical treatment models are not sufficient to address the challenge of the ever-growing demand for cost-effective health services either in a hospital or in other medical care units. Information and Communication Technology (ICT) being the key factor in the development of the latest medical models which has proposed advanced biosensors, efficient patient records management, and digital and computerized medical equipment. These advancements have facilitated bridging the gap between remote and underserved locations and infrastructurally equipped hospitals with the help of seamless information exchange. Such models are termed as *remote healthcare* or *e-health*. In simple words, remote healthcare is to take care of the patient's health from a distant, i.e., remotely. The vision of remote healthcare is to provide healthcare services to patients outside of traditional healthcare establishments, usually at home.

Remote healthcare plays a crucial role in acquisition, management and exchange of personal as well as medical information of the patient, making it easier to follow treatment procedures without moving the patient, as well as the medical staff, here and there. Information exchange is the primary principle of remote healthcare. It should be done in a highly structured manner to support stakeholder interactions in the pathway, both in terms of self-management and lifestyle promotion and disease detection. The exchange of information should be such that, it assures efficient coordination and transition among the various phases to facilitate process continuation.

2.1 "Remote Health Care" Versus "Remote Healthcare"

Here, we would like to make clear the misunderstanding between the two terms used in the literature: "remote health care" and "remote healthcare".

Remote health care: The term "remote health care" can be defined as the use of existing and emerging ITeS and technologies to provide and support remote delivery of health care. It signifies the actions that are carried out to monitor and improve patients' health remotely. Remote health care is referred in the context of remote monitoring, remote diagnosis, e-imaging, remote medication, remote emergency support, remote consulting, etc.

Remote healthcare: The term "remote healthcare" (sometimes referred as telehealthcare) is used in a broader perspective. Remote healthcare refers to the system or industry that supports remote health care and may include other services related to health care, besides the clinical operations, such as logistics, e-commerce and e-marketing of health care, decision support, e-business intelligence in health care, remote home care applications, training and e-learning, online transactional transmissions, etc.

In summary, "remote health care" realizes "remote healthcare" whereas "remote healthcare" facilitates in providing "remote health care".

2.2 Benefits of Remote Healthcare

The key benefits of remote healthcare are as follows:
- Medical services are easily accessible to remote patients.
- Patients don't have to travel long to see doctors. This saves time and money.

- Patients feel less stressed being treated at home or near home, among the family and friends.
- If medical care is easily accessible people are less likely to delay care. Early diagnosis and treatment not only results in better and faster recovery and minimized chance of fatal health consequence but also reduces medical expenses.
- Emergency services and intensive care services, to some extent, can also be delivered remotely.
- Reduction in the need for and better utilization of emergency rooms and ICUs.
- Contagious diseases are managed better by isolating the affected patient and initiating early treatment.
- Shortage of medical professionals and nursing staffs can be addressed by remote healthcare.
- Specialist doctors can attend more patients.
- Remote healthcare curtails the traveling time for accessing medical services, waiting periods at the clinic, contact time with doctors, and duration of hospitalization. This saved time minimizes the productivity loss in organisations which in turn improves the national GDP.
- As an additional benefit, remote healthcare helps the environment by reducing carbon footprint due to less traveling.

2.3 Challenges in Remote Healthcare

Implementing remote healthcare systems are not straight forward. There are several challenges in terms of infrastructure, operations, management, policy, standards, legal, awareness, acceptability, etc. are needed to be addressed. Besides security and privacy which are elaborately discussed in Section 3, below some of the other factors that need to be focused on to achieve the real goal of remote healthcare are mentioned [11].
- Deficiency of adequate infrastructure.
- Lack of proper integration with the legacy and traditional healthcare systems.
- Shortage of specialized and skilled healthcare professionals.
- Lack of awareness and trust. People, by nature, are resistant to changes and reluctant to have faith in new services.
- Inadequate number of service centers for the medical equipment used in remote healthcare.
- Absence of global standards for different equipment and file formats for health data and storage. This complicates the interoperability.
- Absence of a legal framework precisely defining the responsibilities and liabilities of every stakeholder and course of legal actions in case of non-adherence.

- Absence of national e-healthcare policy and regulations.
- Nonexistence of comprehensive and national strategies regarding remote healthcare.

2.4 Remote Health Care Approaches

Remote health care generally has two main approaches: telemedicine (sometimes referred to as e-medicine) and remote health monitoring (sometimes referred to as telecare) [12].

2.4.1 Telemedicine

2.4.1.1 What is telemedicine and how it works?.

Telemedicine is one of the first initiatives toward remote healthcare. It was proposed nearly 30 years back with the purpose of providing affordable medical services, especially for the unprivileged people, in terms of the medical facility, from the places where sufficient healthcare infrastructure is absent. In telemedicine, physicians interact with the patients located in remote and rural areas using information and telecommunication technologies. At the patient's end, at the local health center, the local health technicians act as mediators. They do the preliminary assessment of the patients, conduct basic diagnostic tests, note the information of vital parameters and enter all these relevant information into the electronic health record. These information are sent to the remote physician who assess the patient's health condition, diagnose and sends back the prescription or suggestions to the local health center. The doctors may use video and audio connections for real-time interaction with the patients in order to provide real-time consultations. Doctors may use other sophisticated remotely controlled medical examination equipment such as close-up cameras, microscopes, dermascopes, etc. for better assessment [13]. Modern telemedicine systems make use of IT technologies like high-resolution 4K monitors, high-performance communication networks, telecommunications, responsive websites and pervasive smart devices [7]. Today's telemedicine systems are highly integrated with the cloud and edge computing [14], IoT [15] and sophisticated communication systems.

To build an effective telemedicine system, the system should satisfy the following requirements:

- Efficient to handle emergency medical cases at remote sites and live consultant available with quick response.
- Enhanced intensive healthcare provision by providing telemedicine unit to ICU doctor and medical unit with the in-house telemetry system.

- Facilitates home monitoring by incorporating telemedicine unit at the home of the patient while the other units remain at the main hospital.

2.4.1.2 Infrastructural components of telemedicine.

The following components form the backbone of any telemedicine system:

Local healthcare unit: In a telemedicine system, remote healthcare unit is regarded as a small health center, usually located in a rural or remote area. This is the primary point of contact between patients and healthcare givers. These health centers may comprise a computer with customized medical software connected to a few diagnostic instruments like ECG, EMG, X-ray, etc. for acquiring the health status of the patients. The major components found at a typical local healthcare center are mentioned below:

- Medical peripherals: These digital devices enable the local healthcare attendant to gather patient's vitals, monitor progress, view ultrasounds, check-up lung and heart, and capture images of skin, ears, eyes and other areas. The telemedicine medical kits comprise stethoscopes (interactive or telephonic-based), vital signs monitors, ECGs, spirometers and holters, retinal cameras, ultrasound probes, etc.
- HD Camera: To facilitate remote communication in a smooth manner via video conferencing, a high-quality HD camera is required. It is responsible for exchanging live feeds between patient and doctor from a distance.
- Computer: Health data processing, controlling and overall operations of records management, handling basic medical equipment's, performing communications are done with the help of a computer.

Data communication networks: The records acquired at the local healthcare unit are sent to the doctor remote at a remote hospital. For reliable and robust transmission, a strong data communication network is required to be integrated within the telemedicine system. The audio-visual communication is generally done using public telephone networks, and the recorded digital data are transmitted through IP-based networks using the standard network protocols, e.g., TCP/IP, HTTP, SMTP, etc. The specific network structure of any telemedicine system largely depends on the geographic factors of the area that will be served by the network and the type of local users there. The telemedicine units are equipped with standard network communication units like ADSL, Broadband, 3G/4G/LTE depending on the area to facilitate information exchange between the centers.

Medical database/EHR: The medical database, also called an electronic health record (EHR), is a centralized database that stores patients' data. The medical database in telemedicine is comprised of the following components:

- Subjective description: Refers to the details of patients, their health information with disease history. The information may contain attributes such as symptoms, duration, and description, along with other comments.
- Objective description: Comprise of all the results noted by the doctor in the previous description and comparison with regard to the current situation of the patient.
- Assessment: Details regarding the doctor's diagnosis as well as disease description based on the symptoms of the patient.
- Plan: A complete structure designed and implemented by the doctor regarding diagnosing the patient right from the start to completion of the treatment procedure.
- Application interface: To access the database, it is equipped with a proper web-based application interface which comprises the following modules:
  - Doctor registration: In order to become the part of the telemedicine program, every doctor has to properly register as a new doctor and after registration, the database provides information with regard to only those patients allocated to a specific doctor.
  - Patient registration: To become part of a telemedicine program to take live support from the medical practitioners, every patient has to register the portal with regard to details with valid information. This can be done by the medical staff at the local health center on behalf of the patient.
  - Portal interface: This interface is regarded as the middleware between the doctor at the frontend and database at backend. The doctor can make use of the portal to track all the progress, get notified updates of patients, access all types of images, alter information, interact with the patient live, update the information at the backend.

The database server is tightly integrated into the system to facilitate records storage of patients. The server may be hosted in the cloud for better access and management inexpensively.

Doctor unit/main hospital: The information acquired at the local health center is sent to the doctors at a remote hospital for examining the reports, diagnose, interact with patients and provide appropriate treatment support via tele or video conferencing. The hospital is responsible for collecting the data transmitted by the local healthcare unit. At this end, the doctors sit in the room with ICT-enabled equipment to provide live support to the patients and observe the data transmitted from remote centers. In order to facilitate communication, various audio/video devices (e.g., smartphones, laptops, PCs, etc.) are required. This unit is also connected to the central medical database server.

Online telemedicine portals/websites: To make the telemedicine system interactive, suitable websites and apps, both for PCs and mobiles, are required to be designed and developed. These portals are ideally user-friendly and simple to use with very fewer technicalities so that both patients and medical staffs will be prompted to use the system. With user-friendly online portals, telemedicine provides the following advantages:

- Easy and convenient access to healthcare services.
- Improved QoE (quality of experience) for all the stakeholders.
- Speedy medical follow up.
- Maximizing medical care efficiency.
- User satisfaction attracts new users which leads to market growth.
- Reduced liability because of the completely digital audit trail.

As the use of smart mobile devices (smartphone and tablets) is in continuous upsurge [16], mobile apps are getting more popular. These mobile apps enable users to use telemedicine systems pervasively. The apps are capable of triggering an alarm to alert the doctor in case of an emergency.

2.4.1.3 General architecture of telemedicine and its functionality.
Telemedicine systems, in general, follow a hierarchical tiered structure which includes the following:

- Level 1: Local/remote telemedicine center. These are the local or primary healthcare unit located in rural and remote areas.
- Level 2: City/district hospital. Local/rural health centers are connected to the city/district hospital. The district hospital, optionally, may further be connected to the state hospital.
- Level 3: Speciality center. The city hospital is connected to the speciality centers for disease-specific further assistance.

Fig. 14.1 presents a general architecture of a telemedicine system. A patient requiring medical attention

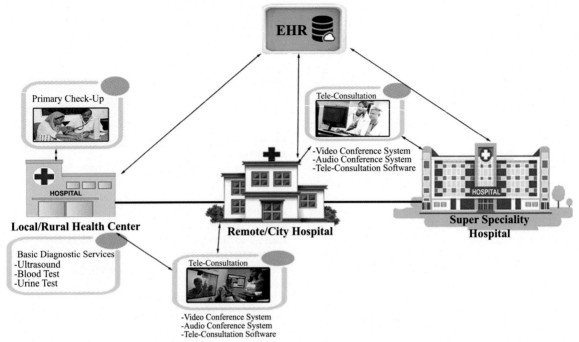

FIG. 14.1 A general telemedicine system.

approaches the nearby local health center where a local health professional (may not be a certified doctor) attends the patient and does the primary health check-up. This unit consists of basic diagnostic equipment and tele-consultation devices linked via PC and Internet to the city hospital. The primary responsibility of the local healthcare unit is to acquire all the vital statistics of the patient in terms of physiological data (e.g., blood, urine, etc.) and images (e.g., ultrasound) and transmits the data to the remote city hospital. After receiving the records, the remote medical practitioner goes through every detail, before proceedings with live Interaction with patients. After carefully examining the basic vital signs, the meeting is booked online between doctor and patient at remote healthcare unit. The doctor makes use of an audio or video conferencing system as well as automation live feeds to have live interaction with the patient. These remote hospitals are connected to a centralized database where all the data of the patient as well as other details and even the recorded audio/video interaction between doctor and patient are also stored. The stored information can be accessed using mobile apps or web-based interface. The main hospitals are also linked to specialist hospitals to provide specialized support to the patients in case of an emergency and these

specialized hospitals have same teleconferencing units enabled to support remote patients.

2.4.1.4 Security vulnerability points in telemedicine.
A telemedicine system requires the data collected through equipment at the local hospitals to be transmitted to the remote super speciality hospital either through live sessions or through web-based applications. Privacy, integrity and authenticity of the collected data must be ensured in a telemedicine system to ensure timely and correct diagnosis and treatment without any abuse of privacy of patient's data. Following are the points of potential security and privacy vulnerabilities in telemedicine systems:

- **Local health center:** Incorrectly calibrated medical equipment is a source of potential failure of the telemedicine system. If the patient data collected by the equipment are incorrect, the diagnosis of the patient suffers serious integrity issues.
- **EHR:** If the doctors in the super-speciality hospital of a telemedicine system can be online, the data is transferred through live streaming or audio/video conferencing. Provided the data is collected using accurate and precise equipment, there is little chance of being modified by any outsider for creating the malicious effect. Assuming that no individual with

malicious intent has access to the personal, stand-alone storage of the doctors in hospitals, the patient data transmitted using standard security protocols over the Internet cannot be accessed by any adversary. However, live streaming is not always feasible because doctors may not be online. In such a case, data is transferred through a web-based application interface and is stored in a database (EHRs) accessible to the doctors. Security of data stored in EHRs is critical as they store data in digital format and Internet hackers do have the potential to access confidential medical information. Another consideration while storing data in EHRs is that the data may be modified by other applications that use the same database. The integrity of data stored on EHRs may be hampered if an unauthorized application modifies the stored data. Identity management of each individual patient is critical because indexing the patient data incorrectly may result in disastrous situations. Any unauthorized access to the EHR may potentially disclose sensitive medical data. Some of the existing popular attacks on databases include SQL injection, denial of service, privilege abuse, unauthorized privilege elevation etc.

- **Remote super-speciality hospital:** The web interface at the remote super speciality hospital is prone to platform attacks as well as privacy abuse by the local staff. If the local storage at the hospital does not have proper security mechanisms, there is a threat of loss of patients' data, breach of privacy, denial of service to the patients etc.
- **Tele-consultation interface:** If the application server used for Tele-consultation does not use secure system design practices and secure communication protocols, each Tele-consultation session is prone to be attacked by an external adversary. The adversary may be interested in aborting the session, learning sensitive medical information about the patients or disrupt the timely and correct diagnosis of the patients. Some of the known attacks on the web interface and include HTTP flood, distributed denial of service attacks (DDoS) etc.
- **The network:** Networking technology used in case of telemedicine is the wired/wireless telephone network or the Internet. The former is easier to eavesdrop than the latter. For Internet communications, complex attacks exist for traffic capture and analysis, external eavesdropping, denial of service, snooping, sniffing etc. Open source tools are available for carrying out these attacks and for analyzing the vulnerability of the network and the server to these attacks.

2.4.2 Remote health monitoring

2.4.2.1 What is remote health monitoring and how it works?.
Remote health monitoring is the latest advancement in remote healthcare which aims to perform automated patient health monitoring from anywhere. Unlike telemedicine, which is basically a reactive approach of healthcare, remote monitoring is generally proactive. Through periodical monitoring of the patient's health status, the health condition is predicted, and the required measures can be taken beforehand to prevent the full manifestation of the disease. The advancement in health sensors [7], wireless sensor network (WSN) [17], and wireless body area network (WBAN) [18] has enabled physicians to check a patient's health in a continuous manner, or whenever required. Patients need not to be hospitalized; rather they are free to be at a place of their choice, generally at home. Different health sensors planted within and on the body sense different physiological data like body temperature, heart and pulse readings, blood pressure, blood sugar, brainwave, the oxygen level in blood, etc. [7]. These data are sent to the concerned health professionals who interpret them to assess a patient's health status and the requirements and recommend suitable medication or treatment accordingly. Remote health monitoring offers real-time patient observations which result in a better diagnosis. This healthcare approach is good, especially for rehabilitating patients, patients suffering from chronic illness, patients who are under mental and physical therapy, and elderly people. The remote health monitoring enables to detect early symptoms of diseases which, in turn, reduces the emergency department visits, hospitalization expenses, duration of stay in hospitals. This increases the overall quality of the patient's life.

2.4.2.2 Infrastructural components of remote health monitoring.
The following components make up a remote health monitoring system in general:

Patient monitoring unit: This unit may be located at a local care unit or at the patient's home. This unit consists of high-end nano-scale bio-sensors for automated reading of a patient's physiological data and overall monitoring of the patient's health. Some of the most common health data collected are oxygen level, heart rate, non-invasive blood pressure, body temperature, respiration, etc. It also may comprise portable medical equipment's to ensure the patient's mobility and flexibility during treatment. Following are the main parts of remote health monitoring that belong to a local care unit:

- WBAN: The primary and major component of a remote health monitoring system is the WBAN which incorporates various biosensors and wearables [7]. These sensors are embedded within or on the patient's body for collecting physiological data continuously. The collected data are assessed for diagnosis, treatment, and overall medical care. With technology enhancements, several medical sensors are proposed which are capable of performing simple as well as some sophisticated jobs such as a temperature reading, skin condition prediction cancer detection, heart attack prediction etc. Following are some examples of sensors used for remote monitoring [12]:
  - Wearable sensors
  - Ingestible sensors
  - Epidermal sensors
  - Blood sampling sensors
  - Tissue embedded sensors

    The sensor to be used depends on the type of disease and vital statistics of patients to be monitored. The sensors, in general, comprised of five subcomponents:

  - Sensor: An embedded chip to detect the physiological data from the patient's body.
  - Microcontroller: Performs very basic level data processing (e.g., data compressions) locally. It also coordinates the functionality of all other components comprising the sensor node.
  - Memory: Facilitates the temporary storage of data.
  - Wireless communication unit: Sends the sensed data to the coordinator to facilitate all communication between nodes to transmit the data wirelessly.
  - Battery Unit: To power on the batteries.
- Coordinator/collector: It is regarded as the central command station of a particular remote health monitoring system. The primary responsibility of a collector is to collect the sensed data from all the sensors, store them provisionally, pre-process (e.g., cleaning and filtering) and send to the remote server through a network gateway.
- WLAN: The communication between sensors and the coordinator is realized through a dedicated wireless LAN (WLAN) or WSN. The most popular WSN technologies used for this purpose are RFID, ZigBee, UWB, BLE, etc. Star topology is most commonly used as the network structure for this purpose.

- Network gateway: To transmit the information at remote locations, all the information is routed via a gateway. The gateway (usually, a dedicated router) acts as the bridge between remote centers as well as remote hospitals for information exchange.

WAN: The sensor data are transmitted to a remote server mainly via the Internet or cellular networks like 3G/4G/LTE. To make the remote health monitoring reliable, a secured and trusted high-performance network infrastructure is required.

Remote medical server/cloud server: The data collected by the coordinator from the sensors are stored in the cloud or in a dedicated medical server. These servers facilitate remote accessing of patient data to medical practitioners. The databases are updated continuously to facilitate precision monitoring of the patient's health.

Management/monitoring unit: Doctors access the patient's data from the server to assess health condition. The doctor sends feedback to the patient and, on a need basis, recommend treatment and medication. Modern remote health servers can trigger alarms in case of abnormality in physiological data to alert doctors, hospital and also the patient. Some of the imperative components of this unit are:

- Pervasive devices: These devices [19] which include laptops, PDAs, smartphones, etc. are used to assist real-time readability of patient health information.
- Application interface: This allows doctors to interact with the remote health monitoring system. The graphically rich interface helps doctors to apprehend and analyze the health statistics easily. Most of these interfaces also allow real-time interactions between patient and physicians.

2.4.2.3 General architecture of remote health monitoring and its functionality. Fig. 14.2 represents the general architecture of a remote health monitoring system. As discussed in Section 2.4.2.2, the primary component of a remote healthcare unit is the WBAN which collects patient's health-related data which are sent through the network gateway to either a cloud or a dedicated medical server from where the doctor gets access of the patient's health record and recommend treatment.

The WBAN consists of different sensors used to read different physiological data. These sensors are connected to a coordinator via WSNs such as ZigBee, Bluetooth, WLAN, etc. The primary role of a coordinator is to maintain communication among the devices and server to facilitate reliable data exchange. Standards like IEEE 802.15.4a [20] and IEEE 802.15.6 [21] are

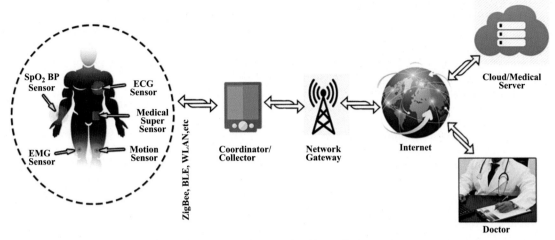

FIG. 14.2 A general remote health monitoring system.

proposed for effective WBAN communication. The coordinator is linked to the gateway whose role is to acquire data from the coordinator and transmit the data to the server over the Internet. The data is stored in a remote medical server database or cloud following the standard security measures. The data can be accessed by the patients and doctors in real time using web-based API or mobile apps to keep live track on the patient's vital signs and treatment progress. The server is also competent enough to provide suitable alerts to recognize serious health anomalies, alert medical staff in case of emergency to take immediate and required actions.

2.4.2.4 Security vulnerability points in remote health monitoring.
Typically, remote health monitoring systems feature all the functionalities of a telemedicine-based remote healthcare system. Additionally, the remote health monitoring systems require the continuous and uninterrupted monitoring of health attributes of patients using a distributed network of sensors. These sensors and the personal gateway or coordinator are potential sources of security and privacy vulnerabilities. Also, with the growing size of patient information, and need to process it at a later point of time, there is a need for a fast and reliable third-party cloud storage. Here, we highlight the security vulnerabilities in these two components of remote healthcare monitoring.

- **Sensors and coordinator:** Sensors are prone to compromise attacks using which an attacker can

potentially collect all the data passing through the sensor or the group it belongs to. Since all the collected data pass through the personal gateway or coordinator, compromising the coordinator can potentially reveal the values of all the health attributes corresponding to the patient. The coordinator directly interacts with the medical server through the network gateway and the Internet. The absence of proper security controls in communication may potentially lead to a privacy breach, obstruction in data delivery and denial of medical services to the patients.

- **Cloud/medical server:** In addition to being prone to failure, a medical server or Cloud can be potentially accessed by unauthorized entities through the web with the intention of producing malicious effects. In the absence of appropriate cryptographic mechanisms, the patient data stored on the medical server can be potentially accessed and/or modified by unauthorized entities including the Cloud service provider.

- **Networking:** The network over which the patient's data is transferred is always a potential source of vulnerability. This is because the nature of the Internet is public and it is exposed to all sorts of users, including the attackers motivated toward creating a malicious effect in the remote healthcare monitoring. Possible attacks on the network in remote healthcare and monitoring include Distributed Denial of Service (DDoS) attacks, session hijacking, spoofing attacks etc.

3 SECURITY & PRIVACY IN REMOTE HEALTHCARE

A great deal of functionality in WBANs for remote healthcare is achieved through the collection of data from the sensors attached to the patients' body. The data collected from various body parts is carried to the medical service provider who can be a human expert or an expert computer system. Data collection serves as a basis for not only the current advice but is also used for predicting possible future problems and can be used for medical research. It is usually the case that data collected by the sensors are transmitted to the healthcare service provider through the personal server called coordinator over a network such as the Internet [18]. Also, the data may have to be stored for a considerable period of time for processing and generating medical advice. For this, the medical service provider may use a dedicated third-party service for storing and processing the data. It is important to preserve privacy and ensure the security of patient's data. Security of patient's data ensures its availability, integrity and non-repudiation. Ideally, the patient who is the owner of the data has discretionary access over his/her health data. This means that the patient can choose to grant access to the collected data to anyone from the outside world. This section defines the security and privacy of data in remote healthcare systems, motivates the need of mechanisms for ensuring data security and privacy in remote healthcare and lists potential security and privacy risks along with their sources. This section also presents widely deployed solutions with security standards and software security protocols. As part of the future of security and privacy in remote and pervasive healthcare through WBANs, this section also briefly introduces patient-customizable security policies and their enforcement.

3.1 Entities in Remote Healthcare: Their Roles, Interests and Impact on Security

Before we highlight the security threats, security requirements, and the security solutions of remote healthcare, it is important to define its various external entities and their potential impacts on the security issues of the remote healthcare system. These entities include:

- **Patient:** Interested in the private, correct, and uninterrupted delivery of healthcare services. A patient provides his/her medical, health, and other related data to the third-party infrastructure/service provider.
- **Government agencies:** They regulate the security and privacy services incorporated in remote healthcare and monitoring by formulating policies. These policies cover the interests of the patients as well as the healthcare service provider. Additionally, these policies also define situations where the security and privacy of an individual patient's data can be overlooked for protecting the interests of a large number of people.
- **Service provider of infrastructural components:** Infrastructure required for remote healthcare is acquired from various vendors who may or may not provide all the hardware and software needed for data collection, its reliable and secure transfer over the local network and through the long-range network like the Internet to the healthcare server. Important infrastructure components include the following:
  - Sensors for data collection
  - Internal (local) short-range networking components
  - Network service provider for long-range (WAN) data communication

These infrastructural components are usually provided by third-party vendors whose services are utilized by the hospital authorities under a contract. All these components deal with the collection and transfer of data between the patient and the remote healthcare server. In the rest of the chapter, we refer to the set of vendors supplying these components as a data collection agency. It is a mandatory requirement that these service providers are auditable under the contract and adhere to the security and privacy policy decided by the healthcare service provider.

It must be ensured that these vendors do not learn or misuse the data collected by the devices provided by them. On the other hand, the interest of the data collection agency against any unwarranted prosecution should also be protected by having proper application-level accountability mechanisms in place.

- **Third-party storage/processing service provider:** Utilizing third-party services for maintaining and processing patient information relieves a hospital or a healthcare service provider of the operational expenditure and the cost of purchase and maintenance of expensive hardware for computing and storage. However, a third-party can be curious about the patients' data and may potentially use it for monetary benefits. If this is the case, the reputation of the hospital can be damaged, and the acceptability of remote healthcare will be limited.

3.2 Security and Privacy Threats in Remote Healthcare

Data security and data privacy are sometimes mistaken to be synonymous because the common goal of both is to protect sensitive data. Data security aims at confronting an adversary who tries to exploit loopholes in the system and steals and/or manipulates a user's sensitive personal data for which the adversary is not authorized. In many application contexts especially health-oriented services, unauthorized manipulation of patient data may potentially result in disastrous situations. This is because data collection without ensuring integrity may potentially lead to incorrect diagnosis and treatment. When a data owner provides the data to any organization, data privacy aims at deterring the organization's intentions, if any, of disclosing the data to anyone and/or misusing the data without data owner's consent. The misuse of data includes activities like unauthorized data publishing, archival for survey purposes without the patient's consent, selling the data to organisations for monetary benefits etc. Any patient who surrenders his/her health data to a hospital has no control afterward on who, apart from the hospital staff, can access the data. If indeed the hospital staff is indulged in selling patient's data, accountability mechanisms must be in place to hold the hospital staff responsible for the breach of security and privacy. This is because if personal details like health data are disclosed to any unauthorized party like the insurer of the data owner (patient), the data owner may suffer serious losses. Consider if the patient is suffering from some major ailment, it is highly likely that the patient may claim the medical insurance in the near future. Learning about the ailment of the patient, his/her insurer may sell the insurance policy to the patient at abruptly high premiums. Other problems with privacy abuse include leading to social prejudices, encouraging unethical business practices and many more. Apart from data disclosure to unauthorized entities, data misuse includes publishing it on public forums, its use for medical research without due consent from the patient etc. Broadly, the security requirements of remote healthcare include:

- Preservation of privacy of patient's data all through the storage duration.
- Prevention of unauthorized access of raw or processed data by any entity other than those authorized by the patient like hospital staff, close relatives, etc.

It is important to highlight the points of security vulnerabilities and activities that may potentially be attacked by an adversary to learn a patient's data.

Also, since availability is another important factor that affects the quality of remote healthcare services, it is important to identify the potential points of failures in remote healthcare that may lead to denial of service to the patient. Sources of security and privacy threats in remote healthcare are classified based on the activities that potentially result in these threats [22]. These activities range from the collection of raw data from the patient's body through various sensors to the complex analyses of the collected data at a storage and computation server.

3.2.1 Distributed data collection and data transmission

Data of a patient can be collected from a patient's body using the sensors planted either inside the patient's body or on the body surface. Since these sensors have a limited lifetime, the sensors have to be replaced quite often over the service duration. Also, due to limited computation power, sensors are unable to perform computationally too expensive operations like asymmetric cryptographic algorithms to achieve both secrecy and access control. However, light-weight symmetric encryption algorithms and elliptic-curve cryptography algorithms are popular for encrypting data by the sensors. Usually, the network of sensors shares a symmetric encryption key under which data collected by the sensors is encrypted before transmission. The main reason for privacy and security breach in data collection phase is disclosure of the shared encryption key. This section discusses potential scenarios that may lead to the disclosure of data encryption key to any adversary. Also, data collection being an important phase demands the availability, integrity and consistency of collected data. The accuracy of the raw data collected from the patient's body determines the quality of diagnosis and treatment. The collected data must not be modified, must be consistent and delivered to the medical service provider in an uninterrupted manner. The participating entities in the data collection phase are the body-attached sensors and the personal coordinator which transmits data collected by sensors to the base station or medical service provider over a large range network like the Internet. What follows is the description of various sources of network and system security threats for the data collection phase of remote healthcare.

3.2.1.1 Sensor node or device compromise. Privacy of data communicated by the sensors nodes that reside in or around the patient's body is ensured through symmetric encryption using an encryption key stored on the local storage of each individual sensor. If any adversary physically captures a sensor node, she can obtain the encryption key. Usually, since each sensor node has a common encryption key on their storage, the adversary not only potentially learns data collected by the compromised sensor, but also by the rest of the sensors in the network. To take control of the monitoring devices, an attacker needs to either attack the coordinator in the remote monitoring system at the patient end over the internet or go to the patient's location and physically get control of the sensors attached to the patient's body. All such methods of device compromise result in privacy breach of the raw data. There are health attributes whose value at first glance reveal no information to humans. However, since the attacker can potentially apply analysis algorithms to retrieve sensitive information, the device compromise problem can lead to the disclosure of the medical condition of the patient. On the other hand, there are health attributes whose values alone are critical, and health condition can be potentially revealed directly by learning the values of those attributes.

3.2.1.2 Dynamics of the network of sensors.

The collection of sensors communicating data related to a particular patient forms a group. This group of sensors is dynamic because the sensors are short-lived and have to be replaced from time to time with the new ones. Also, the sensors may run out of battery power, and they need to be removed temporarily from the group and added again after replenishing the battery power. During the period a sensor is out of the group, any adversary may get physical access to the sensor and obtain the key material stored on it. Since this encryption key is the same for each sensor in the group, the data communication by the group of sensors to the personal coordinator and to the medical server may not be secure anymore. The adversary who has control of the encryption key can decrypt the data transmitted by the group of sensors and obtain values of health parameters of a patient in plaintext form. Adding a new sensor may be required for capturing values of any additional health parameters. If the monitoring of any health parameter required to be stopped, the corresponding sensor is removed from the group. By compromising the removed sensor and obtaining the encryption key stored on it, an adversary can obtain plaintext information communicated by the remaining sensors [23].

To understand this better, consider a group of sensors that share a common encryption key say K. Usually, an encryption key in sensor networks is used by sensors for encrypting the data before transmitting so that only those who share the same encryption key can access the transmitted data. This encryption key is a straightforward solution to the privacy problem in sensor networks. Consider that a sensor is removed from the group and is replaced with a new one. If the key material of the old sensor is not erased properly, any adversary getting possession of the old sensor may get access to the encryption key K stored on it. Now, given that K is not changed after sensor replacement, the adversary who has obtained K through compromise can sense and decrypt all the future transmissions meant for the group of sensors and transmitted to the medical service provider. Thus, keeping the encryption key constant during the course of network operation is a potential source of security issues due to the dynamics of the network of data collecting sensors.

3.2.1.3 External network eavesdropping.

As discussed earlier, the patient's data is communicated to the distant processing or storage server or medical staffs in a hospital over a network like the Internet. An adversary may exploit the security loopholes in the networking technology for intra and inter-domain transmission and gain access to the transmitted data. Attacks on networking technologies are difficult but not rare, especially in the case of heterogeneous networks like in case of remote healthcare [24–27]. Typically, the goal of an external network eavesdropper is to obtain the data exchanged between communicating parties.

Eavesdropping attack can be active or passive depending on the adversary's motivation. For passive eavesdropping, an attacker can simply be sitting somewhere in the network path and capturing all the relevant network traffic for later analysis. The attacker does not need any active connection with the remote healthcare server. For active network eavesdropping, the adversary has to compromise one or more devices in the network path and install the eavesdropping software in the compromised device. This software analyses the traffic, reads the data being exchanged for the attacker, and also lets the attacker potentially modify the data being communicated. One of the popular network eavesdropping attacks is man-in-the-middle (MiTM) attack [28] where the adversary does not obtain the network session keys but establishes authenticated network connections with the communicating parties by masquerading as the authenticated communication party. This way, the adversary obtains all the secret network communications. MiTM attacks are particularly dangerous in remote healthcare systems if the attacker compromises one of the networking components. The main motivation of network eavesdropping is to disrupt ongoing communication and create a malicious effect by reading the data being exchanged.

3.2.2 Data collection and processing

With the continuous collection of a large amount of patients' raw health-related data, arises the need to store, process and manage data. For highly responsive remote healthcare, it is required that the collected data is stored on high-speed storage networks for fast processing and reliable storage. It may be the case that the data collection agency (a hospital) utilizes the services of a third-party storage server for storing patient data. This way of utilizing third-party storage services for storing organisational data is often referred to as storage outsourcing or data outsourcing. Given a semi-trusted or untrusted third-party storage server, storage outsourcing may potentially lead to the disclosure of sensitive information of a patient. A third-party storage and processing server is best described as an honest-but-curious entity that follows the protocol decided upon by the data collection agency but may be interested in learning the data being stored. If data are stored in plaintext form, the third-party storage server can sell data to an unauthorized entity like the patient's insurer. An obvious first choice for dealing with this problem is encrypting the data before storing on a third-party storage server. Encrypting the data is not enough as most encryption algorithms limit the utility of data for processing. The encrypted data looks like a fully random string of bits. So, the results of processing on it do not yield desired results. Homomorphic encryption [29] algorithms exist that encrypt the data such that the results of computations over the original data can be recovered from the results of computations on the encrypted data. However, such encryption algorithms are computationally too expensive to be practical for a large-scale remote healthcare system that stores a huge amount of patient data. Here, we present potential security and privacy threats to data stored on distributed storage servers and processed by the third-party processing servers.

3.2.2.1 Honest-but-curious third-party storage services.
The honest nature of an honest-but-curious storage server compels it to compulsorily store the data provided by its subscriber (healthcare service provider or hospital). However, due to curiosity, it may wish to learn the data stored on it or delegate access to stored data to any unauthorized entity. Consider a scenario where a healthcare service provider stores its patients' data in encrypted form on the server along with the encryption key. Since the

server has been subscribed for storing data, it does so. However, the availability of encryption key means that the server can decrypt the data or potentially delegate access to data by transmitting encrypting keys to its users other than the authorized medical staff. This is a serious threat to the confidentiality and integrity of patient data.

3.2.2.2 Malicious (fully untrusted) third-party storage.
For an honest-but-curious storage service provider, the only motivation is to learn stored data or delegate its access to any unauthorized entity. An even stronger adversary model applicable especially for public storage and computation service provider is that of a fully untrusted or malicious third-party. A malicious service provider not only features capabilities of a semi-trusted server but may also be motivated toward saving its storage cost by not storing the data at all. At a later point of time, when the subscriber wishes to access data, the server responds by sending bogus or modified data. One straightforward solution to this problem is storing a copy of the data locally by the subscriber. However, it is against the real motivation for storage outsourcing. Therefore, in a fully malicious setting, it is an important security challenge to assure the patients that the data being accessed from the server is indeed correct without subjecting the medical healthcare service provider to additional storage and computation overhead.

3.3 Security and Privacy Requirements in Remote Healthcare

This section defines the security requirements of a remote healthcare system using standard security terminology. It is important to identify the concrete security requirements of remote healthcare to address points of possible security vulnerabilities. These requirements encompass all the security services a remote healthcare system must provide in order to ensure the security and privacy protection of both raw and processed data by the data provider (patient and data collection agency).

3.3.1 Requirements for distributed data access security

The sensors in remote health care are distributed all over a patient's body. Accessing values of health attributes of patients should be both error-free and secure. The collection of data using a network of distributed sensors also poses scalability and accountability issues. In the below, data access security requirements of remote healthcare are formally listed with appropriate descriptions.

3.3.1.1 Data access control with revocability. As discussed earlier, selective access of patient data has to be granted to various parties. Patient data must be accessed by the medical staff. Anyone except the concerned medical staff must not be able to access the patient's data. This means that access to patient data must be given based on a policy decided by the patient. The access control policy must be defined such that it distinguishes between the authorized and unauthorized entities. Also, it must be possible to revoke access rights of any authorized entity for any piece of data.

3.3.1.2 Scalability. Ideally, computational and communication cost for accessing a scalable service does not grow linearly with the number of users of the service. In the context of data access security for remote healthcare, ensuring the security of patient data must not impose too much computation and storage cost on the users (hospital staff) or the data collection agency. Setting up and updating the access control policies of a data item should be computationally inexpensive. Security systems for ensuring private data collection and storage must also be scalable. As has been discussed earlier, the network of sensors collecting data from the patient's body and transmitting to the data collection agency is dynamic. In some specific attack scenarios, this addition of sensors to the existing group of sensors or removal of a sensor from the group may cause security problems. It must not be possible for any corrupting adversary to compromise an old removed sensor and obtain the current and/or future transmissions in plaintext form. This is called *forward secrecy*. Similarly, in the event of a new sensor joining the group of sensors, an adversary that corrupts a newly joining sensor node must not be able to access communications carried out by the sensors in the past. This is *backward secrecy*. In a group of sensors, forward and backward secrecy can be achieved by updating the encryption key following every dynamic update operation (sensor node joining or leaving) and securely availing the updated encryption key to only the current valid members of the group of sensors. Scalability of security services also concerns issues due to the growing size of the group of sensors. In case additional sensor nodes are to be added to capture data regarding more health attributes, the growing size of the group of sensors must not affect the time required for availing services to the patients.

3.3.1.3 Flexibility with non-repudiation. Having flexibility in access control mechanisms is practically motivated in remote healthcare systems. Consider a scenario where the patient's medical records have to be shared by a particular department or hospital with another hospital or department for a fixed period of time. Such cases are quite common as they may require to take expert advice from outside the hospital. A temporary access delegation of patient's data is both sufficient and necessary so that after the consultation with the outsider expert, the temporary access rights are revoked again. In such cases, the patient should reserve the rights of specifying access policies of his own and vary them according to various contexts like time, location or other events in a patient's life.

In many cases, the patient may not be in a position to provide the medical staff with his due consent for accessing and sharing data with other medical staff. In such cases, the access policy of patient data must allow someone else on behalf of the patient to delegate data access. In this case, someone is "authorized" on behalf of the patient to take policy decisions. Such temporary authorizations also form an important part of a flexible access control mechanisms for remote healthcare systems.

3.3.1.4 Accountability. An accountable security system has the capability of identifying the medical personnel who try to breach the security and privacy of the patient's data and hold him/her accountable for the same. This is important because it serves as evidence in the court of law in case of any abuse of privacy of patient's data by the hospital staff, data collection agency or the storage and processing server.

3.3.2 Requirements for distributed data storage security

3.3.2.1 Confidentiality. Confidentiality limits the use of data for only specific purposes and by only authorized entities. Data storage servers must keep patient data confidential and must be robust against compromise attacks. Since there may be many users of a storage service, confidentiality of data must be preserved even if one or more users collude with the storage service provider or unauthorized medical staff.

3.3.2.2 Dynamic integrity. Dynamic testing of the integrity of data against unauthorized modifications during storage periods of data on storage servers is very important. Any modifications may potentially lead to disastrous results as it is the degree of correctness of data that determines the correctness and effectiveness of treatment or advice. So, the remote healthcare system must not only detect malicious modifications to data but also generate alarms to the user following an

attempt of unauthorized modifications. All potential modifications during the data storage period must be detected and notified to the user.

Another way the storage server can disrupt the integrity of users' data is by not storing the data at all. Thus, ensuring dynamic integrity also includes ensuring that the data is possessed by the storage server in the first place. A method/scheme used to ensure this is called provable data possession. Similarly, a computation server may also cheat its users by not performing the desired computations at all. To ensure the integrity of the results of computations performed by the third-party computation server, a mechanism is used, known as a result verification mechanism. When result verification is combined with confidentiality of input and output data for computations, it becomes secure computation outsourcing.

3.3.2.3 Dependability.

Unavailability of correct data at the time of requirement may disrupt medical services in emergency cases. This may cause life threats to the patient. All the components of a remote health monitoring system must be dependable and operable in real-time. The network of collecting sensors must be fault tolerant so that failure of a few sensors does not result in disruptions in data collection activity. The data collection server must also be available all the time for data access, data processing and communication of results to the patient. In addition to the availability requirement, dependability also concerns the correctness of the information being stored on servers of remote healthcare systems.

3.4 Privacy Policies and Regulations

Fulfilment of the security and privacy requirements discussed in the previous section depends entirely on the security and privacy policy in place. These security policies may be different for different service providers. Also, the healthcare service providers frame their policy based on the security and privacy policy of the region in which the healthcare service is provided. In remote healthcare, privacy aims at defining who is authorized for accessing a patient's data. A clear specification of the security mechanisms in place for providing quality remote healthcare to its patients enhances the acceptability of remote healthcare.

While there are rising concerns over privacy all over the world, no global definition of the right to privacy exists to date. Right to privacy of one's own body and body related aspects cannot be violated even by the state except in cases of superseding the

right to life of a larger number of people. An efficient and durable structure of laws and policies regulates the protection of privacy in institutions that may not be government agents. An example of one such institution is remote healthcare monitoring systems. In general, systems for ensuring the security and privacy of data are a means of enforcing policies that serve the interests of the stakeholders of the system [30]. A privacy policy gives the users of service the power of self-determination through informed decision making. On the other hand, they also cater to the law of the land and pick and choose what can and what cannot be kept private about any user. In remote healthcare, the stakeholders include the patients, healthcare service provider and the government agencies [31]. The straightforward enforcement of these requirements forms a sound and robust security system for remote healthcare. Security and privacy solutions in remote healthcare systems must take into account all the policy regulations and its provisions in various scenarios. In the below, we discuss important considerations before providing security and privacy solutions in remote healthcare.

3.4.1 Identifying the data owner

In remote healthcare systems, the ownership of the patient's health data changes more frequently than other systems. Reasons that lead to dynamically changing ownership of data ranges from the consensual delegation of ownership by the patients for medical surveys, to government agencies for public interest campaigns etc. Another reason for changing ownership is the occurrence of an event(s) that invoke a clause of the privacy policy designating either the government agencies or the service provider into the owners of patient's health data. Transferring ownership of medical data is a critical activity and must be carried out in an informed manner. It is important before adopting an access policy for medical data to identify its owner. If an individual patient himself is the owner, due consent is required before any access policy comes into force. Upon transferring ownership of patient data, the access policy applicable to using patient data changes and the updated policy may or may not have to be intimated to the patient. In countries like Germany, the patient is the sole owner of his medical data and every time his data are used, a due consent is taken from the patient. This remains true even if the data is properly anonymized. In some other countries with not-so stringent privacy laws, the due consent is required to be taken only at the time of transfer of ownership of patient data.

3.4.2 Which data can be secreted and which cannot be?

A country's data access policies, especially for medical data, clearly specifies what can be hidden or kept in the private domain by the patients and what is meant to be public and what is meant to be produced when asked by competent authorities. These types of requirements concern mainly the security of the region. The outbreak of a contagious disease in a particular part of a region also requires the appropriate authorities other than the medical staff to keep track of the patient's health data. It is the duty of the data collection agency and the medical staff to comply with the access policies of the region and obtain due consent from the patients before data collection.

3.4.3 Who is in-charge in case of emergency situations?

The majority of the discussions presented above focuses the patient as the one who chooses "what and how his health data can be accessed and by whom". There may be cases where the patient himself is not in a position to take policy decisions. This results in a dilemma and may potentially obstruct the delivery of health services to the patient. In such cases, a designated person must be authorized to take such decisions. Now the question — "who authorizes a third person to have discretionary access to a patient's data" is answered in the access policy documents which again, vary from region to region and according to situations.

3.4.4 Policy reviews

There are scenarios where the access policies are reviewed and updated according to legislation from time to time. Recently, in India, the right to privacy has been codified as one of the fundamental human rights of its citizens. This has led to major policy shuffle in most of the organisations in the country. The security and privacy solutions must be durable and sustainable so that changes caused by any policy reviews can be included efficiently without much delay.

3.5 Security Solutions for Remote Healthcare

This section discusses some concrete solutions — cryptographic primitives, frameworks and security standards for all the security requirements highlighted in the previous sections. Table 14.1 summarizes the security and privacy threats in remote healthcare systems and WBAN and also their requirements and solutions.

TABLE 14.1
Summary of Security and Privacy Threats in Remote Healthcare and WBAN, and Their Requirements and Solutions.

Security Threats in WBANs	Security and Privacy Requirements	Security Solutions
Data collection and transmission	Data access control with revocability	Encryption Anonymous authentication Cryptographic access control
	Scalability	Dynamic access control and integrity assurance
	Flexibility with non-repudiation	Anonymous authentication Attribute-based encryption
	Accountability	Digital signature
Data storage and processing	Confidentiality Dynamic integrity	Encryption Message digests and hashing
	Dependability	Provable data possession

3.5.1 Secure and dependable data storage and processing

As discussed earlier, data security requirements include privacy, authentication, non-repudiation and availability. A dependable data storage and processing concern dynamic integrity verification of data stored and computation results. It also involves protecting the secrecy of data and unauthorized modification. Here, we discuss a variety of solutions to suit the requirements discussed above of secure and dependable data storage and processing.

3.5.1.1 Symmetric and asymmetric encryption systems. Encryption is an important tool for preserving the privacy of data. In symmetric encryption, only one key is used for both encryption and decryption. Examples are Data Encryption Standard (DES), Advanced Encryption Standard (AES), Rivest Ciphers (RC1 to RC6) etc. Whereas in asymmetric encryption systems, two keys, namely an encryption key and a decryption are used for encryption and decryption respectively. Examples include RSA (Rivest-Shamir-Adleman) cryptosystem [32], ElGamal cryptosystem

[33], Cramer-Shoup asymmetric encryption algorithm [34] etc. An asymmetric key cryptosystem is usually computationally costlier than the symmetric one. This is because the former involves computations over large primes in a finite field with a sufficiently large order. This makes it unsuitable for use in wireless sensor devices because of their low computing power and limited lifetime. The most important advantage of an asymmetric cryptosystem is that using these, cryptographic keys can be distributed without assuming fully secure communication channels or without having to deliver them physically. For secure delivery of symmetric encryption keys, the symmetric keys can be encrypted using asymmetric key cryptography algorithm under the recipient's public key and transmitted over an insecure channel. Assuming, only the intended recipient has the corresponding secret key, none other than the intended recipient can decrypt and obtain the underlying symmetric key. This technique is also called a digital envelope. Additionally, asymmetric key cryptography algorithms can also be used for digitally signing a document (the procedure is explained in Section 3.5.1.3). The purpose of digital signatures is not data privacy, but non-repudiation and integrity. Digital signatures are based on the principle of reverse encryption. That is, encrypting with the private key and decrypting with the public key. In an asymmetric key encryption setting, a sender encrypts the message using its secret key which can be decrypted by anyone using the corresponding public key which is already published on any centralized public storage. Assuming none other than the sender has its secret key, it can be ascertained that the message has indeed been sent by the sender and is not being repudiated.

To overcome the computational cost of asymmetric cryptosystems over large finite fields, asymmetric cryptosystems have been developed over elliptic curves. This requires comparatively very small numbers to be processed for encryption/decryption for security level equivalent to that achieved in classical asymmetric cryptosystems. Table 14.2 shows the key-sizes of traditional symmetric encryption, Elliptic curve cryptosystems and traditional asymmetric encryption algorithms like the RSA. The majority of the focus on elliptic curve cryptosystems research is about formalizing the security models and building security proofs of these lightweight cryptographic primitives. Though they lack formal proofs of security, there are no attacks known on elliptic curve cryptosystems. Therefore, elliptic curve cryptosystems are popular for asymmetric encryption in sensor networks. Some of the well-known elliptic-curve

TABLE 14.2
Key Sizes for Equivalent Security Levels (in bits) [38].

Symmetric	ECC	RSA
80	163	1024
128	283	3072
192	409	7680
256	571	15,360

cryptosystems are Elliptic-curve Diffie-Hellman (ECDH) key agreement [35], Elliptic Curve Integrated Encryption Scheme (ECIES) [36], Elliptic Curve Digital Signature Algorithm (ECDSA) [37], etc.

Recently, hardware implementations of symmetric ciphers have gained widespread attention for securing data communication using sensor networks [39]. This is because a hardware encrypted device features a fast, portable, and tamper-proof encryption of data traffic through it. A symmetric encryption algorithm is hard-coded into a device's local fast access memory along with the encryption key. An important characteristic of hardware encryption is that it can use biometric features as encryption keys [40]. This makes it more suitable for encryption of data traffic in remote healthcare. Currently, a hardware implementation of AES is widely used as it is the most secure and practical symmetric encryption algorithm. While hardware encryption provides fast and reliable security in communication, this demands more local storage to be available on sensor devices.

3.5.1.2 Anonymous authentication. In security, authentication requires that an activity like read/write/update of data is performed by someone authorized for it. Also, the one performing the activity must be held accountable for the activity. For example, an email server grants access to someone's mail account only if the correct secret credentials are provided. Also, each user is held responsible for the e-mails sent and other activities like deleting the e-mails, saving the e-mails, etc. Anonymity requires that a user should not be uniquely identifiable. For example, the identity of a user visiting a webpage should not be disclosed to anyone, but every time any user visits the webpage, the hit count of the page should increment by one. While anonymity deals with not identifying a user, authentication grants someone access to services based on "who" the user is. The seemingly paradoxical term

Anonymous Authentication [41] is an important feature of remote healthcare for preserving the privacy of the identities of the patients who may store, modify or delete data on the storage server. Designing an anonymous authentication protocol requires building authorized sets of users and devising a method for the server to identify the user as nothing but someone who either belongs or does not belong to the authorized set. No information about the user is revealed that could potentially lead to the disclosure of a user's identity. Some of the important salient features of an anonymous authentication protocol are [42]:

- **Secure authentication:** It requires that no unauthorized user should be able to fool the server into granting it access to services. Anonymity requires that the server should not be able to know the identity of the user it is interacting with.
- **Verifiable anonymity:** It is concerned with providing methods using which the users interact with the server always detects the server's behavior that may potentially lead disclosure of user's identity. This requirement is very difficult to satisfy. Therefore, the relaxed version of it is that if in case there is a possibility of disclosure of the user's identity and the server does so, the user comes to know about it instantly. This is useful because the user, upon detection of the cheating behavior of the server, may disconnect from it without performing any operation. This leads to possible disclosure of the identity of the user who tried connecting to the server, but this information serves as limited motivation for cheating by the server and losing reputation.
- **Revocable anonymity:** It is important in cases where, for example, anonymous users are harassing others. For such behavior of the users, if some action has to be taken, the service provider must reserve the rights to revoke the anonymity of any of its users. Since there is no way of suspending the user's account and there is a risk of losing anonymity of the rest of the users, revocation is carried out using a secure protocol and under the order of the court of law.

3.5.1.3 Dynamic integrity assurance: signature, message digests.

The patient's data in a remote healthcare system should be checked dynamically for integrity. A verifying agency called verifier verifies the dynamic integrity of data. This verifier can be a public entity, the data collector or the data owner. The verifier, given some auxiliary public information and the copy of data being stored by the storage servers, tries to assure that the data stored is consistent and correct. This procedure is executed either periodically or when data is required for processing.

One of the effective ways of assuring the integrity of data is storing cryptographic hash [43] by the verifier. A cryptographic hash function is a one-way function that, given any fixed length input, generates a unique fixed length output. Concretely, a hash function H: $\{0, 1\}^n \rightarrow \{0, 1\}^m$ where n is the input size and m the output block size has the following properties [44]:

i. Deterministic: The same input always results in the same hash output.
ii. One-way: It is computationally infeasible to obtain the input, given the output.
iii. Collision-resistance: Two different inputs do not give the same hash output.

At any point of time, the verifier can verify the integrity of the stored data by computing the hash of the stored data value and compare it with the pre-computed local hash value of the same data. If the data value is unchanged, the two hash values are the same. Otherwise, if even a single bit of data is modified, the two hashes differ on an average by at least half the total number of bits in the two hashes. The problem with this approach is that the storage overhead at the verifier is very large. Indeed, the verifier has to store as many hashes as the number of data items. Fig. 14.3A and B show the hashing and verification procedures, respectively.

To overcome this problem, instead of storing the hash locally, the verifier encrypts the hash under its secret key using an asymmetric key cryptography algorithm and stores it alongside the data item on the storage server. The two algorithms for hashing and encrypting the hash are collectively known as a signature algorithm. Note that the encryption is done using a secret key of the verifier, decryption can be done using its public key. This is the reverse of the conventional encryption process and is done so that no one except the verifier can sign the data. The hash encrypted under the secret key of the verifier is called a digital signature of the message [45]. For verification, the verifier applies a decryption algorithm on the digital signature to get a hash of the message. The verifier also computes the hash of the stored message. Now, the hash obtained by decrypting the signature is compared against the hash computed over the stored message. If the two hashes are equal, the message integrity is verified. Otherwise, the message is considered modified or inconsistent. The algorithms corresponding to decrypting the signature and computing the hash of the data are collectively

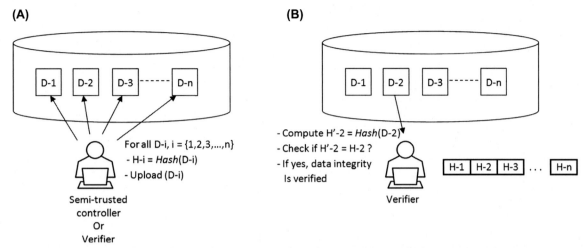

FIG. 14.3 Dynamic integrity verification using a cryptographic hash function. (A) Any semi-trusted entity can compute hash each data item before uploading. (B) Verifier who stores these hashes verifies the integrity of datum D-2.

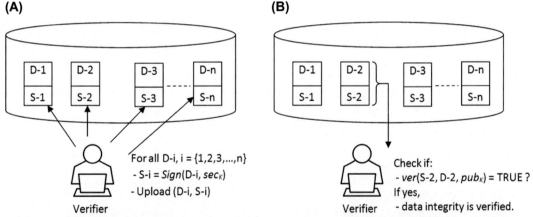

FIG. 14.4 Dynamic Integrity assurance using a digital signature. (A) Verifier signs all the data items using the signature algorithm sign and uploads the data (items, signature)-pair. (B) Verifier verifies the integrity of datum D-2 using the verification algorithm ver.

known as a signature verification algorithm. Since the digital signature is produced using a secret key of the verifier, no one else could have produced it. Therefore, no other value than a hash of the message could be obtained by decryption of the digital signature. Fig. 14.4A and B show how the signing and verification can be used for verifying the integrity of data stored on distributed storage. The advantage of this approach is that nothing has to be stored locally by the verifier. The decryption of the digital signature and computing the hash can be done on the go. However, a shortcoming of this approach is that the verifier cannot be a semi-trusted third party. The verifier must be the data owner or a data collection agency. Indeed, the encryption of the hash of the message requires secret keys, and it is not advisable to share a secret key with a semi-trusted third-party. The approach involving a digital signature is storage efficient, but it is computation and communication intensive.

3.5.2 Cryptographic access control

Access control is for the patients in remote healthcare to grant controlled access of their health data to various parties based on an access control policy. An access control policy specifies access capabilities, according to a party's role. For example, a doctor can access all the health-related data but may not be allowed access to a patient's annual salary or annual premium of the insurance policy of the patient. Also, access to sensitive medical data should not be granted to anyone apart from the hospital staff. An efficient role-based access control mechanism is required for managing access to a patient's health data. Small granularity level and dynamic nature of access control policy make the system robust, efficient and flexible. Fine-grained access control means that access to different pieces of data should be granted to various actors based on the roles they have in the overall system operation. The policy specified by the data owner takes a predicate form which, if satisfied by the role or credentials of a potential user of data, the access is granted. Otherwise, access is denied to the user. Granularity plays an important part in making the access control system efficient and dynamic. A cryptographic primitive for fine-grained role-based access control has been proposed called attribute-based encryption [46−48]. Each data item is assigned a set of attributes which are selected from the set of attributes possessed by the users in the system. An attribute-based encryption grants access to a piece of data to a user only if the attributes possessed by the user are authorized subset of the attributes associated with the data. This set of attributes associated with the data item forms the access control policy of the data item.

As an example, consider a data item which has the attributes "XYZHospital and (CARDIOLOGY or DENTAL)" meaning that anyone from the XYZHospital's cardiology or dental department can access the data item. If anyone from the XYZHospital's orthopedic department tries to access the data, the access is not granted. This is because the set of attributes of the user trying to access the data does not form the authorized subset of the attributes associated with the data. It is possible to revoke or reinstate access rights of any user at any point in time. For example, the access policy of the patient data can be updated as "XYZHospital and (CARDIOLOGY or ORTHOPEDIC)". This means that now, the medical staff belonging to the DENTAL department of the XYZHospital cannot access the data anymore, but the one belonging to ORTHOPEDIC department of XYZHospital can access the data. This dynamic property of access control mechanisms is sometimes referred to as revocability. In the context of attribute-based encryption, it is called attribute revocation. Revoking an attribute from a data item's access control policy reflects in the access rights of multiple parties based on the attribute sets possessed by them.

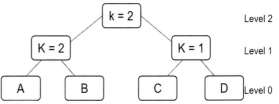

FIG. 14.5 A sample access structure in attribute based encryption.

A set of attributes are associated with a user's secret(s) cryptographically in a way that they cannot be repudiated by anyone else. An access control policy on attribute-based encryption contains several threshold gates connected to form a tree as shown in Fig. 14.5. A threshold gate with threshold K returns a value TRUE only if at least K of its child nodes evaluates to TRUE. For example, in Fig. 14.5, the threshold gates at Level 1 that has K = 2 indicates that it evaluates to TRUE only if both A and B attributes are possessed by a user in question. This gate can also be viewed as an AND gate. Similarly, the threshold gate with K = 1 can be viewed as OR gate. Gates with more general threshold values exist in attribute-based encryption to support expressive access control policies. Access structure of Fig. 14.5 is satisfied by a user having attribute set {A, B, C}. However, it cannot be satisfied by a user with attribute set {B, C, D} or {A, C, D} or {A, B} etc. While this type of access control policies seems to solve most of our problems, the access structures of attribute-based encryption systems are still far from supporting policies as expressive as an SQL statement [49]. The focus of current research if attribute-based encryption is in this direction.

There are two variants of attribute-based encryption, namely key-policy attribute-based encryption (KP-ABE) and ciphertext-policy attribute-based encryption (CP-ABE). The example discussed above is that of a CP-ABE where the policy is associated with the data/ciphertext, and each user has attributes associated with them. In KP-ABE, the policy is associated with the users' secret keys and ciphertext has associated with it the set of attributes. Arguably, CP-ABE is better suitable for scenarios where role-based access control is required.

3.5.3 Security standards

Various application-level solutions for remote healthcare security have been proposed. This section will look at the standardization of security measures in network communication part of the remote healthcare system which mainly consists of WBAN. Here, the

WBAN standard of IEEE 802.15.6 has been considered. For a detail discussion of the WBAN standards, readers are suggested to refer to the article of Pramanik et al. [12]. In this section, only the security aspect of IEEE 802.15.6 has been discussed. The standard defines three levels of WBAN communications:

- **Unsecured communication (level-0):** As the name suggests, in this communication mode, there is no protection against relay attacks, no authentication, and no integrity checks. In this level, data is transmitted in insecure frames.
- **Authentication only (level-1):** In this mode, the data is communicated in an authenticated way. Data privacy and confidentiality are not the concerns in this mode. The authentication only ensures non-repudiation which means that if any changes are made to the data, these changes can be detected and the updating authority can be accounted for the changes such that the updating entity cannot deny that the changes were not done by her.
- **Authentication plus encryption (level-2):** This is the most secure level where both authentication and confidentiality are achieved for WBAN communication. Three keys come into play when a new sensor node joins the network of sensors namely master-key (MK), pairwise temporal key (PTK) and group temporal key (GTK). An MK is either pre-shared between the sensors or is established between the sensors via authenticated association. Using this master key as input, a PTK is established to be used once per session. PTK is obtained by the sensors using a special public polynomial, called a pairwise key generation polynomial. This polynomial takes the secret key of one sensor and the public key of another to generate the session key between the two sensors. This session key can be used for encryption and decryption of data being exchanged between them. The main goal of sensor communication in WBAN is multicast. The overall collection of sensors is divided into activity driven multi-cast groups. For multicast, a GTK is computed based on PTKs of all the group members and the data intended for the whole group is encrypted and decrypted using the GTK.

3.5.4 Wireless, Bluetooth, Zigbee security protocols

The security protocols that are developed as part of wireless security are:

- WEP (wired equivalent privacy)
- WPA (Wi-Fi protected access)
- WPA-2 (Wi-Fi protected access version 2)

WEP was the first security protocol developed for security in wireless communication. WEP was ratified as part of the IEEE 802.11 in 1997. The security services provided by WEP include authentication and weak security. Authentication itself works in two modes:

- **Open system authentication:** In open system authentication, effectively there is no authentication. In fact, anyone can associate with the network. However, by some key distribution mechanism, the symmetric encryption keys are distributed to the intended participants. Messages are sent after encryption using encryption keys. The participant can decrypt only if she has the encryption key.
- **Shared key authentication:** In shared key authentication, the authentication service of WEP called four-step challenge-response handshake comes into play. Basically, the security service of the WEP protocol is provided by encryption using an RC4 stream cipher.

The problem with the RC4 stream cipher used in WEP is that the data encryption key must not be used twice. This means that the synchronization of devices to agree on the same key is required. Moreover, in 2001, the cryptanalysis of WEP was proposed by Scott Fluhrer et al. [50]. This attack broke the WEP protocol using a passive attack on the network. That is why the security service of the WEP protocol is also called Weak Security.

WPA and WPA-2 are developed to overcome the problems of weak security of WEP. Both use a pre-shared pairwise temporal key (PTK) and Group Temporal Key (GTK) for encrypting data for multicast. WPA and WPA-2 include message integrity checks as replacement of the Cyclic Redundancy Codes (CRC) that are designed to prevent an attacker from modifying and resending the data packets. WPA-2 has a major improvement over WPA in that it uses AES-based encryption mode with strong security. In January 2018, WPA-3 was announced as a replacement of WPA-2. WPA-3 uses a 192-bit key which is very strong, and it is expected to overcome security problems due to weak passwords and simplify the processing of joining the network by allowing very limited human intervention.

The highlights of the Bluetooth security protocol are the Link Manager Protocol (LMP) and Logical Link Control and Adaptation (L2CAP). LMP covers services like encryption, authentication, and exchanging the encryption keys. L2CAP supports a higher level of multiplexing and packet reassembly which can help in providing quality of service communication.

The Zigbee standard has a network layer that defines supplementary security services that include a process of authentication and key exchange. All this in addition to the IEEE 802.15.4. The Zigbee protocol also proposes to use the services of a trust center or coordinator, which allows sensor nodes to join the network and distributes encryption keys for encrypting network traffic.

3.6 Challenges and Trade-Offs

3.6.1 Interoperability

Remote healthcare systems are made realizable due to advances in low power communication standards, plug-and-play type device buses, handheld computer systems and internet technology. To increase acceptance of remote and pervasive healthcare, security with device interoperability is an important requirement [51,52]. Sensors attached to the patient's body may be bought by the patient from various vendors. Therefore. It is difficult to share the cryptographic material (e.g., encryption keys) between the sensors from different vendor sources.

As can be understood from the overall network architecture, a remote and pervasive healthcare system consists of several different types of devices with different capabilities. It starts with a sensor node which is least privileged in terms of resources like computing power and battery lifetime. A gateway device with the patient has little more computation and communication power than the sensors. Data transmitted over the internet is collected by the medical service provider which may be operating over a high computing machine to handle multiple concurrent requests. Based on their different computing powers, the devices in remote healthcare systems execute different protocols for achieving security of various aspects of data collection and processing. Devising security standards that address overall security while addressing this interoperability between devices of different computation and storage capabilities is a big concern in remote healthcare.

3.6.2 Security versus efficiency

Efficiency is a measure of the cost of any system. Security protocols for remote healthcare come with a cost associated with them. It is desirable that the security systems for remote healthcare are efficient in terms of storage requirement, communication bandwidth consumption, and computations for achieving the desired level of security. The cost of the security system is usually determined by the cost of assets it protects. If the cost of assets is greater than the cost of security measures, the security system is acceptable. Otherwise, the

security system is considered inefficient. It is advisable that the security measures be deployed only after thorough feasibility and cost analysis.

3.6.3 Security versus usability

The operators of the devices in remote healthcare are either the patients or the medical staff that most probably do not possess any technical expertise. Therefore, it is required that the devices are easy to use and foolproof. Also, the devices should be plug-and-play like and with minimum human intervention required. While omitting human steps from operation is good for usability, it may potentially cause security issues in the remote healthcare system. This is because the minimum involvement also means a low level of understanding and knowledge about the system. There are high chances of a completely ignorant operator of a remote healthcare system falling in the trap of an attacker launching a social engineering attack to learn sensitive information and/or permanent and long-term credentials of the operator.

3.6.4 Security versus availability

Sometimes, security measures adopted to secure any system are too strict that the data cannot be made available in time for processing and generation of advice. Designing a balanced security protocol that does not restrict data flow and still provides the desired level of security is a challenge. This issue is sometimes referred to as "security versus safety".

3.7 Future of Remote Healthcare Security

Data security and privacy policies are dynamic and depend on many factors, including location, time and type of data and its usage. Some countries take the privacy of personal and sensitive data of their citizens very seriously. Whereas, privacy policies in some countries do not demand very stringent privacy protection mechanisms to be in place. Recently, as part of the 2016 General Data Protection Regulation (GDPR) passed by the European Union, new rules have been set up on how companies manage and share personal data. The regulation came into effect on May 25, 2018. In addition to the earlier privacy policy enforced by European Union countries, GDPR requires two more things. Firstly, it requires complete explicit informed consent from the users for accessing and using the personal data of a user. Secondly, there must be a transparent method of revoking that consent. A user can ask for all his data from the organization to verify the consent. Maximum fines per violation are set at 4% of a company's global turnover (or $20 million, whichever is larger). GDPR has affected organizations not only in the EU but also outside the EU because of the global nature of the internet. This shift in privacy policy empowering the users is an indicator of the fact that the organizations are moving from large-scale unified security and privacy rules to more user-centric security services.

The future of remote healthcare security also lies in designing security mechanisms that suit an individual patient. Medical data, once collected, cannot be absolutely deleted from the memory of the data collection agency. However, revoking the consent of sharing the data by a patient would mean that in case of any breach of privacy, a patient can move to the court and get justice. While the users rightly keep getting privacy-aware, existing security mechanisms/frameworks may not be sufficient for proving the required level of flexibility and customization. The future of security of remote healthcare for remote and pervasive healthcare lies in innovative designs to cater to the application-specific and user-specific needs of remote healthcare systems.

4 CONCLUSION

ICT innovations play a key role to promote new medical assistance methodologies, especially in a remote way. Remote healthcare signifies a system that offers healthcare services remotely, which can be provided through telemedicine and remote health monitoring. In remote healthcare, patients and doctors do not get into face-to-face contact because of their geographical distance. Patient's health data are sent to the doctor for remote diagnosis and treatment. Being an internet connected system, any remote health monitoring system is vulnerable to a variety of passive and active security attacks. Breach of privacy, availability, and integrity of patient information can result in disastrous scenarios. The unlawful uses of health data used by an unauthorized person may be fatal. Hence, it is crucial to protect the privacy and integrity of data in remote healthcare. In this chapter, we have presented a wide variety of network and system attacks. Also, potential threats to the security of healthcare systems due to trust issues with various collaborators are also presented. To confront these security threats, standard secured communication protocols, application-level system security solutions and support of regulatory authorities for their enforcement is a must. This chapter has comprehensively described all these security solutions and provided a glimpse of where the security of remote healthcare is heading. Challenges in designing security protocols and application level solutions have also

been highlighted. The future of security in remote healthcare systems, which demands more patient-centric and customizable security solutions, is faced with numerous challenges. The biggest of these challenges is the deletion of data and access revocability. Deletion of already shared data is practically very difficult, and so is the access revocability. Revocability mechanisms while maintaining flexibility of the overall healthcare system are the points of the current focus in the design of cryptographic solutions for remote healthcare. Further works are required to maintain the security and confidentiality of data by introducing advanced encryption-based techniques. Indeed, if remote healthcare is to flourish, along with other technical enhancement, security and privacy issues are needed to be handled with foolproof solutions.

ACKNOWLEDGMENTS
We would like to thank Mr. Shubham Botre, Linux System Administrator, CDAC, Pune, India for his contribution in drawing Figs. 14.1 and 14.2.

REFERENCES
[1] F. Online, Doctor-Population Ratio: In India, One Allopathic Doctor for 11,082 People, Official Data Shows; Bihar, UP Worst Hit, June 20, 2018 [Online]. Available: https://www.financialexpress.com/india-news/doctor-population-ratio-in-india-one-allopathic-doctor-for-11082-people-official-data-shows-bihar-up-worst-hit/1213243/.

[2] S.D. D'Cunha, India's Most Remote Villages Are Getting Better Healthcare with This Cloud-Based Solution, November 21, 2016 [Online]. Available: https://www.forbes.com/sites/suparnadutt/2016/11/21/indias-most-remote-villages-are-getting-better-healthcare-with-this-cloud-based-solution/#201d2d10593b.

[3] S. Singh, S. Badaya, Health care in rural India: a lack between need and feed, South Asian J. Cancer 3 (2) (2014) 143–144.

[4] R. Kumar, Academic institutionalization of community health services: way ahead in medical education reforms, J. Fam. Med. Prim. Care 1 (1) (2012) 10–19.

[5] L. Catarinucci, R. Colella, L. Tarricone, Integration of RFID and sensors for remote healthcare, in: 3rd International Symposium on Applied Sciences in Biomedical and Communication Technologies (ISABEL 2010), Rome, Italy, 2010.

[6] P.K.D. Pramanik, P. Choudhury, IoT data processing: the different archetypes and their security & privacy assessments, in: S.K. Shandilya, S.A. Chun, S. Shandilya, E. Weippl (Eds.), Internet of Things (IoT) Security: Fundamentals, Techniques and Applications, River Publishers, 2018, pp. 37–54.

[7] P.K.D. Pramanik, B.K. Upadhyaya, S. Pal, T. Pal, Internet of things, smart sensors, and pervasive systems: enabling the connected and pervasive health care, in: N. Dey, A. Ashour, S.J. Fong, C. Bhatt (Eds.), Healthcare Data Analytics and Management, Elsevier, 2018, pp. 1–58.

[8] P.K.D. Pramanik, S. Pal, M. Mukhopadhyay, Healthcare big data: a comprehensive overview, in: N. Bouchemal (Ed.), Intelligent Systems for Healthcare Management and Delivery, IGI Global, 2018, pp. 72–100.

[9] I.A. Zriqat, A.M. Altamimi, Security and privacy issues in ehealthcare systems: towards trusted services, Int. J. Adv. Comput. Sci. Appl. 7 (9) (2016) 229–236.

[10] S. Sabnis, D. Charles, Opportunities and challenges: security in ehealth, Bell Labs Tech. J. 17 (3) (2012) 105–111.

[11] S.N. Khalifehsoltani, M.R. Gerami, E-health challenges, opportunities and experiences of developing countries, in: International Conference on e-Education, e-Business, e-Management and e-Learning, Sanya, China, 2010.

[12] P.K.D. Pramanik, A. Nayyar, G. Pareek, WBAN: driving E-healthcare beyond telemedicine to remote health monitoring - architecture and protocols, in: D.J. Hemanth, V.E. Balas (Eds.), Telemedicine Technologies: Big Data, Deep Learning, Robotics, Mobile and Remote Applications for Global Healthcare, Elsevier, 2019.

[13] J. Tan, Drivers of and barriers to E-health care, in: E-health Care Information Systems: An Introduction for Students and Professionals, Jossey-Bass, 2005, pp. 37–51.

[14] P.K.D. Pramanik, S. Pal, A. Brahmachari, P. Choudhury, Processing IoT data: from cloud to fog. It's time to be down-to-earth, in: Applications of Security, Mobile, Analytic and Cloud (SMAC) Technologies for Effective Information Processing and Management, IGI Global, 2018, pp. 124–148.

[15] P.K.D. Pramanik, S. Pal, P. Choudhury, Beyond automation: the cognitive IoT. Artificial intelligence brings sense to the internet of things, in: Cognitive Computing for Big Data Systems over IoT: Frameworks, Tools and Application, Springer, 2018, pp. 1–37.

[16] P.K.D. Pramanik, P. Choudhury, A. Saha, Economical supercomputing thru smartphone crowd computing: an assessment of opportunities, benefits, deterrents, and applications from India's perspective, in: 4th International Conference on Advanced Computing and Communication Systems (ICACCS − 2017), Coimbatore, India, January 2017.

[17] I.F. Akyildiz, W. Su, Y. Sankarasubramaniam, E. Cayirci, Wireless sensor networks: a survey, Comput. Network. 38 (4) (2002) 393–422.

[18] M. Chen, S. Gonzalez, A. Vasilakos, H. Cao, V.C. Leung, Body area networks: a survey, Mobile Network. Appl. 16 (2) (2011) 171–193.

[19] M. Abo-Zahhad, S.M. Ahmed, O. Elnahas, A wireless emergency telemedicine system for patients monitoring and diagnosis, Int. J. Telemed. Appl. 2014 (2014).

[20] IEEE Standards, Part 15.4:Wireless Medium Access Control (MAC) and Physical Layer (PHY) Specifications for Low-Rate Wireless Personal Area Networks (LR-WPANS): Amendment to Add Alternate Phy (Amendment of IEEE Std 802.15.4), IEEE, 2014.

[21] K.S. Kwak, S. Ullah, N. Ullah, An overview of IEEE 802.15.6 standard, in: 3rd International Symposium on in Applied Sciences in Biomedical and Communication Technologies, ISABEL), 2010.

[22] M. Li, W. Lou, K. Ren, Data security and privacy in wireless body area networks, IEEE Wirel. Commun. 17 (1) (2010) 51–58.

[23] J. Camenisch, I. Damgård, Verifiable encryption, group encryption, and their applications to separable group signatures and signature sharing schemes, in: International Conference on the Theory and Application of Cryptology and Information Security, 2000.

[24] N. Zhao, F.R. Yu, Y. Chen, B. Chen, V.C. Leung, Internal collusive eavesdropping of interference alignment networks, in: 85th IEEE International Conference on Vehicular Technology Conference, 2017.

[25] M. Anand, Z. Ives, A.I. Lee, Quantifying eavesdropping vulnerability in sensor networks, in: 2nd International Workshop on Data Management for Sensor Networks, 2005.

[26] N. Zhao, F.R. Yu, M. Li, Q. Yan, V.C. Leung, Physical layer security issues in interference-alignment-based wireless networks, IEEE Commun. Mag. 54 (8) (2016) 162–168.

[27] G.S.S. Geraci, J.G. Andrews, J. Yuan, I.B. Collings, Secrecy rates in broadcast channels with confidential messages and external eavesdroppers, IEEE Trans. Wirel. Commun. 13 (5) (2014) 2931–2943.

[28] F. Callegati, W. Cerroni, M. Ramilli, Man-in-the-middle attack to the HTTPS protocol, IEEE Secur. Priv. 7 (1) (2009) 78–81.

[29] R.L. Rivest, L. Adleman, M.L. Dertouzos, On data banks and privacy homomorphisms, Found. Secure Comput. 4 (11) (1978) 169–180.

[30] M. Blount, V.M. Batra, A.N. Capella, M.R. Ebling, W.F. Jerome, S.M. Martin, M. Nidd, M.R. Niemi, S.P. Wright, Remote health-care monitoring using personal care connect, IBM Syst. J. 46 (1) (2007) 95–113.

[31] A. Cavoukian, A. Fisher, S. Killen, D.A. Hoffman, Remote home health care technologies: how to ensure privacy? Build it in: privacy by design, Ident. Inf. Soc. 3 (2) (2010) 363–378.

[32] R.L. Rivest, A. Shamir, L. Adleman, A method for obtaining digital signatures and public-key cryptosystems, Commun. ACM 21 (2) (1978) 120–126.

[33] T. ElGamal, A public key cryptosystem and a signature scheme based on discrete logarithms, IEEE Trans. Inf. Theory 31 (4) (1985) 469–472.

[34] R. Cramer, V. Shoup, A practical public key cryptosystem provably secure against adaptive chosen ciphertext attack, in: Annual International Cryptology Conference, 1998.

[35] N. Koblitz, Elliptic curve cryptosystems, Math. Comput. 48 (177) (1987) 203–209.

[36] V.S. Miller, Use of elliptic curves in cryptography, in: Conference on the Theory and Application of Cryptographic Techniques, 1985.

[37] D. Johnson, A. Menezes, S. Vanstone, The elliptic curve digital signature algorithm (ECDSA), Int. J. Inf. Secur. 1 (1) (2001) 36–63.

[38] K. Lauter, The advantages of elliptic curve cryptography for wireless security, IEEE Wirel. Commun. 11 (1) (2004) 62–67.

[39] M. Healy, T. Newe, E. Lewis, Analysis of hardware encryption versus software encryption on wireless sensor network motes, in: Smart Sensors and Sensing Technology, 2008.

[40] S. Harper, P. Athanas, A security policy based upon hardware encryption, in: 37th Annual Hawaii International Conference on System Sciences, 2004.

[41] D. He, S. Zeadally, N. Kumar, J.-H. Lee, Anonymous authentication for wireless body area networks with provable security, IEEE Syst. J. 11 (4) (2017) 2590–2601.

[42] Y. Lindell, Anonymous authentication, J. Priv. Confidentiality 2 (2) (2011).

[43] P. Rogaway, T. Shrimpton, Cryptographic hash-function basics: definitions, implications, and separations for preimage resistance, second-preimage resistance, and collision resistance, in: International Workshop on Fast Software Encryption, 2004.

[44] M. Bellare, R. Canetti, H. Krawczyk, Keying hash functions for message authentication, in: Annual International Cryptology Conference, 1996.

[45] S. Goldwasser, S. Micali, R.L. Rivest, A digital signature scheme secure against adaptive chosen-message attacks, SIAM J. Comput. 17 (2) (1988) 281–308.

[46] V. Goyal, O. Pandey, A. Sahai, B. Waters, Attribute-based encryption for fine-grained access control of encrypted data, in: 13th ACM Conference on Computer and Communications Security, 2006.

[47] J. Bethencourt, A. Sahai, B. Waters, Ciphertext-policy attribute-based encryption, in: IEEE Symposium on Security and Privacy (SP '07), 2007.

[48] N. Attrapadung, B. Libert, E.D. Panafieu, Expressive key-policy attribute-based encryption with constant-size ciphertexts, in: International Workshop on Public Key Cryptography, 2011.

[49] R. Bobba, H. Khurana, M. Prabhakaran, Attribute-sets: a practically motivated enhancement to attribute-based encryption, in: European Symposium on Research in Computer Security, 2009.

[50] S. Fluhrer, I. Mantin, A. Shamir, Weaknesses in the key scheduling algorithm of RC4, in: International Workshop on Selected Areas in Cryptography, 2001.

[51] A.-J. Samaher, I. Al-Shourbaji, M. Shojafar, S. Shamshirband, Survey of main challenges (security and privacy) in wireless body area networks for healthcare applications, Egypt. Inf. J. 18 (2) (2017) 113–122.

[52] W. Steve, J. Lebak, J. Yao, J. Creekmore, A. Milenkovic, E. Jovanov, Interoperability and security in wireless body area network infrastructures, in: 27th Annual International Conference of the Engineering in Medicine and Biology Society, 2005.

CHAPTER 15

Virtual Clinic: A CDSS Assisted Telemedicine Framework

ATTA-UR-RAHMAN • MOHAMMED IMRAN BASHEER AHMED

Department of Computer Science, College of Computer Science and Information Technology, Imam Abdulrahman Bin Faisal University, Dammam, Saudi Arabia

1 INTRODUCTION

In the recent times, the healthcare facilities in developing countries are insufficient in rural areas while the population of rural areas, is more than sixty percent. Undoubtedly, the patients in such a region face huge inconvenience in traveling a long distance to the urban areas for their treatment, especially with poor transportation facilities. In this situation, information and communication technology (ICT) can help to extend better healthcare facilities in the rural areas.

Any medical activity involving a factor of distance is defined as telemedicine. Stanberry [1], has proposed the fast and advance development of ICT has changed the world. Telematics applications such as sending images, telephone call is removing the barrier of distance, but it will lead to the loss of the traditional healthcare framework and the comfort one can only have by face to face consultation. Most successful examples of educating medicine at a distance, includes the Mayo clinic, an educational program that was delivered by satellite. Sood et al. [2], has proposed that telemedicine can be defined in different ways by researchers and technology users. Modern telemedicine is a branch of e-health that transfers healthcare services from one location to another by ICT. Many institutions around the world are moving out research activities in the field of telemedicine including Brunel University, University College Cork (Ireland), University of Miami, University of Virginia, Harvard and Boston University, Imperial College London, Harvard and Boston University, the Johns Hopkins University, University of Texas etc. Armstrong and Haston [3], has presented an approach for telemedicine technology. They have proposed a telemedicine project in which connection was setup between the two departments of a remote and urban hospital. Videoconferencing, tele-radiology and

tele-presence were the key features of the link. Connection was based upon ISDN and satellite link. Within a year, near about 120 tele-consultations took place between the hospitals. Results indicated that tele-consultation had enhanced patient care. Estimated cost saving was about sixty-five thousand USD. Carson [4], has described the role of telemedicine in the management of diabetes and chronic diseases. It includes the evaluation of decision support systems (DSS), methodology and framework by using system modeling as a key approach for designing system infrastructure. This paper reviews current evidence and provides clinical applications where telemedicine is beneficial and cost-effective. In Ref. [5], the authors have proposed a wireless telemedicine system named as WISTA offering low cost and easy to deploy systematic framework. It will assist patients during unfavourable conditions. Multimedia consultation occurs between multiple disasters sites to the control center of the disaster. Proposed system supports real-time consultation offering quick treatment procedures. It uses a hierarchical architecture for implementation. For thick disaster areas, a layered structure is functional to support information distribution from large number of patients. Moreover, the performance of the system is computed using OPNET simulation. Other than clinical decision support systems, there exist DSS for other domains [6–11] as well.

In Ref. [9], has telemedicine approach that is proposed in the rural areas of Pakistan. Their proposed idea connects these rural areas with these non-practicing lady doctors through any information and communication technology, whether it is internet, GSM, WiMAX, Satellite etc. For this sake, virtual clinics will be established in the remote and ruler sites, patients will visit these clinics a nursing staff would be available

Telemedicine Technologies. https://doi.org/10.1016/B978-0-12-816948-3.00015-5

to take all necessary information that will be sent by a smart phone to a registered doctor through a central system. Doctor in return would prescribe and prescription will be sent back by smartphone to virtual clinic via same central system. Memon et al. [12], has proposed a telemedicine system that support both store and forward and real-time consultation methods for quick and cost-effective response. Relevant patient information has been sent using ISDN and satellite links. The system approaches toward the mobile doctor in all decisions. The system has various advantages like reduces the cost and time, increases. Pakistan is a thickly populated country where large population resides in rural areas. These areas lack basic health services and specialists. Telemedicine projects in developing countries are introducing new ways of improved health care services. Malik [13], presented an approach for information and communication. It highlights the projects established by Pak-US collaboration, SUPARCO a satellite-based telemedicine centers and establishing global telemedicine network which connect all medical college hospitals and district hospitals. Karim et al. [14], has proposed a virtual medicine approach that is composed of a clinical decision support system deployed at rural area, to provide medical facilities in an active way at reduced cost. This CDSS diagnosis the patient and suggest prescription as well. The cases where it is unable to prescribe, it sends email to the doctor and store response in its database for future use. In this research paper, complete setup of virtual telemedicine approach has been discussed. From the discussion, it is apparent that a comprehensive, practical and cost-effective telemedicine solution equipped with clinical decision support system (CDSS) is still a need of remote areas of Developing Countries. This is addressed in this paper. Rest of the paper is organized as follows. Section 2 highlights the proposed system model, Section 3 describes the research methodology used to carry out the research, Section 4 describes the proposed approach while Section 5 concludes the paper.

2 SYSTEM MODEL

For collecting patients' information from rural communities, the system consists of clinic side interface. The patients from rural areas to the virtual clinic where the health works available in clinic to collect the patient basic data if the patient is new then registered to the system if patient profile already exists then open the patient's profile. Collect the symptoms and vitals of the patients and submit it to the system. Proposed system finds doctor/physician for the patient and sends the data to the doctor along with the suggested medications

and investigations. The submitted data is also stored in our history data base for future use. The doctor side interface is designed for registered doctors where he/she can view their incoming patients requests and corresponding suggestions by the system. The doctor may then prescribe the patient by modifying the system suggestions and/or adding his/her own medications and submit the treatment to the system. The system stores the information in the database as well as sends it back to the corresponding virtual clinic. Virtual clinic staff can collect the suggested treatment and prescription by the doctor and pass it to the patients.

VC-DSS system suggests the doctor to a patient based on the doctor's availability and ranking. Ranking of the registered doctors is done by the proposed system based on their experience, specialization, relevance and response time. The system will store all the information of the registered patients, doctors, investigations and prescription for future use like to maintain patients and doctors' history, decision support, doctor ranking etc. Abstract architecture is shown in the following Fig. 15.1.

The VC-DSS Server is the backbone of whole Decision Support System in Virtual Clinic. This is a main web server which receives patient's information from the virtual clinic passed by health workers serving in virtual clinics in rural areas. The VC-DSS checks the patient's complete information received from virtual clinic for errors and abnormalities and passed it to decision support system as well as store the information in the patient's database. After DSS analysis and processing it is finally delivered to the appropriate doctors for consultation and prescription with suggested medicine and advices. The treatment and prescription given by consultant is then forwarded to virtual clinic side as well as store in the database.

DSS is a decision support tool that decides that the patient is sends to which doctor based on the information, investigation, symptoms of patients and doctor ranking, efficiency and successful recommendation. It also helps for consultants in a way that along with patients it also suggests the medication and treatment. The proposed DSS has performs functions like assigning the suitable doctor, calculating ranking of the doctor and suggesting the medication to the doctor based on his own history as well as system's history.

3 RESEARCH METHODOLOGY
3.1 Apriori Algorithm

Apriori algorithm is used for the problem of association rule mining, the classic algorithm. In the proposed system, it is used to assign consultant to the patients,

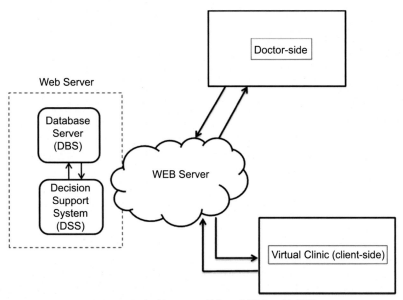

FIG. 15.1 Architecture of Victual Clinic with DSS.

suggest medications using three types of suggestions based on system database history (from all the doctors), doctor's own history and using British National Formulary (BNF). Apriori uses association rule having a support level and a confidence level. Support is the percentage of the population which satisfies the rule. If the percentage of the population in which the attendant is satisfied, then the confidence is that percentage in which the consequent is also satisfied.

3.2 Inductive Learning Algorithm

Inductive Learning algorithm (ILA) is also investigated to assign consultant to the patients, suggest medications using three types of suggestions based on system database history, doctor's own history and using British National Formulary (BNF). In ILA sub tree for each class is established. Rules can be generated from these sub trees. Once the association is selected for rule, it does not consider for future rules. If all are unmarked, go for the combination in sub tree, select maximum group of combination.

3.3 Assigning the Consultant

The goal of Decision support system is to perform analysis, make proper decision and quick response. DSS used an efficient algorithm decide the new patient is refer to which consultant. The algorithm uses the patient information, investigation, vitals, type of disease, and previous history of the database as well as the

doctors ranking and specialty and to refer to the appropriate consultant. DSS also maintain the history of diagnostics, treatment, medication, advice and prescription of the doctors as well as their response time and effectiveness of prescription and medication feedback for future use i.e. patients' history and ranking of doctors maintained. DSS monitors the patients and doctor's important data to make the better decision in future. Apriori algorithm and Inductive Learning algorithm are used to establish the rules to assign doctors to the patients by using prior knowledge and history from database systems. Patient's important data for example patient's basic information, history, blood pressure, temperature, symptoms list, images, reports history, diabetes and health situation etc. The doctors' important data include the ranking, specialty, diagnostics, prescription, and effectiveness of the treatment and medication. Ranking of registered doctors is one key role of the DSS system.

3.4 Proposed FRBS Assisted Ranking

FRBS are becoming popular for sake of recommendation, ranking etc. [15–18]. The ranks assign to the doctors are excellent, good, average, low, very low based on various factors like their response time, effectiveness of the treatment and experience. When a patient visits VC, DSS dynamically allocates the doctor with the highest rank depending on his daily limit. In case the limit is over then second highest rank holder is approached etc.

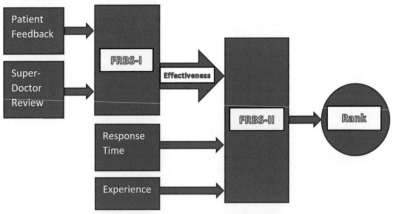

FIG. 15.2 FRBS conceptual model.

This helps in control the workload and increases the system efficiency by using knowledge driven decision-making system. In this regard, a fuzzy rule-based system (FRBS) for efficient ranking calculation is proposed. The FRBS is two folds. First the effectiveness of the doctor's previous prescriptions is calculated that depends on the patients' feedback and super-doctor's level of satisfaction toward the doctor's prescription for a specific case. Once the effectiveness is calculated, two further factors are involved to calculate the final rank. Those factors are doctor's experience and response time. Fig. 15.2 shows the model of the proposed FRBS.

3.4.1 Components of the FRBS

In first level of FRBS, the fuzzy there are two input variables namely *patient-feedback* and *super-doctor-review*, which results in fuzzy output variable *effectiveness*. In turn, the output of FRBS-I (effectiveness), along with *experience* and *response-time* become inputs to FRBS-II, which eventually generates the fuzzy output variable *rank*. Brief description of components used in both FRBS is enlisted in Table 15.1.

Fig. 15.3A and B show the input variables of FRBS-1 and Fig. 15.3C shows the output variable of FRBS-1. Similarly, Fig. 15.3D and E show the input variables of FRBS-2 and Fig. 15.3F shows the final output of the both FRBS.

Fig. 15.4 shows the rule surfaces of FRBS-1 and FRBS-2. For instance, Fig. 15.4A shows rule surface for FRBS-1 which shows that Effectiveness is directly proportional to patient-feedback and super-doctor-review. Moreover, both have equal impact on output variable (effectiveness). Fig. 15.4B shows the impact of

TABLE 15.1
Parameters Used in FRBS.

#	Parameter	Value
1	Number of input variable and membership functions in FRBS1	2 Each with three (3) membership functions.
2	Number of output variable and membership functions in FRBS1	1 With five (5) membership functions
3	Number of rules in FRBS1	9
4	Number of input variable and membership functions in FRBS2	3 With three, three and five membership functions respectively
5	Number of output variable and membership functions in FRBS2	1 Seven (7) membership functions
6	Number of rules in FRBS2	45
7	Type of fuzzifier	Triangular
8	Type of Fuzzy Inference Engine used	Mamdani
9	Type of de-fuzzifier	Center average de-fuzzifier (CAD)
10	Rule generation principle	Heuristic

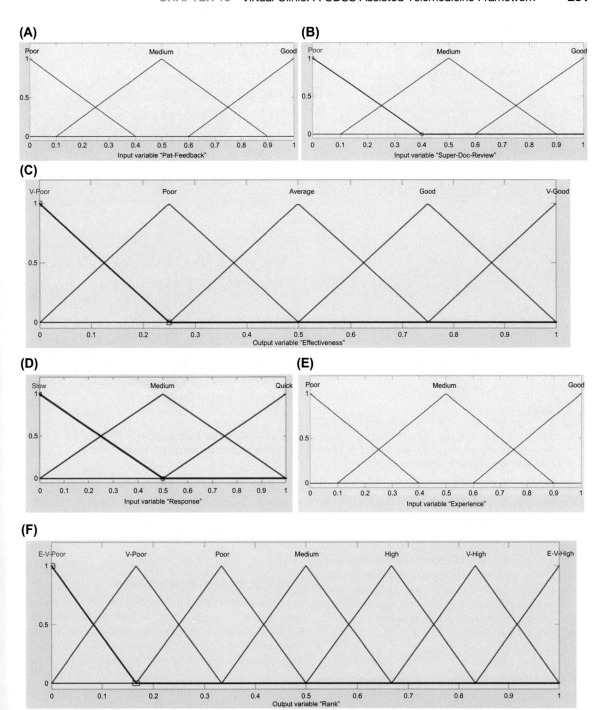

FIG. 15.3 Input and output variables used in both FRBS. (A) Input variable "Patient Feedback", (B) Input variable "Super-doctor reivew", (C) Output variable "Effectiveness" which is also input to FRBS2, (D) Input variable "Response", (E) Input variable "Experience", (F) Output variable "Rank".

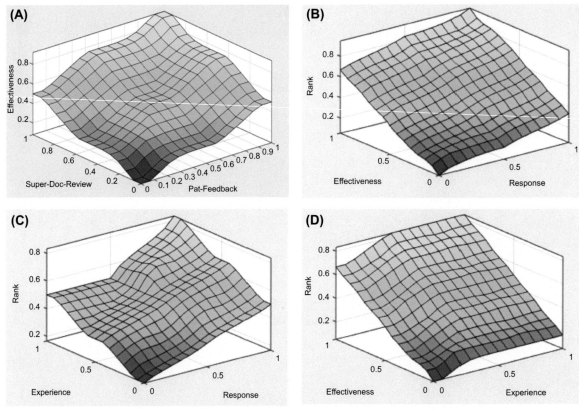

FIG. 15.4 Rule surfaces of both FRBS. (A) Rule surface of FRBS-1, (B) Rule surface of FRBS-2 for effectiveness and response, (C) Rule surface of FRBS-2 for experience and response, (D) Rule surface of FRBS-2 for effectiveness and experience.

effectiveness and response on the rank of the doctor. In this regard, both have the impact on rank, but the effectiveness has more impact compared to response time. Similarly, Fig. 15.4C and D shows the joint impacts of experience-response and effectiveness-experience on the rank. From this, we can see that experience and response have almost same effect while in effectiveness-experience case, effectiveness is given more weightage compared to the experience on the rank of doctor, respectively.

3.5 Proposed CDSS

One of the facilities provided by this system is to suggest the medicine to the doctors, while he is prescribing the patients. This system provides three suggestions to the doctors by using the overall history in the system, the doctor's own history and the BNF database. Apriori and Inductive Learning algorithms are investigated in this regard. The doctor can use any of the suggestions provided by the system or can modify an existing

suggestion based on current symptoms of the patient and effectiveness of the medication. The modifications will also be stored in the system for future use as a new instance.

4 RESULTS AND DISCUSSION

A sample data-set given below and the DSS results are given below by using Apriori algorithm and Inductive Learning algorithm. Apriori algorithm is the mining tool to find the frequent item-sets for association rules. This algorithm uses the prior knowledge to find the frequent item-sets to establish the rules. Here a sample dataset is used to establish the rules examples. This dataset is given in Table 15.2.

4.1 Suggest the Doctor
4.1.1 Using Apriori algorithm

➣ Min Confidence = 50%
➣ Min Support = 20%

TABLE 15.2 Sample Dataset From Database.						
Pts.	**Gender**	**Age**	**Disease**	**Symptoms**	**Medication**	**Dr.**
Pt1	M	55	Asthma	Shortness of breath, chest pain	Theophylline	A
Pt2	M	35	Diabetes	Hunger, increased urination, thirst	Chlorpropamide	B
Pt3	M	60	Heart attack	Heartburn, arm pain, back pain	Aspirin	D
Pt4	M	21	Dengue	Fever, headache, muscle pain	Acetaminophen	E
Pt5	F	18	Pneumonia	Cough, shortness of breath, fever	Amoxicillin, augmentin	F
Pt5	M	3	Dengue	Muscle pain, fever	Tylenol	G
Pt7	M	5	Pneumonia	Fever, chest pain	Augmentin	G
Pt8	M	35	Diabetes	Hunger, thirst	Chlorpropamide	B
Pt9	F	26	Dengue	Headache, muscle pain	Acetaminophen	E
Pt10	F	2	Dengue	Headache, muscle pain	Tylenol	G
Pt1	M	1	Pneumonia	Fever, chest pain	Augmentin	G
Pt12	M	19	Pneumonia	Fever, chest pain	Amoxicillin, augmentin	F
Pt13	M	4	Pneumonia	Chest pain	Augmentin	G
Pt14	M	15	Diabetes	Hunger, thirst	Chlorpropamide	B
Pt15	F	18	Pneumonia	Fever, Chest pain	Amoxicillin	F

➢ Age grouping
  ➢ 0−10 → 1
  ➢ 10−20 → 2
  ➢ 20−30 → 3
  ➢ 30−40 → 4
  ➢ 40−50 → 5
  ➢ 50−60 → 6
  ➢ >60 → 7

Rules examples

➢ Rules (1, Pneumonia → Dr. G) = (3/15)/(5/15) = 0.6
  ⇨ 60% Confidence
➢ Rules (1 → Dr. G) = (5/15)/(5/15) = 1
  ⇨ 100% Confidence
➢ Rules (Heart Attack → Dr. D) = (1/15)/(1/15) = 1
  ⇨ 100% Confidence
➢ Rules (Diabetes → Dr. B) = (3/15)/(3/15) = 1
  ⇨ 100% Confidence

4.1.2 Using Inductive Learning algorithm
Rules generation

➢ Rules (Disease = Asthma) → Dr. A
➢ Rules (Disease = Diabetes) → Dr. B
➢ Rules (Disease = Heart Attack) → Dr. D
➢ Rules (Age < 10) → Dr. G

➢ Rules (Disease = Dengue and Age > 10) → Dr. E
➢ Rules (Disease = Pneumonia and Age > 10) → Dr. F

Comparison. Table 15.3 shows the ILA subtrees against each doctor. Table 15.4 provides the comparison between the both schemes. The common suggestions got the highest priority (0) and appear top of the list, while the individual suggestion got the medium priority (0.5). In case where, there is no suggestion for a scenario, which will be treated as lowest priority (1) and this will appear in the lower part of the list. Comparison shows that ILA performs better in terms of number of rules.

4.2 Suggest Medications by Local History
4.2.1 Using Apriori algorithm

➢ Min Confidence = 50%
➢ Min Support = 20%
➢ Age grouping
  ➢ 0−10 → 1
  ➢ 10−20 → 2
  ➢ 20−30 → 3
  ➢ 30−40 → 4
  ➢ 40−50 → 5
  ➢ 50−60 → 6
  ➢ >60 → 7

TABLE 15.3
Inductive Learning Subtrees-Assign Doctor.

Pts.	Gender	Age	Disease	Symptoms	Medication	Dr.
DR. A – SUB TREE						
Pt1	M	55	Asthma	Shortness of breath, chest pain	Theophylline,	A
DR. B – SUB TREE						
Pt2	M	35	Diabetes	Hunger, increased urination, thirst	Chlorpropamide	B
Pt8	M	35	Diabetes	Hunger, thirst	Chlorpropamide	B
Pt14	M	15	Diabetes	Hunger, thirst	Chlorpropamide	B
DR. D – SUB TREE						
Pt3	M	60	Heart attack	Heartburn, arm pain, back pain	Aspirin	D
DR. E – SUB TREE						
Pt4	M	21	Dengue	Fever, headache, muscle pain	Acetaminophen	E
DR. F – SUB TREE						
Pt5	F	18	Pneumonia	Cough, shortness of breath, fever	Amoxicillin, augmentin	F
Pt12	M	19	Pneumonia	Fever, chest pain	Amoxicillin, augmentin	F
Pt15	F	18	Pneumonia	Fever, chest pain	Amoxicillin	F
DR. G – SUB TREE						
Pt5	M	3	Dengue	Muscle pain, fever	Tylenol	G
Pt7	M	5	Pneumonia	Fever, chest pain	Augmentin	G
Pt9	F	6	Dengue	Headache, muscle pain	Acetaminophen	G
Pt10	F	2	Dengue	Headache, muscle pain	Tylenol	G
Pt1	M	1	Pneumonia	Fever, chest pain	Augmentin	G
Pt13	M	4	Pneumonia	Chest pain	Augmentin	G

TABLE 15.4
Comparison.

Rule	ILA	Apriori Algorithm	Priority
Disease = Asthma	A		0.5
Disease = Diabetes	B	B	0
Disease = Heart attack	D	D	0
Age < 10	G	G	0
Disease = Dengue and age > 10	E		0.5
Disease = Pneumonia and age > 10	F		0.5
Disease = Pneumonia and age < 10		G	0.5

Rules examples

➢ Rules (M,1, Chest Pain → Pneumonia) = (3/15)/ (6/15) = 0.6

   ⇨ 60% Chance of Pneumonia

➢ Rules (M,1, Chest Pain → Augmentin) = (3/15)/ (5/15) = 0.53

   ⇨ 53% Chance of Augmentin

➢ Rules (M,1, Chest Pain → G) = (3/15)/(5/12) = 0.55

   ⇨ 53% Chance of Dr. G

➢ Rules (2, Fever → Pneumonia) = (3/15)/(5/15) = 0.6

   ⇨ 60% Chance of Pneumonia

4.2.2 Using Inductive Learning algorithm

Table 15.5 shows the ILA sub trees to suggest medication from the local history.

TABLE 15.5
Inductive Learning Subtrees-Suggest Medicine From Local History.

Pts.	Gender	Age	Disease	Symptoms	Medication
THEOPHYLLINE – SUB TREE					
Pt1	M	55	Asthma	Shortness of breath, chest pain	Theophylline
CHLORPROPAMIDE – SUB TREE					
Pt2	M	35	Diabetes	Hunger, increased urination, thirst	Chlorpropamide
Pt8	M	35	Diabetes	Hunger, thirst	Chlorpropamide
Pt14	M	15	Diabetes	Hunger, thirst	Chlorpropamide
ASPIRIN – SUB TREE					
Pt3	M	60	Heart Attack	Heartburn, arm pain, back pain	Aspirin
ACETAMINOPHEN – SUB TREE					
Pt4	M	21	Dengue	Fever, headache, muscular pain	Acetaminophen
Pt9	F	26	Dengue	Headache, muscle pain	Acetaminophen
AMOXICILLIN – SUB TREE					
Pt5	F	18	Pneumonia	Cough, shortness of breath, fever	Amoxicillin, augmentin
Pt12	M	19	Pneumonia	Shortness of breath, fever	Amoxicillin, augmentin
Pt15	F	18	Pneumonia	Shortness of breath, fever	Amoxicillin
AUGMENTIN – SUB TREE					
Pt5	F	18	Pneumonia	Cough, shortness of breath, fever	Amoxicillin, augmentin
Pt12	M	19	Pneumonia	Shortness of breath, fever	Amoxicillin, augmentin
Pt7	M	5	Pneumonia	Fever, chest pain	Augmentin
Pt1	M	1	Pneumonia	Fever, chest Pain	Augmentin
Pt13	M	4	Pneumonia	Chest pain	Augmentin
TYLENOL – SUB TREE					
Pt5	M	3	Dengue	Muscle pain, fever	Tylenol
Pt10	F	2	Dengue	Headache, muscle pain	Tylenol

Rules generation

➢ Rules (Disease = Asthma) → Theophylline
➢ Rules (Disease = Diabetes) → Chlorpropamide
➢ Rules (Disease = Heart Attack) → Aspirin
➢ Rules (Disease = Dengue and Age < 10) → Tylenol
➢ Rules   (Disease = Dengue   and   Age > 10) Acetaminophen
➢ Rules   (Disease = Pneumonia   and   Symptom = Shortness of Breath) → Amoxicillin
➢ Rules   (Disease = Pneumonia   and   Symptom = Chest Pain) → Augmentin

Comparison. Same behavior can be observed here, ILA performs better in terms of number of rules due to high number of sub-trees that are based on different medicines. Still those suggestions will appear on top of the list in the doctor graphical user interface (GUI) where both algorithms agree and then there comes the suggestions with at least one of the algorithms agrees and eventually the remaining suggestions that are usually due to a smaller number of instances of that type.

4.3 Suggest Medications by Doctors History
Table 15.6 shows sample dataset for doctor's history.

4.3.1 Rules examples using Apriori algorithm
➢ Rules $(Dengue \rightarrow Tylenol) = (4/10)/(4/10) = 1$
    ⇨ 100% chance of *Tylenol*

TABLE 15.6
Data Set Doctors History.

Pts.	Gender	Age	Disease	Symptoms	Medication
Pt1	M	5	Flu	Fever	Arinac
Pt2	M	3	Diarrhea	Loose stools	Navidate
Pt3	M	6	Flu	Fever	Arinac
Pt4	M	2	Dengue	Fever, headache, muscular pain	Tylenol
Pt5	F	8	Pneumonia	Chest pain, fever	Augmentin
Pt5	M	3	Dengue	Muscle pain, fever	Tylenol
Pt7	M	5	Pneumonia	Fever, chest pain	Augmentin
Pt8	M	5	Diarrhea	Loose stools	Navidate
Pt9	F	6	Dengue	Headache, muscular pain	Tylenol
Pt10	F	2	Dengue	Headache, muscle pain	Tylenol

TABLE 15.7
Inductive Learning Subtrees.

Pts.	Gender	Age	Disease	Symptoms	Medication
ARENAC – SUB TREE					
Pt1	M	5	Flu	Fever	Arinac
Pt3	M	6	Flu	Fever	Arinac
NAVIDATE – SUB TREE					
Pt2	M	3	Diarrhea	Loose stools	Navidate
Pt8	M	5	Diarrhea	Loose stools	Navidate
TYLENOL – SUB TREE					
Pt4	M	2	Dengue	Fever, headache, muscle pain	Tylenol
Pt9	F	6	Dengue	Headache, muscle pain	Tylenol
Pt5	M	3	Dengue	Muscle pain, fever	Tylenol
Pt10	F	2	Dengue	Headache, muscle pain	Tylenol
AUGMENTIN – SUB TREE					
Pt5	F	8	Pneumonia	Cough, shortness of breath, fever	Augmentin
Pt7	M	5	Pneumonia	Fever, chest pain	Augmentin

➢ Rules (Chest Pain → Augmentin) = $(2/10)/(2/10)$ = 1
  ⇨ 100% chance of Augmentin
➢ Rules (Pneumonia → *Augmentin*) = $(2/10)/(2/10)$ = 1
  ⇨ 100% chance of *Augmentin*
➢ Rules (*Fever* → *Arenac*) = $(2/10)/(4/10) = 0.5$
  ⇨ 50% chance of *Arenac*

4.3.2 Using Inductive Learning algorithm
Table 15.7 shows the ILA subtrees to suggest medicine from the doctor's history.

Rules generation
➢ Rules (Disease = Flu) → Arenac
➢ Rules (Disease = Diarrhea) → Navidate
➢ Rules (Disease = Dengue) → Tylenol
➢ Rules (Disease = Pneumonia) → Augmentin

Table 15.8 shows the comparison of Apriori algorithm and Inductive Learning algorithm in terms of suggesting medicines from doctor's history. Highest priority (0) is given to the common suggestions and lower is assigned to distinct suggestions that is either by Apriori or Inductive Learning algorithm not both. In case there a disease with no suggestions by both algorithms is assigned a lowest priority. However, they will appear as an instance.

4.4 Suggest Medications Using BNF Dataset

Table 15.9 shows the dataset extracted from British National Formulary (BNF).

TABLE 15.8
Comparison.

Disease	ILA	Apriori Algorithm	Priority
Flu	Arenac		0.5
Diarrhea	Navidate		0.5
Dengue	Tylenol	Tylenol	0
Pneumonia	Augmentin	Augmentin	0

TABLE 15.9
Data Set for BNF.

Disease	Symptoms	Medication
Diabetes	Shortness of breath, chest pain	Chlorpropamide
Diabetes	Hunger, increased urination, thirst	Chlorpropamide
Heart attack	Heartburn, arm pain, back pain	Aspirin
Dengue	Fever, headache, muscle pain	Acetaminophen
Pneumonia	Cough, shortness of breath, fever	Amoxicillin, augmentin
Dengue	Muscle pain, fever	Tylenol
Pneumonia	Fever, chest pain	Augmentin
Diabetes	Hunger, thirst	Chlorpropamide
Dengue	Headache, muscle pain	Acetaminophen
Dengue	Headache, muscle pain	Tylenol

4.4.1 Rules examples using Apriori algorithm

➤ Rules for (Dengue, *Tylenol*)
  ⇨ Rules ($Tylenol \rightarrow Dengue$) = $(2/10)/(4/10) = 0.5$
  ⇨ 50% Chance of *Tylenol*
  ⇨ Rules ($Dengue \rightarrow Tylenol$) = $(2/10)/(2/10) = 1$
  ⇨ 100% Chance of Dengue
➤ Rules (M,1, Chest Pain $\rightarrow$ Augmentin) = $(1/10)/(2/10) = 0.5$
  ⇨ 50 Chance of Augmentin

4.4.2 Using Inductive Learning algorithm

Table 15.10 shows the ILA subtrees from the dataset given in Table 15.9. The dataset belongs to BNF databased that contains knowledge about the medications against the diseases already.

TABLE 15.10
Inductive Learning Sub Trees to Suggest Medicine Using BNF.

Disease	Symptoms	Medication
CHLORPROPAMIDE – SUB TREE		
Diabetes	Shortness of breath, chest pain	Chlorpropamide
Diabetes	Hunger, increased urination, thirst	Chlorpropamide
Diabetes	Hunger, thirst	Chlorpropamide
ASPIRIN – SUB TREE		
Heart Attack	Heartburn, arm pain, back pain	Aspirin
ACETAMINOPHEN – SUB TREE		
Dengue	Headache, muscle pain	Acetaminophen
Dengue	Headache, muscle pain	Acetaminophen
AUGMENTIN – SUB TREE		
Pneumonia	Cough, shortness of breath, fever	Amoxicillin, augmentin
Pneumonia	Fever, chest pain	Augmentin
TYLENOL – SUB TREE		
Dengue	Muscle pain, fever	Tylenol
Dengue	Fever, muscle pain	Tylenol
AMOXICILLIN – SUB TREE		
Pneumonia	Cough, shortness of breath, fever	Amoxicillin, augmentin

4.4.3 Rules generation

➢ Rules (Disease = Diabetes) → Chlorpropamide
➢ Rules (Disease = Heart Attack) → Aspirin
➢ Rules (Disease = Dengue and Symptom = Headache) → Acetaminophen
➢ Rules (Disease = Pneumonia and symptoms = Shortness of Breath) → Amoxicillin
➢ Rules (Disease = Pneumonia and symptoms = Chest Pain) → Augmentin
➢ Rules (Disease = Pneumonia and symptoms = Fever) → Tylenol

It is apparent that ILA provides more range of results compared to Apriori algorithm in terms of suggestions.

5 CONCLUSION

This paper investigates on clinical decision support system for a virtual clinic, a telemedicine proposal of healthcare service provision in remote areas of developing countries. The idea is to connect the remote area patents with urban area hospitals/doctors using information and communication technologies. In addition, the system is equipped with clinical decision support system (CDSS), which helps the system to choose the appropriate doctor for a patient based on his previous prescriptions and a Fuzzy Rule Based System is augmented in the system to rank an appropriate doctor. CDSS also helps doctors in terms of suggestion of the medication for the patient based on his/her symptoms. This is accomplished by applying data mining techniques (Apriori algorithm, Inductive Learning algorithm) on system history (by the same doctor, by all the doctors) and online BNF database. Results shows that inductive learning has outperforms Apriori algorithm in many ways. In future, it is intended to investigate further improvements in the DSS.

REFERENCES

[1] Stanberry, Telemedicine: barriers and opportunities in the 21st century (internal medicine in the 21st century), J. Intern. Med. 247 (2000) 615–628.

[2] S. Sood, V. Mbarika, S. Jugoo, R. Dookhy, C.R. Doarn, N. Prakash, R.C. Merrell, What is telemedicine? A collection of 104 peer-reviewed perspectives and theoretical underpinnings, Telemed. J. E-Health 13 (5) (October 2007) 573–590, https://doi.org/10.1089/tmj.2006.0073.

[3] I.J. Armstrong, W.S. Haston, Medical decision support for remote general practitioners using telemedicine, J. Telemed. Telecare 3 (1) (1997) 27–34.

[4] E.R. Carson, Clinical decision support, systems methodology, and telemedicine: their role in the management of chronic disease, IEEE Trans. IT Biomed. 2 (2) (1998) 80–88.

[5] C. Yuechun, A. Ganz, WISTA: a wireless telemedicine system for disaster patient care, Mobile Network. Appl. 12 (2–3) (2007) 201–214.

[6] Atta-ur-Rahman, S. Das, Big data analysis for teacher recommendation using data mining techniques, Int. J. Control Theory Appl. 10 (18) (2017) 95–105.

[7] Atta-ur-Rahman, S. Das, Data mining for students' trends analysis using Apriori algorithm, Int. J. Control Theory Appl. 10 (18) (2017) 107–115.

[8] Atta-ur-Rahman, S.A. Alrashed, A. Abraham, User behavior classification and prediction using FRBS and linear regression, J. Inf. Assur. Secur. 12 (3) (2017) 86–93.

[9] Atta-ur-Rahman, M.H. Salam, S. Jamil, Virtual clinic: a telemedicine proposal for remote areas of Pakistan, in: 3rd World Congress on Information and Communication Technologies (WICT'13), 2013, pp. 46–50. December 15–18, Vietnam.

[10] S.G. Cho, S.B. Kim, A data-driven text similarity measure based on classification algorithms, Int. J. Industr. Eng. 24 (3) (2017) 328–339.

[11] Atta-ur-Rahman, A. Bakry, K. Sultan, M.A.A. Khan, M. Farooqui, D. Musleh, Clinical decision support system in virtual clinic, J. Comput. Theor. Nanosci. 15 (6) (2018) 1795–1804.

[12] T.D. Memon, B.S. Chowdhry, M.S. Memon, The potential of telemedicine system. An approach towards a mobile doctor, in: National Conference on Emerging Technologies, 2004.

[13] A.Z. Malik, Telemedicine country report-Pakistan, in: E-health Networking, Application and Services, 9th International Conference, 2007, pp. 90–94.

[14] S. Karim, Clinical decision support system based virtual telemedicine, Intell. Hum. Mach. Syst. Cybernet. 1 (2011).

[15] Atta-ur-Rahman, I.M. Qureshi, A.N. Malik, M.T. Naseem, Dynamic resource allocation for OFDM systems using DE and fuzzy rule base system, J. Intell. Fuzzy Syst. 26 (4) (2014) 2035–2046.

[16] Atta-ur-Rahman, I.M. Qureshi, A.N. Malik, M.T. Naseem, A real time adaptive resource allocation scheme for OFDM systems using GRBF-neural networks and fuzzy rule base system, Int. Arab J. Inf. Technol. 11 (6) (2014) 593–601.

[17] Atta-ur-Rahman, I.M. Qureshi, A.N. Malik, M.T. Naseem, QoS and rate enhancement in DVB-S2 using fuzzy rule base system, J. Intell. Fuzzy Syst. 30 (1) (2016) 801–810.

[18] A. Alqahtani, F.A. Alhaidari, Atta-ur-Rahman, M. Mahmud, K. Sultan, Decision support system assisted E-recruiting system, J. Comput. Theor. Nanosci. 16 (2019) 1–6.

Index

Note: Page numbers followed by "f" indicate figures, "t" indicate tables and "b" indicate boxes.

Printed in the United States
By Bookmasters